T0139924

Value Based and Intelligent Asset Management

Adolfo Crespo Márquez · Marco Macchi ·
Ajith Kumar Parlikad
Editors

Value Based and Intelligent Asset Management

Mastering the Asset Management Transformation in Industrial Plants and Infrastructures

Editors
Adolfo Crespo Márquez
Department of Industrial Management
School of Engineering of the
University of Seville
Camino de los Descubrimientos
Seville, Spain

Marco Macchi
Department of Management,
Economics and Industrial Engineering
Politecnico di Milano
Milan, Italy

Ajith Kumar Parlikad
Department of Engineering
University of Cambridge
Cambridge, UK

ISBN 978-3-030-20706-9 ISBN 978-3-030-20704-5 (eBook)
https://doi.org/10.1007/978-3-030-20704-5

This Springer imprint is published by the registered company Springer Nature Switzerland AG
The registered company address is: Gewerbestrasse 11, 6330 Cham, Switzerland

I dedicate this book to my friend and
companion Pedro Moreu de León
as sincere gratitude to his many life lessons

Adolfo Crespo Márquez

To my parents, to whom I owe the possibility of doing an amazing job; and to my mentor Marco Garetti, with whom I have been growing professionally, learning dedication and passion to research

Marco Macchi

To Priya and Sid, for enduring my extended absences during my secondments which made this book possible

Ajith Kumar Parlikad

Foreword

Asset management has been happening throughout human history. Early humans knew how to make, utilize, and discard tools that provided the means for their hunter-gatherer livelihood. Over 250 years ago or so, modern humans created and utilized assets that provided the means to harness water and steam to power the mechanized production of the First Industrial Revolution. The scope, range, and variety of engineered assets continued to increase as man harnessed electricity to power the Second Industrial Revolution, and then information and communications technologies as the driver of the Third Industrial Revolution. By the time, this book is published, the Fourth Industrial Revolution (4IR) will be underway, and the management of engineered assets toward our seemingly insatiable desire for instant gratification will continue to be a challenge.

The human desire for instant gratification is a daunting value proposition given the scope, range, and variety of engineered assets that will be required. The 4IR technologies will not only enable us to develop smarter and intelligent assets but also the concomitant fusing of the biological, digital, and physical worlds implies that we must seek new ways to manage such assets. Thus, the concept of value takes on new significance for the management of increasingly complex and more sophisticated engineered assets like rail and road infrastructure, industrial facilities, and cyber-physical systems.

This book includes seventeen chapters contributed by several scholars carrying out research on the management of smart and intelligent assets. The scope covered in the book ranges from technical issues like failure prediction, reliability, condition monitoring, diagnostics and prognostics, and digital twins, to models for decision making based on the value ethos. Many case studies are presented in the book, and such case studies constitute an invaluable resource for academia, as well as practitioners and policy makers.

Seville, Spain
February 2019

Joe Amadi-Echendu, D.Phil. CEng. PrEng. FISEAM
Professor, Engineering and Technology Management
University of Pretoria

Formerly Editor-in-Chief
Engineering Asset Management Review Series, Springer

Director and Chair of Board
International Society of Engineering Asset Management (ISEAM)

Director and Board Treasurer
International Association for the Management of Technology (IAMOT)

Preface

The fundamental motivation of this book is to contribute to the future advancement of asset management in the context of industrial plants and infrastructures. The book aims to foster a future perspective that takes advantage of value-based and intelligent asset management in order to make a step forward with respect to the evolution observed nowadays. Indeed, the current understanding of asset management is primarily supported by well-known standards. Nonetheless, asset management is still a young discipline and the knowledge developed by industry and academia is not set in stone yet. Furthermore, current trends—in new organizational concepts and technologies—lead to an evolutionary path in the field. Therefore, this book aims to discuss this evolutionary path, starting first of all from the consolidated theory, then moving forward to discuss:

- The strategic understanding of value-based asset management in a company;
- An operational definition of value, as a concept on the background of value-based asset management;
- The identification of intelligent asset management, with the aim to frame a set of "tools" recommended to support the asset-related decision-making process over the asset lifecycle.

The book compiles information gathered from interesting research and innovation efforts in projects that were relevant to this scope, especially considering the evidences from state of the art and current research trends of Physical Asset Management (PAM) and Operations and Maintenance (O&M) of industrial plants and infrastructures. Among the new trends, digitalization is enabling new capabilities for asset management, by means of the appearance of cyber-physical systems (CPSs), and the subsequent issues resulting from building the digital twins of the physical assets. This may lead to a new era of intelligent asset management systems. At the same time, basic principles of asset management will continue to be relevant in the new era, helping to guide the development of digitalization programs in assets intensive companies, and being transformed along the evolutionary path toward the achievement of a more digitized and intelligent management.

Relevant Topics

One of the main challenges in the field of physical asset management is to enhance the identification and quantification of cost and value to evaluate the total cost and value of industrial assets throughout their lifecycle. These concepts have been widely discussed in the literature, by offering different perspectives and also using plenty of terms partially overlapping or providing slightly different interpretations. Terms, such as total cost of ownership (TCO), lifecycle cost (LCC), whole-life cost (WLC), cost of ownership (COO) and, if extending to values, total value of ownership (TVO) and whole-life value (WLV), are widely cited. If one surfs the Internet, a myriad of definitions and references can be found. This does not mean that the terms are well understood and widely adopted in practice.

Considering the industrial applications of TCO and TVO, it is worth remarking that their benefits are clearly envisioned (e.g., the benefits of TCO can be considered cost control support, management strategy selection, quality optimization, and best cost-effectiveness management). However, in practice, some missing links can be pointed out with regard to their use: Even though the need and desire to implement lifecycle costing is very much talked about, there are a number of difficulties that limit a widespread adoption by industry. This is even more challenging when extending to value and, thus, to the whole-life value, which is a more recent concept.

Another relevant challenge addressed by physical asset management is the assurance of the cost and value along the asset life cycle. Henceforth, appropriate "tools" are required in order to assure that the value delivery from industrial assets (at reasonable cost) is effectively achieved and, when not, that proper decisions are activated with the aim to guarantee value delivery. In particular, proper "tools" should be used when planning in advance, and when monitoring and controlling the effective outcomes, to eventually activate re-planning in case of extant discrepancies with respect to expectations, thus leading to a continuous improvement of what is decided over the asset life cycle. Identification and quantification of value delivered by the assets are essential in all the cases.

Structure of the Book

The book is divided into four parts. In Part I, the first chapter introduces fundamental concepts used in this book and presents a generalized framework providing relevant dimensions of value-based and intelligent asset management. The rest of the chapters in this part offer a long-term perspective of asset management, dealing with topics like societal impact of investments in infrastructure assets, performance and economic impacts of investments in manufacturing plants, and long-term deterioration and renewal of assets.

In Part II, the value-based decision-making approach is stressed as an overall perspective for management of the assets over their life cycle and also exemplified in real-world specific cases. The concept of value, understood as presented in the first Chapter of this book, is *operationalized* to drive day-to-day management decisions and activities.

Part III is dedicated to different advanced developments at the operational level. Different tools are presented to predict and/or to determine properly assets conditions leading to the release and execution of the maintenance activities. Predictive analytics are used to make predictions about assets' future behavior. Many techniques from data mining, statistics, modeling, machine learning and artificial intelligence can be applied to analyze current data to make predictions about future. The scalability of these emerging models, in this new scenario of individualized asset prognostics, is another topic discussed in this part of the book, trying to find a compromise between accuracy and computational power of these tools.

Part IV is devoted to new emerging processes and new ideas that can be implemented by exploiting the power of new technologies such as cyber-physical systems that can certainly embed more intelligence and orientation to value in existing asset management systems.

European Project and Worldwide Collaboration

This book results from a collaboration of the authors, strengthened within the context of Sustain Owner, "Sustainable Design and Management of Industrial Assets through Total Value and Cost of Ownership," a project sponsored by the EU Framework Program Horizon 2020 and based on a knowledge sharing scheme involving many universities worldwide, from the Americas, Asia, and Africa.

Chapters Including Previously Published Research Results

This book compiles a set of chapters that were previously published as journal papers by the research groups involved in the Sustain Owner project. The editors would like to identify the correspondence between each chapter and the original research paper. According to Springer policy, the publishers were asked to provide their permissions for this work to be presented in its current form. The editors thank the publishers for their cooperation making this book possible. The referred chapters are:

- Chapter 2: Heaton, J., Parlikad, A.K., "A conceptual framework for the alignment of infrastructure assets to citizen requirements within a smart cities framework," *Cities*, Volume 90, pp 32–41, 2019.

- Chapter 3: Roda I., Garetti M., “Application of a Performance-driven Total Cost of Ownership (TCO) Evaluation Model for Physical Asset Management”. In: Amadi-Echendu J., Hoohlo C., Mathew J. (eds) 9th WCEAM Research Papers. *Lecture Notes in Mechanical Engineering*. Springer, Cham, 2015, © Springer International Publishing Switzerland 2015, https://doi.org/10.1007/978-3-319-15536-4.
- Chapter 5: Roda, I., and M Macchi. “A framework to embed Asset Management in production companies.” *Proceedings of the Institution of Mechanical Engineers, Part O: Journal of Risk and Reliability* 232, no. 4: 368–378, 2018, © IMechE 2018, https://doi.org/10.1177/1748006x17753501.
- Chapter 6: Srinivasan, R., Parlikad, A.K., “An approach to value-based infrastructure asset management,” *Infrastructure Asset Management*, Volume 4, Issue 3, pp 87–95, 2017.
- Chapter 9: Olivencia Polo F.A, Ferrero Bermejo J. Gómez Fernández JF., Crespo Márquez A., “Failure mode prediction and energy forecasting of PV plants to assist dynamic maintenance tasks by ANN based models”. *Renewable Energy,* Volume 81, pp 227–238. 2015.
- Chapter 10: Liu, B., Liang, Z., Parlikad, A.K., Xie, M., Kuo, W., “Condition-based maintenance for systems with aging and cumulative damage based on proportional hazards model,” *Reliability Engineering & System Safety*, Volume 168, pp 200–209, 2017.
- Chapter 11: C. Colace, L. Fumagalli, S. Pala, M. Macchi, N. R. Matarazzo, M. Rondi., “Implementation of a condition monitoring system on an electric arc furnace through a risk-based methodology.” *Proceedings of the Institution of Mechanical Engineers, Part O: Journal of Risk and Reliability*, Volume 229, Issue 4, August 2015, 327–342, 2015, © IMechE 2015, https://doi.org/10.1177/1748006x15576441.
- Chapter 12: Erguido A., Crespo Márquez A.. Castellano E., Gómez Fernández JF., “A dynamic opportunistic maintenance model to maximize energy- based availability while reducing the life cycle cost of wind farms”. *Renewable Energy,* Volume 114, pp 843–856. 2017.
- Chapter 13: Negri E., L. Fumagalli, M. Macchi, “A Review of the Roles of Digital Twin in CPS-based Production Systems”, in *Proceedings 27th International Conference on Flexible Automation and Intelligent Manufacturing, FAIM2017*, Volume 11, 939–948, 27–30 June 2017, Modena, Italy, (Eds.) Marcello Pellicciari, Margherita Peruzzini, 2017, 2351-9789, © 2017 The Authors. Published by Elsevier B.V., https://doi.org/10.1016/j.promfg.2017.07.198.
- Chapter 14: Li, H., Salvador-Palau, A., Parlikad, A.K., “A Social Network of Collaborating Industrial Assets,” *Proceedings of the IMechE Part O: Journal of Risk & Reliability*, Volume 232, Issue 4, pp. 389–400, 2018, © IMechE 2018, https://doi.org/10.1177/1748006x18754975.
- Chapter 15: Salvador-Palau, A., Liang, Z., Lutgehetmann, D., Parlikad, A.K., “Collaborative Prognostics in Social Asset Networks,” *Future Generation Computer Systems*, Volume 92, pp 987-995, 2019.

– Chapter 16: Chekurov S, Metsä-Kortelainen S, Salmi M, Roda I, Jussila A., "The perceived value of additively manufactured digital spare parts in industry: an empirical investigation". *International Journal of Production Economics,* 2015, 87–97, 2018, 0925-5273 © 2018 The Authors. Published by Elsevier B.V. T., https://doi.org/10.1016/j.ijpe.2018.09.008.

Seville, Spain — Adolfo Crespo Márquez
Milan, Italy — Marco Macchi
Cambridge, UK — Ajith Kumar Parlikad

Acknowledgements

The authors wish to thank specific people and institutions for providing their help during the years 2016–2019, making the publication of this book possible.

This research work was performed within the context of Sustain Owner ("Sustainable Design and Management of Industrial Assets through Total Value and Cost of Ownership"), a project sponsored by the EU FP Horizon 2020, MSCA-RISE-2014: Marie Skłodowska-Curie Research and Innovation Staff Exchange (RISE) (grant agreement number 645733—Sustain-Owner—H2020-MSCA-RISE-2014).

We offer our sincere gratitude to the project partners of Sustain Owner—Prof. Joe Amadi-Echendu at the University of Pretoria, Prof. Makarand Kulkarni at the Indian Institute of Technology (Mumbai), Prof. Ashish Darpe at the Indian Institute of Technology (Delhi), Prof. Mohsen Jafari at Rutgers University, Prof. Jay Lee at the University of Cincinnati, and Prof. Julio Canales and Prof. Adolfo Arata Andreani at Pontificia Universidad Católica de Valparaíso. This book would not be possible without the secondments carried out by our researchers with them and their input during their secondments with us.

We would like to thank our colleagues in the Department of Industrial Management of the University of Seville, in the Department of Engineering at the University of Cambridge and in the Department of Management, Economics and Industrial Engineering (DIG) at the Politecnico di Milano. They did provide their help and support to this work; many of them also co-authored some of the papers serving as basis for different chapters of the book. We fully appreciate the amicable and friendly working atmosphere in our research groups over these years, where the area of asset management could importantly develop.

We would like to specially thank Dr. Irene Roda as she was a cornerstone, coordinating the Sustain Owner project during its scientific and technical activities, besides being part of the knowledge exchange and, thus, bringing valuable concepts as a follow-up of her doctoral work.

Likewise, during this time we had the opportunity of interaction with engineers and research fellows from universities around the world contributing to this book: Mondragon University, City University of Hong Kong, Aalto University, and VTT Technical Research Centre of Finland. We received precious benefits of their knowledge, which in turn have influenced the development of this book. At the same time, we would like to thank them for their friendship, enriching our personal life.

We also want to give special thanks and our warm recognition to companies and associations providing their exceptional and very valuable professional experience to the book. Within these companies, many colleagues gave their support offering a practical view helping us to understand different aspects of asset management. In order of appearance in the list of contributors, these institutions are: Tenaris Dalmine, IK4-Ikerlan, Magtel Systems, Viesgo Distribución Eléctrica and the Institut für Mathematik, AG Topologie.

Last but not least, we wish to express our gratitude to our families for their unconditional trust, encouragement, and unconditional support enabling this work to be accomplished.

To all of them, Thanks.

Contents

Editors and Contributors

About the Editors

Adolfo Crespo Márquez is currently Full Professor at the School of Engineering of the University of Seville and Head of the Department of Industrial Management. He holds a Ph.D. with Honors in Industrial Engineering from this same University. His research works have been published in journals such as *Reliability Engineering and System Safety*, *International Journal of Production Research*, *International Journal of Production Economics*, *European Journal of Operations Research*, *Omega*, *Decision Support Systems and Computers in Industry*, among others. He is the author of nine books, the last six with Springer Verlag (2007, 2010, 2012, 2014, and 2017) and Aenor (2016) about maintenance, warranty, supply chain, and asset management. He is Fellow of ISEAM (International Society of Engineering Asset Management) and leads the Spanish Research Network on Asset management and the Spanish Committee for Maintenance Standardization (1995–2003). He also leads the SIM (Sistemas Inteligentes de Mantenimiento) research group related to maintenance and dependability management and has extensively participated in many engineering and consulting projects for different companies, for the Spanish Departments of Defense, Science and Education as well as for the European Commission (IPTS). He is President of INGEMAN (a National Association for the Development of Maintenance Engineering in Spain) since 2002.

Marco Macchi is currently Full Professor at Department of Management, Economics and Industrial Engineering of Politecnico di Milano. His teaching activity is in the area of production systems, industrial technologies, asset lifecycle management, operations and maintenance management, both to undergraduate and postgraduate students; he also regularly teaches for the doctorate program in Management Engineering, in the area of modeling and simulation of complex systems. His research interests are focused on several topics concerning asset and operations management within engineered systems. He is author or co-author of 4 books and more than 100 papers at national and international levels. Serving to the scientific community, he is currently Vice-chair of the IFAC (International Federation for Automatic Control) TC (technical committee) 5.1 Manufacturing Plant Control, and Chair of the IFAC Working Group A-MEST (Advanced Maintenance Engineering, Services and Technology), WG affiliated to the IFAC TC 5.1 Manufacturing Plant Control; he is Member of the IFIP (International Federation for Information Processing) WG 5.7 Advances in Production Management Systems and the IFAC TC 5.3 Enterprise Integration and Networking; besides, he is Book Reviews Editor and Editorial Board Member of the *International Journal Production Planning & Control: The Management of Operations*, Taylor & Francis.

Ajith Kumar Parlikad is Senior Lecturer in Industrial Systems at Cambridge University Engineering Department (CUED). He is Head of the Asset Management Group at the Institute for Manufacturing and a Co-investigator for the Centre for Smart Infrastructure and Construction and the Centre for Digital Built Britain. His research focusses on examining how asset information can effectively be managed and used to improve asset investment and maintenance decision making. He is Author or Co-author of 2 books and has published more than 100 papers in peer-reviewed international journals and conference proceedings. His research has been funded by the UK Government, EU H2020, and Industry. He sits on the steering committee of the IFAC Working Group on Advanced Maintenance Engineering, Services and Technology and is a member of the Management Advisory Board for the Centre for Digital Built Britain. His current research focusses on the development of infrastructure digital twins and how data from disparate sources such as BIM models and condition-monitoring and sensing technologies can be integrated and brought to bear to improve asset management.

Contributors

Eduardo Castellano MIK Research Centre, Mondragon University, Gipuzkoa, Spain

Sergei Chekurov Department of Mechanical Engineering, School of Engineering, Production Technology, Aalto University, Espoo, Finland

Cristian Colace Tenaris Dalmine S.p.A., Dalmine, BG, Italy

Adolfo Crespo Márquez Department of Industrial Management, School of Engineering, University of Seville, Seville, Spain;
Intelligent Maintenance System Research Group (SIM), Department of Industrial Management, School of Engineering, University of Seville, Seville, Spain

Antonio de la Fuente Carmona Intelligent Maintenance System Research Group (SIM), Department of Industrial Management, School of Engineering, University of Seville, Seville, Spain

Antonio González Diego Viesgo Electrical Distribution. PCTCAN, Santander, Spain

Asier Erguido Ruiz Ikerlan Technology Research Centre, Operations and Maintenance Technologies Area, Gipuzkoa, Spain;
Department of Industrial Management, School of Engineering, University of Seville, Seville, Spain

Eduardo Candón Fernández Intelligent Maintenance System Research Group (SIM), Department of Industrial Management, School of Engineering, University of Seville, Seville, Spain

Jesús Ferrero Bermejo Magtel Operaciones, Seville, Spain

Luca Fumagalli Department of Management, Economics and Industrial Engineering, Politecnico di Milano, Milan, Italy

Marco Garetti Department of Management, Economics and Industrial Engineering, Politecnico di Milano, Milan, Italy

Juan F. Gómez Fernández Intelligent Maintenance System Research Group (SIM), Department of Industrial Management, School of Engineering, University of Seville, Seville, Spain;
Department of Industrial Management, School of Engineering, University of Seville, Seville, Spain

Vicente González-Prida Díaz Intelligent Maintenance System Research Group (SIM), Department of Industrial Management, School of Engineering, University of Seville, Seville, Spain

Antonio J. Guillén López Intelligent Maintenance System Research Group (SIM), Department of Industrial Management, School of Engineering, University of Seville, Seville, Spain

James Heaton Institute for Manufacturing, University of Cambridge, Cambridge, UK

Juan Izquierdo Ik4-Ikerlan Technology Centre, Operations and Maintenance Technologies Area, Gipuzkoa, Spain

Ari Jussila VTT Technical Research Centre of Finland Ltd., Espoo, Finland

Way Kuo Department of Systems Engineering and Engineering Management, City University of Hong Kong, Kowloon, Hong Kong

Hao Li Institute for Manufacturing, Cambridge, UK

Zhenglin Liang Department of Engineering, Institute for Manufacturing, University of Cambridge, Cambridge, UK

Bin Liu Department of Systems Engineering and Engineering Management, City University of Hong Kong, Kowloon, Hong Kong

Daniel Lütgehetmann Institut für Mathematik, AG Topologie, Berlin, Germany

Marco Macchi Department of Management, Economics and Industrial Engineering (DIG), Politecnico di Milano, Milan, Italy

Pablo Martínez-Galán Fernández Intelligent Maintenance System Research Group (SIM), Department of Industrial Management, School of Engineering, University of Seville, Seville, Spain

Nelson R. Matarazzo Tenaris Dalmine S.p.A., Dalmine, BG, Italy

Sini Metsä-Kortelainen VTT Technical Research Centre of Finland Ltd., Espoo, Finland

Pedro Moreu de León Intelligent Maintenance System Research Group (SIM), Department of Industrial Management, School of Engineering, University of Seville, Seville, Spain

Elisa Negri Department of Management, Economics and Industrial Engineering, Politecnico di Milano, Milan, Italy

Fernando Olivencia Polo Magtel Operaciones, Seville, Spain

Simone Pala Department of Management, Economics and Industrial Engineering, Politecnico di Milano, Milan, Italy

Adrià Salvador Palau Institute for Manufacturing, Cambridge, UK

Ajith Kumar Parlikad Department of Engineering, Institute for Manufacturing, University of Cambridge, Cambridge, UK

Irene Roda Department of Management, Economics and Industrial Engineering, Politecnico di Milano, Milan, Italy

Maurizio Rondi Tenaris Dalmine S.p.A., Dalmine, BG, Italy

Antonio Sola Rosique INGEMAN. Association for the Development of Maintenance Engineering, School of Engineering, Seville, Spain

Mika Salmi Department of Mechanical Engineering, School of Engineering, Production Technology, Aalto University, Espoo, Finland

Javier Serra Parajes Intelligent Maintenance System Research Group (SIM), Department of Industrial Management, School of Engineering, University of Seville, Seville, Spain

Rengarajan Srinivasan Department of Engineering, University of Cambridge, Cambridge, UK

Min Xie Department of Systems Engineering and Engineering Management, City University of Hong Kong, Kowloon, Hong Kong

Part I
Long-Term Vision for Proper Asset Management

Chapter 1
Fundamental Concepts and Framework

Adolfo Crespo Márquez, Marco Macchi and Ajith Kumar Parlikad

Abstract This chapter introduces those terms and concepts that we consider fundamental for the reader to understand the rest of the book. It provides a background and introduces a generalized framework providing relevant dimensions of value-based and intelligent asset management. Firstly, understanding the value that an asset can provide and how value-based asset management can be implemented is fundamental. Secondly, it is essential to understand that the realization of the value that an asset provides to an organization can be also done at a different indenture level to the one where asset operation and maintenance is managed. Indeed, understanding the systemic dimension of the problem is therefore a fundamental aspect too. Equally important is the emphasis that asset management places on an asset life cycle approach, to deal properly with many strategic decisions regarding investment and reinvestment in new capacity, extension of the useful life, assets health analysis, identification of possible major maintenance needs, etc. Thirdly, in a world subject to a sweeping digital transformation, we must also make use of better methods, skills and abilities that help us improve our levels of intelligence in management and allow us to take advantage of the data and information at our disposal to reach levels of unprecedented asset management. The last part of this Chapter is dedicated to present the generalized framework, to ease the understanding of these asset management dimensions, and to deal with each one of them in a proper manner for the long-term vision.

A. Crespo Márquez (✉)
Department of Industrial Management, School of Engineering, University of Seville, Camino de los Descubrimientos s/n, 41092 Seville, Spain
e-mail: adolfo@us.es

M. Macchi
Department of Management, Economics and Industrial Engineering, Politecnico di Milano, Piazza Leonardo da Vinci, 32, 20133 Milan, Italy
e-mail: marco.macchi@polimi.it

A. K. Parlikad
Department of Engineering, Institute for Manufacturing, University of Cambridge, Cambridge CB3 0FS, UK
e-mail: aknp2@cam.ac.uk

A. Crespo Márquez et al. (eds.), *Value Based and Intelligent Asset Management*,
https://doi.org/10.1007/978-3-030-20704-5_1

Keywords Asset management · Value-based management · Whole-life assessment · Asset system and network level · Decision-making

1 The Definition of Asset Management

The current understanding of Asset Management (AM) is primarily supported by well-known standards provides a good reference to define what is an 'Asset' and what is 'Asset Management'.

The BSI PAS 55-1 standard specifications [1] defines an asset as "*plant, machinery, property, buildings, vehicles and other items and related systems that have a distinct and quantifiable business function or service*". The ISO 55000 standard on Asset Management [2] generalizes the definition, stating that an asset is "*an item, thing or entity that has potential or actual value to an organization*".

Correspondingly, some literature distinguishes the assets a company has, and classifies them as:

- Intangible assets, including designs, knowledge, software, intellectual property, and processes;
- Tangible assets, including liquid assets (cash or inventories) and fixed assets (consisting of physical assets such as the land, buildings and infrastructures, IT equipment, machineries, hardware, and product and service equipment).

Amongst them, we are primarily concerned with Physical Assets such as buildings, industrial plants or infrastructure. However, it is evident that nowadays intangible assets are also growing in their presence and importance; especially, Intelligent Asset Management will be increasingly relying on them, through the close connection of Cyber and Physical spaces.

BSI PAS 55-1 defines asset management as the "*systematic and coordinated activities and practices through which an organisation optimally and sustainably manages its assets, and their associated performance, risks and expenditures over their lifecycle for the purpose of achieving its organisational strategic plan*". A rather simplified, yet powerful definition as put forward by the ISO 55000 standard, states that Asset Management is the "*coordinated activity of an organisation to realise value from assets*" [2].

The critical and new term in the definition of Asset Management is value. Value is considered to be one of the fundamental pillars of Asset Management. It can be "tangible or intangible, financial or non-financial, and includes consideration of risks and liabilities" [2], and is the contribution of AM as a value-adding process to the core business of an organisation [2–5]. Therefore, it is worth clarifying what means value as a core element of AM.

2 Defining Value

Value is obtained by acquiring assets that allow an organization to fulfil its strategic objectives [5], and ensuring that the assets keep fulfilling those objectives throughout their life.

Notwithstanding this general understanding, no single detailed definition of value delivered by the assets can be found in the literature. In fact, its specific definition is very much dependent on the company's purpose, the nature of its assets (i.e., tangible and intangible assets), its strategic objectives and the expectations of its stakeholders.

What is generally agreed is that the realization of value involves balancing costs, risks, opportunities and benefits arising from the way assets are specified, procured, deployed, used, maintained and disposed. Hence, the assertion that the whole-life needs to be considered in making any asset management decision, is inherently important to value delivery.

Three remarks can be outlined as primary assumptions when dealing with value realization from assets:

1. Each organisation has to define its own conception of value, given the specific business context in which it operates;
2. When the management and utilisation of an asset encompass more than one organisation (it is normally the case especially for infrastructure assets), managing value requires to deal with conflicting objectives, asymmetric sharing of costs and risks, opportunities and benefits, etc.;
3. A whole-life assessment of value should be considered in order to deal with the fulfilment of company's objectives and the stakeholders' expectations throughout the whole-life of the assets.

Given these assumptions, we rely on some criteria derived from the literature and the experience shared during the SustainOwner project, to introduce a more operational definition of value. To this end, we consider a first requirement for proper value-based asset management implementation: asset management objectives should be clearly identified and should be SMART (Specific, Measurable, Achievable, Realistic and Timely), so to effectively drive operations and, thus, the actions performed towards the operations excellence. Our definition of stakeholders' requirements and value metrics aims at complying with this implementation requirement; the steps to this end are summarized in Table 1.

The definition of value metrics is aligned with the usual practices of performance measurement. Indeed, the organization (or its asset management function) needs to identify the desired performance targets and measures for each asset as well as for the asset management system [6]. Doing so, the asset management system performance measures will need to encompass technical (at system/network level and individual asset level), economic and organizational dimensions, reflecting the holistic characteristic of asset management; an even larger perspective, inclusive of the theories of sustainability, may also recommend the use of economic, environmental, social dimensions to frame the performance measures. All in all, the defined performances

Table 1 Definition of stakeholders' requirements and value metrics

#	Step	Description
1	Identification of the key stakeholders	The stakeholders as interested party that "can affect, be affected by, or perceive themselves to be affected by a decision or activity", are the starting point to define value [2]. Thus, the key stakeholders are firstly identified. Their identification is influenced by the business context, and should take into account the need to consider the *whole-life* assessment of value
2	Definition of the stakeholders' requirements	The stakeholders' requirements—expression of their objectives, needs and expectations—should be understood. We consider this understanding after the operational context is properly identified. Henceforth, the specific assets, assets systems or networks are firstly set within the boundaries of the operational context considered for the *whole-life* assessment; afterwards, the stakeholders' requirements are defined
3	Conversion of the stakeholders' requirements into value drivers and metrics	Requirements are converted into the main attributes assigned by the stakeholders to the assets, asset systems or networks under assessment. The attributes will be generally named as value elements, that is the elements worth of deserving proper control because they are influent on the value realization (i.e., on the balance of costs, risks, opportunities and benefits). In particular, the value elements are defined in two terms, as value drivers and value metrics: (i) the value drivers are the characteristics of assets, asset systems or networks, relevant to their business function or service as perceived by the stakeholder; (ii) the value metrics are the performance measures used in order to assess the value drivers

should be used to explicitly measure, through proper metrics, the achievement of satisfaction (or not) of the different value drivers set by the stakeholders.

Tables 2 and 3 show some cases of dependencies of value drivers and metrics from the stakeholders and their requirements. The cases are an extract taken from interviews to stakeholders in different contexts, that is a networked infrastructure and a production plant, while the asset studied is respectively the asset network (i.e. metro network) and the asset system (i.e., a complex production plant).

Generalizing the evidences shown in the table, the stakeholders' requirements result from the stakeholders identified amongst the interested parties internal or external to a company; value drivers and metrics are subsequently defined.

Some examples of stakeholders' requirements are herein summarized as derived from different contexts:

- For Customers: High% of Reliable delivery of products/services, High% of products Availability, 100% Quality of products, 100% Safety;

Table 2 Stakeholders' requirements and value metrics for a company in the food and beverage sector

Sector and asset	Stakeholders	Stakeholder type	Stakeholder requirements	Value drivers	Value metrics
Food and Beverage—production plants	Production department	Internal	Meeting production planning target	Production	Production volume Number of satisfied (production) orders
			Meeting quality standards	Quality	Scraps rate
	Maintenance department	Internal	Maximizing asset availability	Availability	Plant availability
	Energy department	Internal	Optimizing energy consumption	Energy	Energy efficiency
	Safety department	Internal	Granting 100% safety	Safety	Number of near missed/safety un-compliance
	Customers	External	Buying good product High service level	Customer satisfaction	Selling volume Fill rate

Table 3 Stakeholders' requirements and value metrics for a company in the transport sector

Sector and asset	Stakeholders	Stakeholder type	Stakeholder requirements	Value drivers	Value metrics
Transport—metro rail network	Customers	External	Reliable service	Service	Lost customer hours
					Reliability risk
			Cheap service	Cost	CAPEX
					OPEX
			Safe service	Safety	Frequency of accidents
					Number of passenger fatalities and casualties
			Good ambiance	Service	Complaints from users
				Reputation	Column inches of bad press
	Unions	Internal	Better pay	Cost	OPEX
				Well-being at work	Number of workplace accidents
			Safe environment	Safety	Number of workplace accidents
			Low impact on the environment	Sustainability	Carbon footprint
					Sustainability costs
					Penalties
	Metro	Internal	Safe workplace	Safety	
	Maintenance team			Well-being at work	Ambience at workplace

(continued)

Table 3 (continued)

Sector and asset	Stakeholders	Stakeholder type	Stakeholder requirements	Value drivers	Value metrics
			Efficient interventions	Cost	OPEX
			Ability to do things quickly	Cost	OPEX
				Service	
	Metro	Internal	Reliable, on-time service	Service	Lost customer hours
	Operations team		Cheaper to run	Cost	OPEX
			Minimizing disruptions to service	Service	Column inches of bad press
				Reputation	
	Metro	Internal	Safety	Safety	Frequency of accidents Number of passenger fatalities Number of workplace accidents
	Profession heads		Legal compliance	Compliance	Penalties Column inches of bad press
			Quality of works	Cost	OPEX

(continued)

Table 3 (continued)

Sector and asset	Stakeholders	Stakeholder type	Stakeholder requirements	Value drivers	Value metrics
				Service	
			Standards compliance	Compliance	Penalties Column inches of bad press
			Efficient interventions	Cost	OPEX
				Service	
			Efficient technologies	Cost	OPEX
			Low impact on the environment	Sustainability	Emissions
	Regulator	External	Effective technologies	Cost	OPEX
				Service	
			Wise investments (value for money) Low impact on the environment	Cost	OPEX, CAPEX Emissions
				Service	
				Sustainability	
	Mayor of City	External	Image of the city	Reputation	Ambience for users
					Column inches of bad press

- For Asset Owners: Optimized Availability of equipment, Meeting Quality standards of products, Optimized Cost-effectiveness, Maximized Efficiency, Minimized Operational expenditures (as, e.g., Cost of operation), 100% Safety;
- For Asset Maintainers/Managers: Minimized Operational expenditures (as, e.g., Cost of maintenance), Maximized Reliability of equipment, Achievement of service level agreements;
- For Regulators/Investors: Optimized Cost-effectiveness, Guarantee of long-term sustainability, Low Stock price, Low Environmental impact, 100% Safety.

The case studies developed during SustainOwner further revealed the following key aspects of value:

- Different concepts of value exist, depending on the stakeholders;
- Value drivers and metrics can be in conflict with each other;
- Value features both a quantitative and a qualitative nature.

Differing concepts of value. It was important to observe that each of the stakeholders had different ways in which they derived value from the assets, asset systems or networks. For instance, the key requirement for customers of a rail network is the ability to complete the journey on time, resulting in the value driver of reliability (of the journey time completion), whereas the key requirement for employees of the network is the guarantee of a safe working environment, resulting in a value driver of safety at work. Moreover, for the same value driver (e.g., cost), the metrics also differ depending on the stakeholder. For instance, for the operator of highways, the critical component of expenditure was the maintenance cost whereas, for the user, the critical component of expenditure arose from decreasing fuel efficiency due to poor road surfaces or congestion. As another example, in the context of a production case, to measure the driver efficiency there could be metrics expressed as yields accounting energy, quality and availability (losses) of a production line; indeed, energy efficiency, quality rate and availability are typical yields of interest of different stakeholders internal to the company, i.e. respectively the energy manager, the quality manager and the maintenance manager.

The implication of this evidence is that the decision-makers have to take a multi-objective approach to driving their decisions, and have to not only weigh different value drivers, but also weigh the preferences of different stakeholders during their decision-making process. Besides, it should be noted that each of the value drivers is often measured using metrics that have different units and scales. In order to cope with these issues, we found three approaches taken by the companies we examined:

- To keep the metrics and units separate, and include them in a multi-objective decision model;
- To convert all the value metrics into a single unit, often monetary; and
- To convert all the value metrics into an ordinal unit-less scale (e.g., 1–10).

Conflicting value drivers and metrics. In an ideal world, the different value drivers and metrics should align with each other, so that value realization can be performed without any compromise. In reality, these value drivers and metrics can be in conflict with each other. For instance, minimizing disruption to road users to address their value driver of journey time reliability conflicts with their value driver of comfortable and economic travel, which requires timely maintenance (which disrupts traffic). Also in a production system, as each production asset is a shared resource, at least between production and maintenance operations, there is always a kind of conflicting objective: this is influent on the usage time of the asset itself, to balance the achievement of conflicting value drivers such as, e.g., availability, cost and production continuity.

This confirms the implication already remarked: the decision-makers have to take a multi-objective approach to driving their decisions; therefore, more operationally speaking, the decision-makers have to identify Pareto-optimal fronts where multiple solutions that balance the conflicting objectives are found. The best solution is then taken that addresses some of the practical and/or socio-political considerations influent for the asset management strategy.

Quantitative and qualitative nature of value. Decisions are driven by hard quantitative data and information. However, it could be seen that some value drivers and metrics are qualitative in nature (e.g., impact on brand image/reputation). In our case studies we found two possible and equally valid alternatives to cope with this:

- Some organisations consider such qualitative values separate to the quantitative ones, and bring them into the consideration while choosing a solution from the Pareto-optimal front;
- Other organizations attempt to convert the qualitative values into some form of quantitative figure, bringing them into the optimisation problem directly; for instance, some organisations choose to quantify the impact on brand image by estimating the number of customers they would lose, and hence the impact on revenue/profit.

3 Value-Based Asset Management

Putting value at the core of asset management implies a subsequent understanding: that asset management does not necessarily mean the "management of assets"; instead, it means "management of value" provided by the organisation's assets. The core characteristic of asset management is therefore its focus on the complete portfolio of assets that a company adopts to fulfil its objectives—not the focus just on single assets. Doing so, asset management is essentially a strategic activity.

This strategic role is maintained based on planning and control of the asset-related activities [5]. Then, it is essential that any asset management decision—driving the asset-related activities—should assess how the decision outcome will affect, positively and/or negatively, the value provided by the portfolio of assets. To this end, we consider a further requirement for proper value-based asset management implementation: groupings of assets specifying the portfolio in more operational concepts, are needed in order to effectively assess decisions aimed at the value realization; in fact, individual assets seldom provide value on their own, while they contribute to the value generated in a group, i.e. at the system or network level; therefore, asset management objectives should be strategically focused on the asset system or the network level, and decisions should be evaluated, for the value delivered, primarily at this level. This complies with a general recommendation provided by the ISO 55000: "an organization may choose to manage its assets as a group, rather than individually" [2].

Considering SMART (Specific, Measurable, Achievable, Realistic and Timely) asset management objectives, effectively driving operations towards the realization of value, involves balancing costs, risks, opportunities and benefits against the desired performance of the assets. Indeed, performance is a key concept moving to a change in the management philosophy: value is generated by an asset system or network when it delivers the intended functionality at a required level of performance—i.e., required as defined by the stakeholders through their value metrics—so to achieve the organisation's and the stakeholders' objectives.

Therefore, when shifting to value-based asset management, the core focus is clearly changing, i.e. from a singular consideration of costs to a combined balanced consideration of performance, costs and risks associated with the assets. It is a relevant change, also induced by the wider scope targeted by multiple stakeholders.

Table 4 shows the key differences between a traditional cost-based approach, and the ideal value-based asset management approach for making decisions [7].

When migrating to value-based asset management, the asset management system should be enhanced by embedding a value-based decision-making approach, with the purpose of entailing a systematic approach to identify and quantify how value is affected by the decisions concerned within asset management. More specifically, a value model, capable to quantify the value delivered by the assets, is required to focus on how any decision outcome may affect the value provided by the assets.

Figure 1 shows how the value model sits within the Asset Management system. Its primary functions are:

Table 4 Key differences between cost-based and value-based asset management (adapted from [7])

	Cost-based	Value-based
Core focus	Cost	Cost, risk and performance
Management philosophy	Minimise expenditure while satisfying performance requirements	Maximise performance while satisfying budgetary constraints
Stakeholder focus	Decision maker or asset owner	All stakeholders of the asset
Impact on service	Maintain minimum service levels	Explore innovative approaches to improving service levels
Decision focus	Generally focuses on asset-specific issues	Focuses on system/network level dependencies and business value

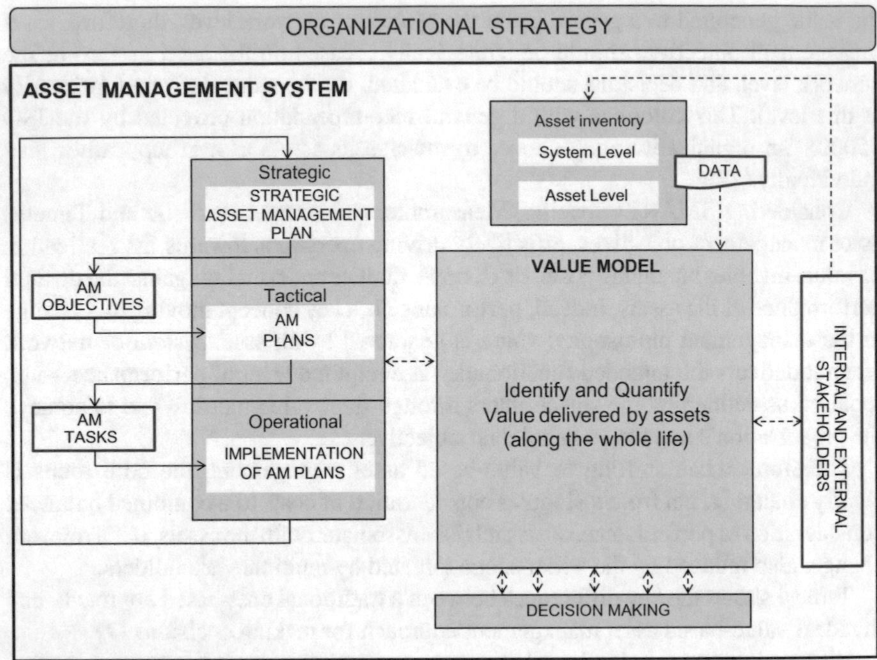

Fig. 1 The value model integrated in the AM system (adapted from [8])

- To enable and control AM planning and operative implementation, by supporting decision-making and options/scenario analysis, both when planning in advance, and when monitoring and controlling the actual operations;
- To control alignment between the AM plans and operations, and the AM strategy;
- To support communication of assets' value contribution internally and externally to the organization (i.e. including target stakeholders internal and external to a company), with the primary purpose of enhancing commitment and continuous improvement.

Similar to the different concepts of value, there is no value model that fits all needs and expectations, thus no single model is appropriate for all situations. The actual model to be used is dependent on the particular scenario and decision under concern. This does not mean that some general criteria, helpful to drive building and using a value model, cannot be identified. Indeed, the role of a value model to support asset-related decision-making in the Asset Management system from strategic to operational level, is fundamentally leading to its building and usage. To this regard, the following three factors that places the key requirements on the role of the value model to support decision-making, are considered:

- The value model is driven by stakeholders' requirements and value metrics;
- The value model primarily focuses on the value realization at system or network level;

- The value model is concerned with a whole-life assessment of the value realization.

The first requirement was previously discussed in Sect. 2 of this Chapter, and it is a relevant background for value-based asset management implementation. The other two requirements—also previously cited due to their close link with the concept of value and value-based management—are illustrated with more details in the next subsections.

4 Value Realization at Asset System and Network Level

It is commonly understood that value realization is placed, from a business perspective, at the asset system or network level, as the individual assets seldom provide value on their own. Nevertheless, decisions locally taken for the individual assets are relevant for their consequence at the system or network level: individual assets have the ability to affect the value delivered by the system or network, depending on their criticality to the business function or service. The impact of a single asset could also have an influence on the operations of other assets, because of load sharing between alternative assets as well as other sharing mechanisms such as, e.g., on resources.

The focus at system or network level has led to the emergence of the term "multi-unit system" advocated in the reliability engineering research field [9]. This is dealing with the presence of various interactions and dependences among the components (thus, the assets) within a system or network. Various classifications are generally based on the types of dependence among the members of a system. Nicolai and Dekker [10] laid the foundations for three main categories, that is: economic, stochastic, and structural dependence. Structural dependence occurs when an intervention for a component requires that other components be intervened (either replaced or dismantled) at the same time. Stochastic dependence is referred to as failure interactions among the components. Economic dependence implies that a joint maintenance of several components either negatively or positively affects the overall expense. Olde Keizer et al. [11] established also the concept of resource dependence in which components in a system share common resources such as budget or spare parts. These authors also extended the definition of structural dependence by further classifying this into technical and performance dependence. A system is considered technically dependent when intervention activities for certain components are restricted by those of other components; the negligence of this dependence inevitably results in an infeasible maintenance strategy. Performance dependence [12] is applied when an intervention activity, deterioration, and failure of a component affects the performance of other components and the overall system; this dependence is also referred to as system configuration in the survey by Wang and Chen [13].

Compared to traditional reliability engineering, despite a similar aim to enhance the system reliability, asset management extends its focus beyond, as it aims at improving system safety and reducing cost to delivering the intended functionality at a required level of performance, with the end objective to reach the benefits and opportunities addressed by an organisation and its stakeholders. In this integrated view, a decision on which asset and how it should be improved is determined by multiple performance measures—that is, more value metrics, so that the utility of system reliability is maximised. On one hand, the intricate dependences require substantial computational resources in order to deal with operations and maintenance problems. However, in most cases, a subset of interdependencies is modelled considering the practical problem at hand, and the resources the organization wants to spend on modelling. This is essential to strike the right balance between modelling complexity and accuracy. Indeed, it is essential to use a reasonable model of the system to control and, then, influence the value's output. Using such a kind of model, adequately fitted to the problem at hand, opens up new opportunities for devising innovative policies that, compared to optimisation strategies focused on individual assets, enhance the efficiency and effectiveness of the overall system or network.

Given the relevance of the asset system/network model to control the value realization, in the reminder we focus respectively on its building and use for AM decision-making.

Building an asset system/network model. When building a system/network model, the dependences need to be understood to make effective asset management decisions driven by value: therefore, starting from the objectives and the asset management performances due to value drivers and metrics, specific kinds of interdependencies are to be considered in order to evaluate the value contribution of a local decision (at a single asset) at the system or network level. For example, a bridge on its own does not deliver value, while the associated road network generates value for the users and owners; hence, the bridge should be considered as an asset within the network, evaluating the effect of any intervention activity (planned or unplanned) on the value delivered by the road network wherein it is placed. The evaluation aims at setting the criticality of the bridge with respect to the intended functionality of the road network due to many reasons (as the traffic volume, types of affected user, the need of compliancy with some legal or normative rules, etc.). A similar consideration can be transferred to factories and production systems within the factories: a production system is not just a single asset but a composition of assets; therefore, the value of a single asset, i.e. a production equipment or machinery, is not only evaluated as a single asset by itself; it is evaluated for its criticality on the functionality delivered at the asset system level. Therefore, a critical machinery is typically designated as bottleneck of a production line, as it deserves attention in intervention activities due to the fact that it constrains the production capacity of the line, one of the targeted value drivers.

Using an asset system/network model. An organisation is in need of a modeling technique for prioritizing assets in a system or network when their performance contributes unequally to the overall system/network value. Indeed, in a context with limited resources (i.e. budgetary constraints), asset managers have to give priority to assets that are more critical to the system/network, so that appropriate intervention activities (i.e. appropriate decisions) can be planned and operated accordingly. This problem entails the incorporation of criticality analysis into a Value-based decision-making approach. It results from a founding principle of Asset Management, i.e.

system orientation, where criticality of the individual assets at the system (equivalently, at the network) level is considered an essential aspect in order to ensure focusing efforts and resources on the right activities [14].

As criticality is a relevant concept to habilitate the system orientation of asset-related decisions, it is worth remarking three features of criticality analysis that we consider essential. Such features are illustrated in the reminder, jointly with selected examples provided to suggest some linkage to specific models which are to be considered as forms of value models dealing with the assessment of value realization at asset or system level.

- The concept of criticality analysis is typically employed with the aim to establish the relationship between asset reliability and system performance; this leads to a hierarchy of assets prioritized in accordance with a given set of value drivers/metrics and related parameters. For example, see [15] where criticality analysis is based on two parameters, that is: failure frequency levels and impacts on business objectives in terms of functional loss severity (this expresses the loss with respect to more value drivers such as reliability of service). These are adopted to provide a prioritization of assets in a power plant (i.e. the asset system) based on the risks incurring with failures. A resulting risk matrix is correspondingly used in order to visualize the results of the criticality analysis, which is helpful for the communication of the assets' value contribution within the plant.
- While carrying out criticality analysis, it is important to understand the dynamic nature of criticality and the influence of changes, happening in criticality, on asset management decisions. Typically, the changes within an extant asset system/network may regard the operating environment as well as the assets conditions, and they may lead to such criticality changes. To this regard, the case of Adams et al. [16] is worth of a mention: they propose the adoption of a dynamic criticality-based model to which the inputs are updated as the operating environment changes. This considers a portfolio in which the risk profile and the consequence of asset failures (thus, the impact on value) change over time.
- An asset system, such as a production plant or a networked infrastructure, must be viewed as a collection of assets that interact, when interdependencies between the component assets can affect the systemic value contribution. Concerning the risk to incur in failures, interdependencies imply that failures (or even deteriorations) of any asset within the system or network can have knock-on effects throughout the system/network, and might create additional costs or risks not located in the single asset: this effect might not be clear when managing each asset independently. Therefore, a criticality analysis, aware of system interdependencies, is essential. Focusing on production systems, for instance, while individual asset-level analysis estimates performances with respect to a single component as a machinery or a production equipment (Overall Equipment Effectiveness, shortly OEE, is the performance measure best representing this concept), system-level analysis allows considering systemic value contribution arising from components' interdependencies [17, 18]. To this regard, Roda and Macchi [19] propose a system-level analysis aimed at performance evaluation model for production systems, consid-

ering the need of a factory-level performance metric that enables to track system effectiveness. To this end, they combine the OEE calculation approach and system reliability analysis methods, to guide the improvement of a buffered multi-state production system based on the most important criticalities of single components (either production equipment or buffers) assessed for their impact at system level.

5 Whole-Life Assessment of Value Realization

A main challenge in the field of Asset Management is to enhance the identification and quantification of cost and value to evaluate the total cost and value of assets throughout their lifecycle. In particular, a value model, to quantify the value delivered by the assets, is required specially to focus on how any decision outcome affect the value provided by the assets throughout the lifecycle. The value model is then concerned with the whole-life assessment of the value realization and, in some way, can be seen as an extended view enriching what already theorized when focusing the quantification of the total cost throughout the lifecycle.

Notably, a variety of terms has been commonly used in different industries, all of which traditionally aim to support decisions that consider the costs incurred across the complete life of an asset, instead of focusing simply on purchase prices or acquisition costs. Terms such as TCO (Total Cost of Ownership), LCC (Life Cycle Cost), WLC (Whole Life Cost), COO (Cost of Ownership) are a sample of a myriad of definitions and references that can be found in Internet. Their common objective is to promote thinking of the potential whole-life cost arising through the assets' use; this pushes to go beyond the only consideration of capital expenditures (CAPEX) with the ultimate goal to properly evaluate the operational expenditures (OPEX). Whole-life thinking is a general concept that can be transferred to the evaluation of value realisation throughout the assets' life. Therefore, TVO (Total Value of Ownership) and WLV (Whole Life Value) can be used to support decisions that consider the values delivered across the complete life of an asset. However, these terminologies are to be intended more as forthcoming rather than actual practice. Notably, practical understanding and use is in general far from being consolidated in industrial practice: on the one hand, life cycle costing and all the related terminologies were very much talked about, even if there are a number of difficulties that limit a widespread adoption by industry; on the other hand, this becomes even more challenging when extending to value, thus, to the whole life value, that is a more recent concept.

This evidence does not mean that the terminologies do not build on promising theories to exploit. Indeed, in a Value-based Asset Management, Whole-Life Value is an essential concept. It can be defined as extension of the definition of the value concept already discussed in previous Section. Building on the literature and our interpretative summary, we consider that WLV consists of the "benefits and opportunities, costs and risks associated with an asset over its whole-life, set by taking into account the interests of all stakeholders affected by its design, manufacture, existence and usage, and assessed considering economic, social and environmental impacts". This working definition is introduced with the following aims:

- To frame the search for the best trade-offs induced by the conflicting value drivers and metrics due to the key stakeholders;
- To relate the search to all the decisions that can be taken along the whole life of the asset;
- To remark the need to assess the impacts of Asset Management in the wide scope of sustainability, thus leading to the triple bottom line assessment of economic, social and environmental impacts.

All in all, the definition sets the context where conflicts of short-term and long-term interests and objectives rise to the attention of the decision makers. In order to better seize the whole-life as a framing concept, it is worth defining it in more detail. Two related terms are illustrated to this end, that is the lifecycle and the life of an asset.

What is the lifecycle of an asset? As noted by [2], "*an asset management system provides a structured approach for the development, coordination and control of activities undertaken on assets by the organization over different life cycle stages*". Therefore, the lifecycle is a founding term of an asset management system.

Notably, the whole-life or lifecycle of an asset consists of a number of stages. However, naming and number of the stages usually vary: different industries or companies may determine different naming according to different conventions and even norms.

Figure 2 provides a lifecycle model adapted from literature (adapted from Badurdeen et al. [20]). Therein, three stages are identified—Beginning of Life, Middle of Life and End of Life; lifecycle activities and related decisions undertaken on assets are then allocated at each stage.

Figure 3 illustrates the impact due to a range of activities that form the asset lifecycle, such as design, purchase and construction, commissioning, operation and

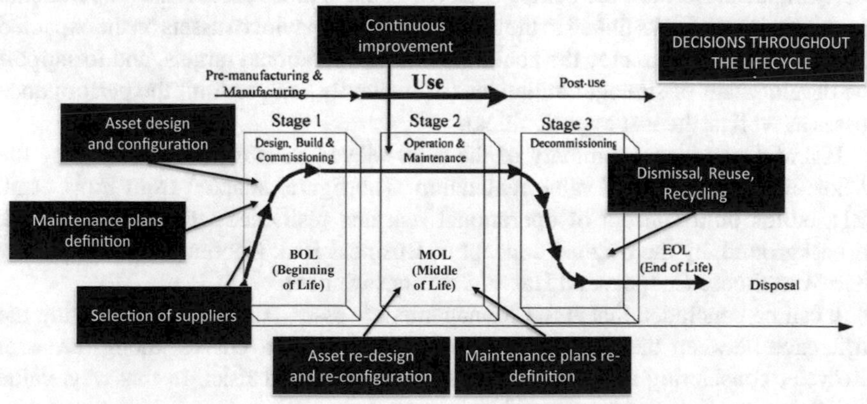

Fig. 2 A three-staged lifecycle model (adapted from Badurdeen et al. [20])

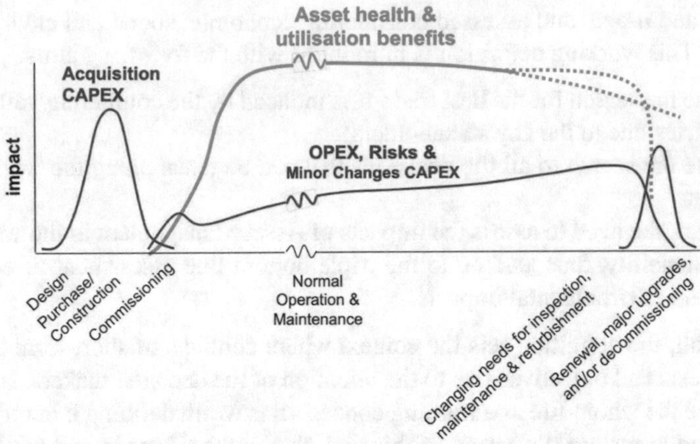

Fig. 3 Lifecycle activities and impacts (Courtesy: The Woodhouse Partnership)

maintenance, refurbishments, renewals, major upgrades as well as decommissioning. Impacts correspond to the value realization throughout the lifecycle of the asset. In fact, costs, risks, benefits/opportunities are evaluated in relationships to the lifecycle activities. As seen in the figure, they are quite varied. Benefits/opportunities are received throughout the operational life of the asset, whereas expenditures are incurred throughout the whole lifecycle, from CAPEX to OPEX.

Risks are also considered in the operational life of the assets. They may include not only hazards leading to safety-related risks, but also uncertainties of assets' behavior as well as opportunities leading to risks of losses due to a lack or partial delivery of the intended assets' functionality. It makes up the risk orientation of Asset Management: when making decisions related to assets, every company should apply all the measures required in the legislative/normative framework of the specific sector to reach the level of compliance for the management of critical risks related to safety impacts. Besides, the company has to manage those asset-related operational risks—such as the risks linked to the effect of future behavior of assets on the expected performance—to guarantee the achievements of operational targets, and to support the development of strategic initiatives (equivalently said, to limit the performance losses as well as the lost opportunities).

Figure 4 provides a summary of the asset-related risks to be considered by the Whole-life assessment of value realization. The figure, adapted from Frost et al. [21], builds on a concept of operational risk and resilience which is influenced, as background, by the original concept of Business Risk Continuum developed by PriceWaterhouse Coopers and Harvard Business School.

It can be concluded that Asset Management is essentially about maximising the difference between the areas under the cost and benefit 'curves' along the asset lifecycle, considering also the need to limit asset-related risks. In this way, value realization can be properly planned and controlled.

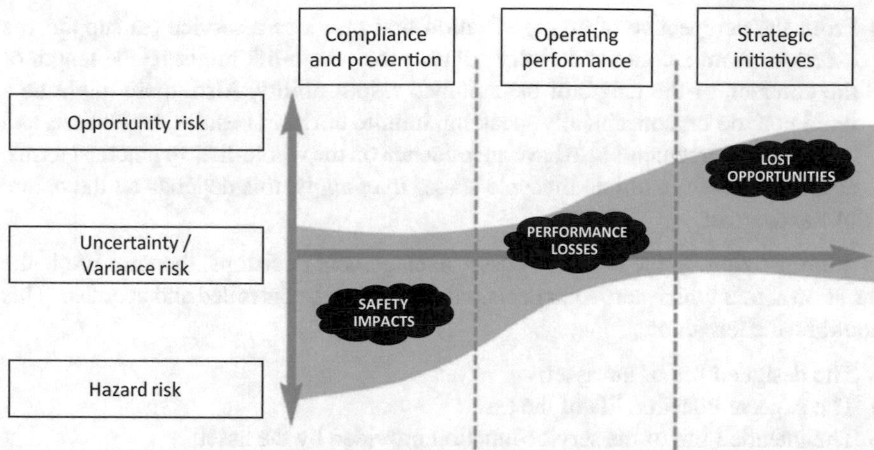

Fig. 4 Different perspectives on asset-related risks (adapted from Frost et al. [21])

What do we mean by life of an asset? The most commonly understood or used definition of life is the designed life of the asset, most often as defined by the manufacturer or designer of the asset. However, most often it is found that assets are used for a duration that is quite different from the designed life. Many assets are kept in service for far longer than their designed life, by means of life extension activities (such as, e.g., refurbishments, renewals and major upgrades), especially if replacing these assets involve heavy capital investment. Instead, some assets last for far less than their designed life as well, due to changes in functional requirements, or through obsolescence when OEMs exploit technological development to design and offer assets featuring higher cost-effectiveness with respect to the assets installed in customers' premises, or when OEMs forces obsolescence by stopping to make spare parts.

In order to manage the different meanings of asset life, it is essential to have a realistic understanding of the expected lifetime of the asset when calculating the whole-life cost or value. Two scenarios can be considered.

- From the perspective of an organization that uses the asset, and based on the fact that the asset exists to deliver some particular business function or service, the whole-life can consider the lifetime of the functionality or service provided, which in some cases might be, conceptually speaking, infinite. In practical terms, this is equivalent to state that, in such circumstances, the whole-life may include multiple lifecycles consisting of different activities, such as design, build, procure, operation and maintenance, and disposal activities. Additionally, when considering asset systems or networks, or making asset management decisions from a system/network perspective, decision-makers inherently focus on the functionality or service requirement, and are therefore dependent on long-term multiple lifecycle stages.

- From the perspective of an organization that provides a service (to support the asset), or from a contract delivery context, the whole-life might be the length of the contract, or the length of the assumed responsibility. Also, these might be a fixed horizon or, conceptually speaking, infinite horizon; besides, regulations and contractual agreements also have an influence on the whole-life. In practical terms, also in this case, multiple lifecycle stages may apply, this depends on the nature of the contract.

Summarizing, in the context of asset management decisions, the term life is the duration across which particular decisions are planned, controlled and operated. This could be the length of:

- The designed life of the asset;
- The expected/desired life of the asset;
- The intended life of the service/function provided by the asset;
- The designed/expected/desired life of the longest lasting component of a complex asset;
- The planning horizon;
- The regulatory cycle;
- The contract duration.

It is important that organisations are consistent in their use of the terminology, so that decisions made across their assets are aligned as far as possible. It is then essential to understand the term whole-life considering its dependence on the nature of the decision, the business driver of the organisation, regulatory or service contract requirements, and the type of the asset.

6 Intelligent Asset Management

The value model sits within the asset management system, endowed with all the characteristics discussed in previous sections in order to respect the key requirements on its role to support decision-making—i.e., being driven by stakeholders' requirements and value metrics, primarily focusing on the value realization at system or network level, and concerned with a whole-life assessment of the value realization. It is now worth to focus the attention on the Asset Management system itself and, in particular, its capabilities owed to the resources that constitute it.

Asset Management resources typically include:

- Organization, overall business process and related asset-control activities;
- Skills and competences;
- Information, as well as technologies and engineering methodologies to manage and process it.

They are intended as a set of "tools" to enable the effective and efficient development, coordination and control of AM plans and assets' lifecycle activities, and their continual improvement.

Indeed, AM "tools" are the means that support the execution of the value model and, more in general, the actuation of the AM decision-making process. Notably, proper "tools" should be used when planning in advance, and when monitoring and controlling the effective outcomes, to eventually activate re-planning in case of extant discrepancies with respect to expectations, thus leading to a continuous improvement of what is decided over the asset life cycle. It enables to assure cost and value along the asset lifecycle; indeed, it is remarkable that such "tools" are required to assure that the value delivery from assets (at reasonable cost) is effectively achieved and, when not, that proper decisions are activated to guarantee the value delivery.

The "tools" are all essential to move towards an Intelligent Asset Management, devised as part of the vision towards which the present book is addressed. In particular, among the new trends, it is clearly evident that digitalisation is enabling new capabilities for Asset Management, by means of the appearance of Cyber Physical Systems (CPS), and the subsequent issues resulting from building 'digital twins' of physical assets. This may lead to a new era of intelligent asset management systems, and is directly related to building new capabilities required for data/information management and processing. Nevertheless, at the same time, organization, business process and related activities, as well as skills and competences, are also fundamental to guide the development towards value-based asset management; besides, they will be transforming along the evolutionary path towards the achievement of a more digitized and intelligent management. Therefore, the next sub-sections are developed to illustrate the capabilities of a value-based and intelligent asset management system, with the aim to frame the "tools" recommended to support the asset-related decision-making process over the asset lifecycle.

6.1 Organization, Business Process and Asset-Control Activities

The AM resources are firstly presented in separated sub-sections before the concluding remarks on their future developments.

<u>*Organization*</u>

The application of Value-based Asset Management in a company calls for an organizational architecture and culture that promote the concept of whole-life, value-based and system/network-wide AM, and its reception as an effective business process [3]. To this end, different organizational functions have to be involved in the AM process, with their specific role in the organization's structure. Notably, in order to integrate AM in a company as a single process, two fundamental requirements should be complied:

- Integration and coordination across different organizational functions taking part in the AM value chain;
- Clear definition of AM related roles, responsibilities and authorities.

The first thing to be considered is the interdisciplinary and collaborative nature of the AM system that can be gleaned from the definition of asset life cycle from a user viewpoint itself [5]. At each life cycle stage of an asset, different disciplines and hence different organizational functions are needed in order to support decision-making. Indeed, the success of a capital-intensive organisation often depends on its ability to develop, coordinate and control the asset-related activities efficiently and effectively, by improving the integration across different disciplines and the cross-functional coordination [2, 5].

Secondly, as also stated in [2], leadership and workplace culture are "determinants of realization of value". This includes clearly defined roles, responsibilities and authorities. In particular, it is advocated that extant organization roles, involved in the AM value chain, are modified in order to include AM responsibilities and accountabilities. A subsequent development for this is that "conventional process control rooms may need to be converted into ownership, management, and utilisation centres" [22]. In fact, keeping an integrated view of AM, many decisions at the strategic level depends on how the asset-related activities (and decisions) are planned, controlled and operated through effective ownership, management and utilisation processes, which assures creating, or at least, sustaining the value delivery as defined in the expectations of the stakeholders. Under this perspective, the whole organization contributes to leadership and workplace culture, covering different responsibilities. As noted by [2], "top management is responsible for developing the asset management policy and asset management objectives and for aligning them with the organizational objectives. Leaders at all levels are involved in the planning, implementation and operation of the asset management system".

Aligned with the two above mentioned requirements, research has found evidences for a successful AM.

- First of all, there is a need for an integrated approach to governance, and a company's board plays an important role on asset management effectiveness since the board members are often responsible and accountable for the tussles between CAPEX and OPEX management.
- Secondly, senior management can also play a crucial role in influencing the employee attitude and mental preferences on asset management practices: as employee involvement and commitment to purpose is essential for asset management, an effective leadership helps towards changing the culture and, then, enhancing the employee commitment.
- Last but not least, it is evident that an effective leadership and top-level support is essential in order to deal with different matters required by decision-making, in particular: the alignment between stakeholder objectives, current AM plans and interventions; the formulation of asset management objectives; the management of multiple actors with different interests and objectives. It is also relevant for successful implementation of asset management practices. Finally, a visionary and innovative leadership is closely tied in with the development of proper business models that distribute the costs, risks and benefits/opportunities of asset management amongst the different stakeholders.

Value-based Asset Management is deeply rooted into these evidences; indeed, these could be seen as a kind of organisational prerequisite to future developments.

Business process and asset-control activities

Complementary to the organisational part of AM, there is a strong need to embed the AM process to properly run it, in a systematic way, within a company. Roda and Macchi [14] provide advice on how to embed the AM process in production companies (where AM is still scarcely adopted, as a business process).

In particular, nowadays, it is evident that a general consensus has been reached about AM as an essential process contributing to an organization's objectives (also thanks to the publication of the ISO 5500X standards on AM systems). Nonetheless, there is still a need of clear indications on how to tailor the AM process in the specific context of a company in a given sector. Indeed, managers aim at understanding how to implement AM in their own reality. To this end, the framework, shown in Fig. 5, was defined based on literature analysis and focus groups findings arising from production companies; therein, the fundamentals to guide the integration of AM are systematized. Two dimensions are identified—the asset life cycle and the hierarchical level of the asset-control activities; and four founding principles—life cycle, system, risk and asset-centric orientation—as levers to integrate asset management within an industrial organization. In particular, the following framework's issues can be pointed out:

- Activities within the AM business process (also referred to as asset-control activities, in the reminder) should be established at different asset-control levels, i.e. From strategic to operational level; then, the company has to be able to position any asset-related decision in terms of control level (strategic, tactical or operational), by simultaneously checking the alignment with the other levels;
- Asset-control activities within the AM business process should be related to the asset lifecycle stages, wherein asset-related decision-making is required;
- The founding AM principles should be the primary basis to establish the AM as whole business process; indeed, the asset-control activities should entail such principles for successful AM implementation.

It is worth providing a short description of the principles:

- Lifecycle orientation leads to incorporate in the AM process long-term objectives and performances, to drive decision-making and, thus, the management of conflicts of short term and long-term interests and objectives;
- System orientation makes criticality of the assets at the system/network level an essential aspect, in order to ensure focusing efforts and resources on the right activities;
- Risk orientation influences the establishment of a risk culture, empowering the leaders and employees with the information to be predictive decision-makers, which is considered to be instrumental to achieve "Best-in-Class performance";
- Asset-centric orientation remarks the essentiality to have a common asset database providing basic reference to information regarding assets' properties, and usable for strategic, tactical and operational decisions.

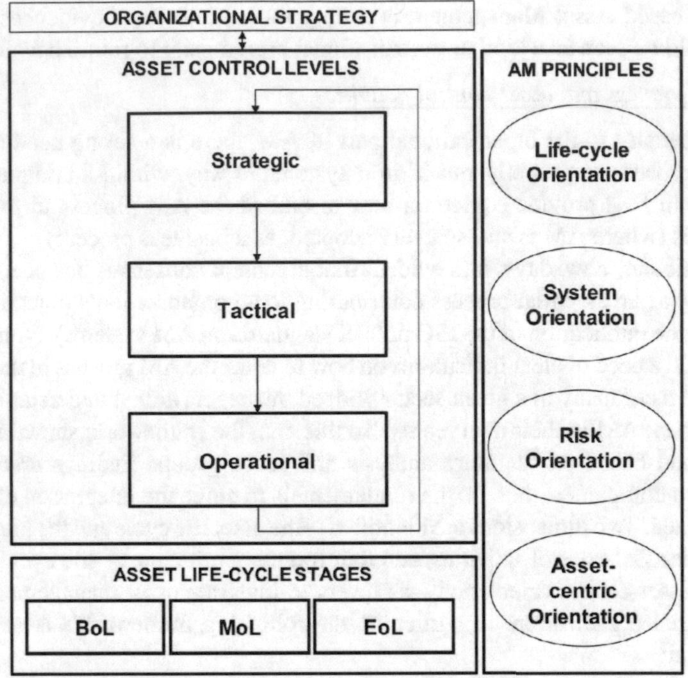

Fig. 5 Framework for embedding AM in production companies [14]

When the decision-making is provided with proper value models, the principles of Lifecycle, System and Risk orientation are inherently included in the AM process. Instead, asset-centric orientation refers to the key role of information and the related technologies and engineering methodologies to manage and process it, with the purpose to support the execution of the decision-making process. This framework is further explained in Chap. 5.

Future developments

Looking at the future developments, we may expect a change in the actual establishment of the AM business process and asset-control activities, as the process will cross different organizations while the activities will be correspondingly spread along the AM value chain, also enhancing the role of providers—both OEMs and service providers. To this regard, there is a need to explore the increasing popularity of product-service systems (PSSs), and how the asset management frameworks are needed in order to support them.

Extending their traditional business into the service domain and offering bundles of products and services (PS) was in fact a natural response for many firms [23, 24] to secure additional sources of revenue and profits [25–27], and to answer the pressures from customers and environmentalists on sustainability [28]. This phenomenon, seen in manufacturing, was called servitization [29], and represents business models that have evolved from a "pure product" orientation toward an integrated Product-Service System (PSS). PSSs are where both products and services conform a single offer and the business model is no longer "selling only product" but selling an "integration of product and services". The concept of PSSs brings along an interesting debate about the issue of asset ownership. In this new business model, the ownership of the product and its utilization processes are decoupled, thus customers (a more suitable term for the users of these products) will not typically buy the product, but its availability or capability.

Therefore, PSSs are a hot-topic amongst increasing number of industries where the manufacturer retains the ownership (hence the responsibility for management) of the asset, and the customer simply pays for the service/business function/output of the asset. A classic example of this is Rolls Royce "power by the hour" model. Variants of this are found in the public sector as well with Public Private Partnership (PPP) and Private Finance Initiative (PFI) models with varying degrees of success. In this way such business models can be made more effective, and asset management practices and systems are suitable for such approaches.

It is notable that PSSs are well aligned with the perspective promoted by Value-based Asset Management: amongst the dominant elements of PS business models agreed by many authors [see e.g. 30, 31, 32, 33], the Value Proposition, also referred to as PS offering, concerning the bundle of products and services offered, represent the benefit/opportunity for which the customer is willing to pay. It is, therefore, clear that PSSs will be able to affect the organizational part of Value-based Asset Management; in particular, we may expect that the AM business process and the related asset-control activities will be reshaped by such servitization trend, leading to a distribution of the activities across the AM value chain.

On the whole, there is a general feature that currently emerges: implementing a good asset management system is a major organisational and cultural shift—requiring a change in processes, methods, techniques, and, in some cases, even organisational structures that go beyond the single company. In order to support the change, breaking down of organisational silos that have been created over long periods of time requires top management support and motivation, as well as leaders as "change agents" in different organizational levels.

6.2 Skills and Competences

Competences are amongst the resources required to enable the effective and efficient development and delivery of AM plans and assets' lifecycle activities, and their continual improvement [2]. In fact, organisations seeking to improve their Asset Management systems and performance need to ensure they have suitably competent people in place. Therefore, it is important for a company to put effort to this end. This is required to enable the effective transition to Value-based Asset Management.

Regarding required AM competencies, given by the ability to apply knowledge and skills to achieve intended results, details can be found, as a meaningful example,

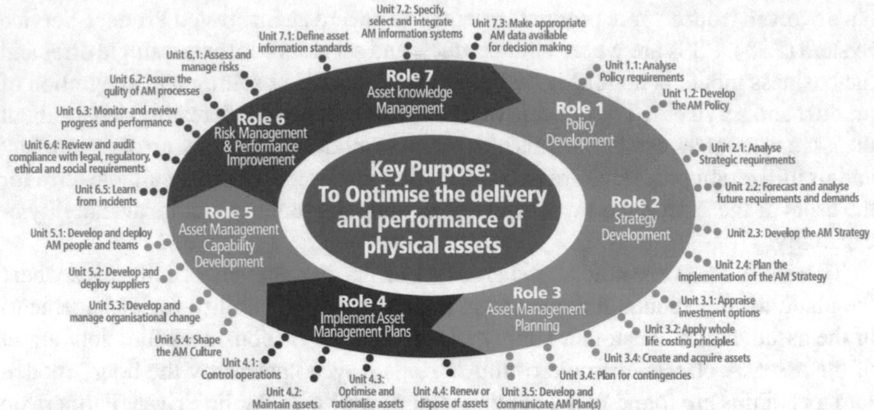

Fig. 6 Asset management competences and skills (Competence Framework, [34])

in the IAM report (Competence Framework, [34]). Figure 6 provides a summary of the 27 Units of Competence in the Framework and their distribution across 7 Key Roles within Asset Management.

Having maps similar to the one proposed by the IAM report is important, as such maps play the role of a reference model to cover the wide scope of abilities to be developed according to what is needed by the organisational part of AM and, in particular, by the needs of the AM process to be embedded within a company.

Future developments

Nowadays, and in the future, people are a key asset of an Asset management system, thus developing competence and skills is essential. Specifically, appointing competent people in Key Roles within the Asset Management system, helps supporting the proper leadership as well as the capability to master the asset-control activities within the AM business process, the asset-related decision-making, and the promotion of the founding AM principles.

Competence and skills are closely linked with the need of tailoring the AM process in the specific context of a company in a given sector: even if a map of competence and skills is relevant, it requires adaptation to fit the specific context where the AM business process, and the different roles involved in the process, are implemented. It is a relevant interest for future works, as competence and skills may be actually deployed in different ways according to the characteristics of the context.

Besides, the current development due to digitization is causing a rethink to balance the tasks required by humans and machines. In particular, different scenarios are expected to occur: besides the effects of automation, as similarly experienced in the past (i.e. with the replacement of humans with machines in order to perform specific tasks), it is worth remarking the possibility to enable an assisted proceduralization of tasks assigned to humans, as well as an augmented intelligence to support their decisions. This is a trend that is not yet consolidated, thus, the overall expectation is that the future will reshape the maps of competence and skills in a digitized environment.

6.3 *Information, Technologies and Engineering Methodologies*

The AM resources are firstly presented in separated sub-sections before the concluding remarks on their future developments.

Information

Data is a critical asset in today's organizations [35]. The availability of valuable (and big) data is paramount to making the best decision in AM. Notably, with the complexity of AM (resulting from management needs due to asset portfolios, fleets, asset systems or networks), the exploitation of interconnected data and data-driven decision-making approaches [36] to generate useful information, as well as the information management and integration within suited architectures (also capable to integrate and exploit the big data available from assets) [37], are still challenges, determining some limits to a proper decision support: bringing all of the disparate data together into a widespread asset-centric information is yet a key problem nowadays faced by asset owners, users and providers, as well as impacting the communication to different stakeholders. Nevertheless, an integrated asset information management strategy is required [38] in order to effectively habilitate the AM decision-making. Indeed, asset intensive organisations rely on asset data, information and even asset knowledge as key enablers in undertaking both strategic asset management activities and operational activities.

When considering which data are needed for supporting informed AM decision-making, it must be taken into account that the data that are required can vary from case to case depending on the definition given to value. What is generally recognized is that both technical and financial data are needed. The major challenge in obtaining the required data is that they are heterogeneous, hence they are usually scattered among separate information systems (such as administrative IT systems, industrial IT systems, and non-automatic sources) [39, 40]. May it be at IT based or not, it is widely agreed that integration must facilitate the bi-directional flow of data and information into the decision-making at all levels [39] of asset-control. The core idea is that, a common asset database should be developed where for each asset, at different aggregation level, different data can be stored all together. The asset database would provide basic reference to information regarding assets' properties for supporting decisions [40].

Due to the widespread use of different information systems and tools, interoperability has been claimed in order to manage data, information and knowledge along the asset lifecycle. The publication of standards such as, e.g., ISO 10303 [41] and ISO 15926 [42] provided the momentum for a flurry of academic activity in this area.

In particular, research on asset information management took off in the scope of the product lifecycle management (PLM) community in the early 2000s, developing concepts as product data and knowledge management (PDKM) [43]; research continued, leading to remark nowadays the importance of ontologies in manufacturing systems, for different reasons, such as interoperability, knowledge sharing and reuse, as well as support reasoning [44]. Recently, with the publication of PAS 1192 [45] suite of guidance documents, researchers have started showing interest in the development of asset information models and BIM (Building Information Modelling); thus, in addition to the complexities involved in the effective organisation and transfer of data to clients at the end of construction projects, challenges associated with supporting whole-life asset management have also been an emerging focus of interest. Indeed, the requirement of long-term availability of asset information for large-scale infrastructure e.g. during operation and maintenance phases or beyond, is still a challenge. The importance of long-term availability can be understood by the consequences of its lack: non-availability of critical information may cause severe impacts on decisions. In some cases, non-availability may result in closure of business, or delays, or disruptions in infrastructure operations. Even if this is partly due to the nature of multi-organisational management of infrastructure assets, it can be considered as a generalized problem; in fact, industrial plants also require such an asset information model, to guarantee the long-term availability throughout the factory and assets' life cycle. Identifying the risks associated with the information loss and deterioration is critical for sustainable whole-life asset management [46].

Besides the long-term availability, the quality of decisions made is often constrained by the quality of data available to the decision-maker. Thus, data quality is another interesting and recent topic for research. Indeed, data and information quality has been a research area in its own right, but this has also been a specific focus within the asset management community. The first step towards high quality data for any organisation is the Data Quality (DQ) assessment [47]. Further on, in addition to exploring and classifying the critical data quality problems faced by asset management practitioners, research has also focused on developing frameworks for assessing and improving data quality.

Technologies

A key development to improve data quality and long-term availability is the emergence of digital technologies and their adoption in asset management. In this regard, Internet of Things (IoT) and Cyber-Physical System (CPS) are currently leading concepts of the transformative process pushed by a blend of mature technologies.

- Most commonly talked about is the concept of IoT, which aims to integrate the physical world with the digital world. It envisions a variety of things around us, connected, interacting and cooperating with each other to reach common goals. Such things allow to collect and exchange data by means of Internet [48, 49] and through networks of devices [50, 51]. The integration of several enabling technologies—such as identification, sensing and communication technologies—allows the implementation of the IoT concept [50, 51].

- A CPS—an extension of the IoT concept—is a mechanism that is controlled or monitored by computer-based algorithms, tightly integrated with the Internet and its users [52]. In CPSs, physical and software components are deeply intertwined, each operating on different spatial and temporal scales, exhibiting multiple and distinct behavioral modalities, and interacting with each other in a myriad of ways that change with context. As such, CPSs are "systems of collaborating computational entities which are in intensive connection with the surrounding physical world and its on-going processes, providing and using, at the same time, data-accessing and data-processing services available on the Internet" [52].

This results from a number of organised research activities spun out in many countries around the world, the most popular one being Industrie 4.0 (aka worldwide as Industry 4.0), which is a national strategic initiative from the German government through the Ministry of Education and Research and the Ministry for Economic Affairs and Energy. It aims to drive digital manufacturing forward by increasing digitisation and the interconnection of products, value chains and business models. Similar initiatives at the national level exists in many other countries under different names, however Industry 4.0 is perhaps the most popular initiative and thus has become a de facto terminology used for digital manufacturing.

This evolution will also affect the asset management, expected to rely in the future on cyber-physical systems as intelligent resources; in this regard, we may even name them also as cyber-physical assets. The evolution will enable managing—within the Cyber space built on top of the IoT infrastructure and other technologies, such as cloud computing—the Big Data resulting from the interconnectivity of machines and sensors, in order to finally reach the goal of intelligent data management and computation [53]. Big data analytics, that access the data and promise to provide fast decision-making with the use of smart analytics tools, combining richer real-time analytics, fusion and interpretation, to build greater intelligence [54, 55], are in fact going to complement the IoT with further features useful for decision-making support. Advanced simulation will also play a key role, amongst the tools relevant for prediction of the stochastic processes that occur in the physical world [56].

In such a technology landscape, an important concept is often remarked, that is the Digital Twin (DT), meant as the digital counterpart of the physical system. In particular, the DT is considered as a virtual entity, relying on the sensed and transmitted data of the IoT infrastructure, as well as on the capability to elaborate data through data analytics and simulation technologies, with the purpose to allow optimizations and informed decision-making. The potentials for decision support along the asset lifecycle offered by DT modeling are highly promising [14], leading to think of future applications built on the DTs of Physical Assets.

The following list aims at showing the potential of the CPS-based transformation, inclusive also of DT modeling. The list does not pretend to be complete; nevertheless, it could not be complete, as the use of CPSs and DTs for asset management is still at its initial development stage, and cannot actually cover all the asset-control levels, lifecycle stages, and related activities and decisions:

- Davies et al. [57] present the role of the DT in a smart factory, with the purpose to allow intelligent algorithms processing the data generated by the physical system, to yield information on its performance, condition and health in real time.
- Penas et al. [58] consider the need for methods and tools to create a smart CPS aimed at improving effectiveness, safety, productivity and reliability of production systems.
- Lee et al. [55] adopt smart analytic tools with the purpose to improve prognostics and health management of production equipment and machineries.
- Monostori et al. [52] remark that an appropriate prognosis, in a digitized manufacturing, can improve accuracy and reliability in predicting resource needs and allocation, maintenance scheduling and remaining service life of equipment.
- Liu and Xu [59] highlight that big data and associated analytics play a significant role in optimizing production quality and improving equipment service; they also exploit the virtual world, by means of models of the production resources for which simulation plays a role, to support real-time decision-making.

Engineering methodologies

Companies also need to use specific engineering methodologies, integrated to provide a decision support. In fact, such methodologies are rooted in the AM application domain and they should serve with the purpose to set the value model and to elaborate the required information for asset-related decision-making. Different methodologies are needed, depending on the specific case and value model required by the case.

The methodologies are an additional resource, besides the technologies offered in the Industry 4.0 umbrella, but complementary. Cyber-physical systems/assets will in fact enable to empower the adoption of such methodologies in a digitized environment, through the enhanced capabilities owed to the data quality and long-term availability of asset information. This will also allow to improve the extant methodological theories, primarily resulting from reliability modeling and condition-based maintenance modeling communities.

In particular, the research done on reliability and condition-based maintenance modelling underpins much of our understanding of how assets and systems/networks behave, how to model their failures, and how to quantify the deterioration of their condition and performance throughout their life. One of the most important contributions was made in 1972 by Sir David Cox with the Proportional Hazards Model, which allowed us to analyse equipment survival probability in terms of several factors simultaneously [60]. This led to advancements in the field of condition monitoring and condition-based maintenance. More recently, the development observed under the term of Prognostics and Health Management (PHM) is remarkable: a body of knowledge is included that nowadays is considered as an engineering discipline [61, 62]. PHM aims to assess the reliability of a product/asset in its actual life cycle conditions to determine the advent of failures, and mitigate system risks [63], which creates opportunities for a subsequent use of the resulting information (in maintenance, logistics, life cycle control, asset management, etc.). Indeed, PHM is seen as a cornerstone of advanced maintenance systems, within new digitized environments, while being connected with reliability modeling methodologies such as Reliability Centered Maintenance (RCM) [64].

This, coupled with the renewal theory, underpins most of the mathematical models developed for optimising maintenance and replacement decisions [65]. Today, modelling techniques range from simple approaches such as Reliability Block Diagrams to more complex models using Markov Chains and Petri Nets [66]. Besides, maintenance, inspection, and renewal decision models are typically driven by the development of cost functions based on reliability models and renewal theory. Also, more complex models involve Continuous Time Markov Chains and Markov Decision Processes (with complete or partial observability) [67]. Eventually, it is worth observing that, over the last couple of decades, increasing attention is being given to multi-component systems where the dependency between the constituent components (of the system) are explicitly modelled. Optimisation techniques used for these complex decision models lean towards evolutionary computing techniques such as Genetic Algorithms and Simulated Annealing because such problems are very hard to solve analytically [9].

Future developments

Overall, information, as well as technologies and engineering methodologies to manage and process it, are key resources to support the asset-related decision-making process over the asset lifecycle. Moreover, they directly lead towards building the foundation of an Intelligent Asset Management system. Nevertheless, this scenario is not actually achieved, but a future vision. A lot of challenges are open; the major ones are herein summarized as a result of previous discussions.

- Information management and integration are still requiring a major research effort in order to enable interoperability amongst heterogeneous information systems and tools and, subsequently, an effective decision support. Herein, the development of asset information models is extremely relevant, with the aim to assure the long-term availability of information across the assets' lifecycle.
- Cyber-physical assets and Digital Twins of Physical Assets are still in the early stage development; more experiences and use cases will be required, aimed to support the whole spectrum of relevant decisions in the asset-control activities.
- Traditional reliability and condition-based maintenance modeling methodologies should take advantage of the new capabilities made available by the Cyber-physical assets and Digital Twins, to go beyond the current potentials for decision support.

The evolutionary path is any way set towards an asset-centric orientation, where data are transformed and managed to properly provide information and knowledge regarding assets' properties, usable for strategic, tactical and operational decisions. This is the expected trend for the coming years.

7 Framework

This final section of the chapter firstly aims to present a generalized framework providing a perspective of the relevant dimensions that we suggest for developing a Value-based and Intelligent Asset Management. It is the main outcome from capturing the learnings we achieved during the SustainOwner Project.

The framework dimensions are a summary of what previously discussed along the chapter.

- The value model sitting within the Asset Management system with its role to support decision-making is the first dimension. The model is endowed with three characteristics to comply with the role—i.e., being driven by stakeholders' requirements and value metrics, primarily focusing on the value realization at asset system or network level, and being concerned with a whole-life assessment of the value realization.
- Other dimensions result from the capabilities of the Asset Management System owed to the resources constituting it. These AM resources act as "tools" that support the execution of the value model and, in general, the actuation of the AM decision-making process. They are: the organization, business process and asset-control activities, the skills and competences, and the information, as well as technologies and engineering methodologies to manage and process it.

The AM "tools" should be used when planning in advance, and when monitoring and controlling the effective outcomes, to eventually activate re-planning in case of extant discrepancies with respect to expectations, thus leading to a continuous improvement of what is decided over the asset life cycle. It enables to assure cost and value along the asset lifecycle. Whereas, the value model is at the basis of the assurance: it is in fact the core element of the Value-based decision-making approach embedded within the AM system, aimed to identify and quantify how value is affected by the decisions concerned within Asset Management; as such, it enables to control the decision-making process to finally deliver value from assets.

The framework dimensions are summarized in the following Fig. 7, according to IDEF0 formalism [67].

The framework is proposed as a conceptual structure aimed at outlining the primary dimensions (or factors) to leverage in order to master the transformation towards Value-based and Intelligent Asset Management. Through it, the book aims to foster a future perspective, making a step forward with respect to the current understanding of Asset Management primarily geared with well-known standards.

8 Conclusions

In this Chapter the notion "framework" was used to refer to a conceptual structure to transmit or address complex issues about some area of knowledge, asset management in our case, through a generic outline or approach. The framework also includes a

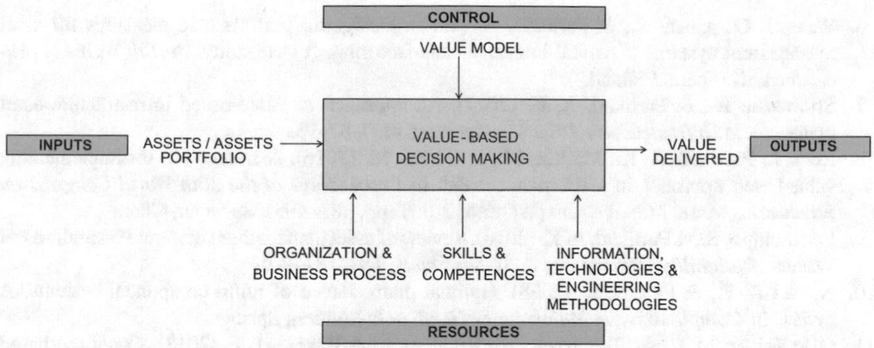

Fig. 7 Framework for value-based and intelligent asset management

so called “value model”. The term “model” refers to an abstract representation of entities and their relationships involved in the scope of different asset management problems, that can be implemented, for instance, by means of a computer language or any engineering formalism. In this sense, both the value model and the framework serve as templates for the development and comparison of specific business and operational scenarios within the asset management scope. Overall, the framework proposed consists of processes, as well as control factors—as the value model, and resources—as ICT tools, engineering methods and techniques, all required for supporting industrial asset and maintenance management continuous improvement toward excellent asset’s contribution in value delivered.

In the next Chapters (that is, from Chaps. 2 to 16) different issues within the asset management domain are addressed, taken at different hierarchical level of the asset-control activities, while considering different problem settings; in the last Chapter (Chap. 17) the framework is used to capture the dimensions of the asset management problems under a similar structure.

References

1. British Standards Institute, PAS55:2008-1:2008. Specification for the optimized management of physical assets, 2008.
2. International Standards Organisation, ISO 55000:2014(E) Asset management—Overview, principles and terminology, 2014.
3. Liyanage, J. P., & Kumar, U. (2003). Towards a value-based view on operations and maintenance performance management. *Journal of Quality in Maintenance Engineering, 9*(4), 333–350.
4. Amadi-Echendu, J., Willett, R., & Brown, K. (2010). What is engineering asset management? *Definitions, concepts and scope of engineering asset management* (pp. 3–16). London: Springer.
5. El-Akruti, K., Dwight, R., & Zhang, T. (2013). The strategic role of engineering asset management. *International Journal of Production Economics, 146*(1), 227–239.

6. Wang, J. Q., Chen, Z., & Parlikad, A. (2015). Designing performance measures for asset management systems in capital-intensive manufacturing: A case study. In *10th WCEAM proceedings,* Tampere, Finland.
7. Srinivasan, R., & Parlikad, A. K. (2017). An approach to value-based infrastructure asset management. *Infrastructure Asset Management, 4*(3), 87–95.
8. Roda, I., Parlikad, A. K., Macchi, M., & Garetti, M. (2016). A framework for implementing value-based approach in asset management. In *Proceedings of the 10th World Congress on Engineering Asset Management (WCEAM 2015)* (pp. 487–495). Springer, Cham.
9. Petchrompo, S., & Parlikad, A. K. (2018). Review of asset management literature on multi-asset systems. *Reliability Engineering & System Safety, 181,* 181–201.
10. Nicolai R. P., & Dekker R. (2008). Optimal maintenance of multi-component systems: A review. In *Complex System Maintenance Handbook*, London, Springer
11. Olde Keizer, M. C. A., Teunter, R. H., Veldman, J., & Babai, M. Z. (2018). Condition-based maintenance for systems with economic dependence and load sharing. *International Journal of Production Economics, 195*, 319–327. https://doi.org/10.1016/j.ijpe.2017.10.030
12. Rasmekomen, N., & Parlikad, A. K. (2013). Maintenance optimization for asset systems with dependent performance degradation. *IEEE Transactions on Reliability, 62*(2), 362–367.
13. Wang, R., & Chen, N. (2016, December). A survey of condition-based maintenance modeling of multi-component systems. In *2016 IEEE International Conference on Industrial Engineering and Engineering Management (IEEM)* (Vol. 2016, pp. 1664–1668). https://doi.org/10.1109/ieem.2016.7798160.
14. Roda, I., & Macchi, M. (2018). A framework to embed asset management in production companies. *Proceedings of the Institution of Mechanical Engineers, Part O: Journal of Risk and Reliability, 232*(4), 368–378.
15. Crespo Márquez, A., Moreu De Leõn, P., Sola Rosique, A., & Gõmez Fernández, J. F. (2016). Criticality analysis for maintenance purposes: A study for complex in-service engineering assets. *Quality and Reliability Engineering International, 32*, 519–533.
16. Adams, J., Srinivasan, R., Parlikad, A. K., González-Prida, V., & Crespo, A. M. (2016). Towards dynamic criticality-based maintenance strategy for industrial assets. *IFAC-Papers, 49,* 103–107.
17. Xu, Y., Elgh, F., & Erkoyuncu, J. (2013). Cost Engineering for manufacturing: Current and future research. *International Journal of Computer Integrated Manufacturing*, 37–41.
18. Liang, Z., Parlikad, A. K. (2015). A condition-based maintenance model for as-sets with accelerated deterioration due to fault propagation. *IEEE Transactions on Reliability*.
19. Roda, I., & Macchi, M. (2019). Factory-level performance evaluation of buffered multi-state production systems. *Journal of Manufacturing Systems, 50,* 226–235.
20. Badurdeen, F., Shuaib, M., & Liyanage, J. P. (2012). Risk modeling and analysis for sustainable asset management. In J. Mathew, L. Ma, A. Tan, M. Weijnen, & J. Lee (Eds.), *Engineering asset management and infrastructure sustainability* (pp. 61–75). London: Springer.
21. Frost, C., Allen, D., Porter, J., & Bloodworth, P. (2001). *Operational risk and resilience*. Butterworth Heinemann.
22. Amadi-Echendu, J. (2004). Managing physical assets is a paradigm shift from maintenance. In *Engineering Management Conference* (pp. 1156–1160).
23. Wise, R., & Baumgartner, P. (1999). Go downstream: The new profit imperative in manufacturing. *Harvard Business Review, 77*(5), 133–141.
24. Pawar, K. S., Beltagui, A., & Riedel, J. C. K. H. (2009). The PSO triangle: Designing product, service and organization to create value. *International Journal of Operations & Production Management, 29*(5), 468–493.
25. Mathieu, V. (2001). Service strategies within the manufacturing sector: Benefits, costs and partnership. *International Journal of Service Industry Management, 12*(5), 451–475.
26. Gebauer, H., & Fleisch, E. (2007). An investigation of the relationship between behavioural processes, motivation, investments in the service business and service revenue. *Industrial Marketing Management, 36*(3), 337–348.
27. Neely, A. (2009). Exploring the financial consequences of the servitization of manufacturing. *Operations Management Research, 1*(2), 103–118.

28. Mont, O. K. (2002). Clarifying the concept of product-service system. *Journal of Cleaner Production, 10*(3), 237–245.
29. Vandermerwe, S., & Rada, J. (1988). Servitization of business: Adding value by adding services. *European Management Journal, 6*(4), 314–324.
30. Kindström, D. (2010). Towards a service-based business model-key aspects for future competitive advantage. *European Management Journal, 28*(6), 479–490.
31. Meier, H., Roy, R., & Seliger, G. (2010). Industrial product-service systems. *IPS2, CIRP Annals—Manufacturing Technology, 59*(2), 607–627.
32. Schuh, G., Klotzbach, C., & Gaus, F. (2008). Service provision as a sub-model of modern business models. *Production Engineering: Research and Development, 2*(1), 79–84.
33. Gaiardelli, P., Resta, B., Martinez, V., Pinto, R., & Albores, P. (2014). A classification model for product-service offerings. *Journal of Cleaner Production, 66,* 507–519.
34. IAM. (2008). The IAM competences framework (November).
35. Borek, A., et al. (2014). A risk based model for quantifying the impact of information quality. *Computers in Industry, 65*(2), 354–366.
36. Petchrompo, S., & Parlikad, A. K. (2019). A review of asset management literature on multi-asset systems. *Reliability Engineering and System Safety, 181,* 181–201.
37. Campos, J., Sharma, P., Gabiria, U. G., Jantunen, E., & Baglee, D. (2017). A big data analytical architecture for the asset management. *Procedia CIRP, 64,* 369–374.
38. Ouertani, M., Parlikad, A., & McFarlane, D. (2008). Towards an approach to select an asset information management strategy. *International Journal of Computer Science and Applications, 5*(3), 25–44.
39. Moore, W. J., & Starr, A. G. (2006). An intelligent maintenance system for continuous cost-based prioritisation of maintenance activities. *Computers in Industry, 57*(6), 595–606.
40. Kans, M., & Ingwald, A. (2008). Common database for cost-effective improvement of maintenance performance. *International Journal of Production Economics, 113,* 734–747.
41. ISO 10303-1:1994. Automation systems and integration—Product data representation and exchange. Part 1: Overview and fundamental principles.
42. ISO 15926:2003. Industrial automation systems and integration—Integration of life-cycle data for process plants including oil and gas production facilities.
43. Kiritsis, D. (2011). Closed-loop PLM for intelligent products in the era of the Internet of things. *Computer-Aided Design, 43*(5), 479–501.
44. Negri, E., Fumagalli, L., Garetti, M., & Tanca, L. (2016). Requirements and languages for the semantic representation of manufacturing systems. *Computers in Industry, 81,* 55–66.
45. PAS 1192-3:2014. Specification for information management for the operational phase of assets using building information modelling (BIM).
46. Masood, T., Cuthbert, R., Parlikad, A. K., & McFarlane, D. C. (2013). Information futureproofing for large-scale infrastructure. In *Proceedings of 3rd IET Asset Management Conference* (6 pages), London, November 2013.
47. Woodall, P., Borek, A., & Parlikad, A. K. (2013). Data quality assessment: The hybrid approach. *Information & Management, 50*(7), 369–382.
48. Ashton, Kevin. (2009). That 'Internet of Things' thing. *RFiD Journal, 22*(7), 97–114.
49. Sarma, S., Brock, D. L., & Ashton, K. (2000). *The networked physical world.* Auto-ID Center White Paper MIT-AUTOID-WH-001.
50. Atzori, L., Iera, A., & Morabito, G. (2010). The Internet of Things : A survey. *Computer Networks, 54*(15), 2787–2805.
51. Iera, A., Floerkemeier, C., Mitsugi, J., & Morabito, G. (2010). The Internet of things [Guest Editorial]. *Wireless Communications, IEEE, 17*(6), 8–9.
52. Monostori, L., et al. (2016). Cyber-physical systems in manufacturing. *CIRP Annals—Manufacturing Technology, 65,* 621–641.
53. Lee, J., Bagheri, B., & Kao, H. A. (2015). A Cyber-Physical Systems architecture for Industry 4.0-based manufacturing systems. *Manufacturing Letters, 3,* 18–23.
54. Davis, J., Edgar, T., Porter, J., Bernaden, J., & Sarli, M. (2012). Smart manufacturing, manufacturing intelligence and demand-dynamic performance. *Computers & Chemical Engineering, 47,* 145–156.

55. Lee, J., Kao, H. A., & Yang, S. (2014). Service innovation and smart analytics for industry 4.0 and big data environment. *Procedia CIRP, 16*, 3–8.
56. Negri, E., Fumagalli, L., & Macchi, M. (2017). A review of the roles of Digital Twin in CPS-based production systems. *Procedia Manufacturing, 11,* 939–948.
57. Davies, R., Coole, T., & Smith, A. (2017). Review of socio-technical considerations to ensure successful implementation of industry 4.0. *Procedia Manufacturing, 11,* 1288–1295.
58. Penas, O., Plateaux, R., Patalano, S., & Hammadi, M. (2017). Multi-scale approach from mechatronic to Cyber-Physical Systems for the design of manufacturing systems. *Computers in Industry, 86,* 52–69.
59. Liu, Y., & Xu, X. (2017). Industry 4.0 and cloud manufacturing: A comparative analysis. *Journal of Manufacturing Science and Engineering, 139*(3).
60. Cox, D. R. (1972). Regression models and life-tables. *Journal of the Royal Statistical Society: Series B, 34*(2), 187–220.
61. Cheng, S., Azarian, M. H., & Pecht, M. G. (2010). Sensor systems for prognostics and health management. *Sensors, 10*(6), 5774–5797.
62. Pecht, M. G. (2008). *Prognostics and health management of electronics*. Wiley.
63. Haddad, G., Sandborn, P., & Pecht, M. G. (2012). An options approach for decision support of systems with prognostic capabilities. *IEEE Transactions on Reliablity, 61*(4), 872–883.
64. Guillén, A. J., Crespo, A., Macchi, M., & Gómez, J. (2016). On the role of prognostics and health management in advanced maintenance systems. *Production Planning and Control: The Management of Operations, 27*(12), 991–1004.
65. Dekker, R., & Scarf, P. A. (1998). On the impact of optimisation models in maintenance decision Making: The state of the art. *Reliability Engineering and System Safety, 60,* 111–119.
66. Li, Y., Zio, E., & Lin, Y. (2012). A multistate physics model of component degradation based on stochastic petri nets and simulation. *IEEE Transactions on Reliability, 61*(4), 921–931.
67. NIST. IDEF Ø: 1993. Integration Definition for Function Modeling. FIPS Publication 183.

Chapter 2
A Conceptual Framework for the Alignment of Infrastructure Assets to Citizen Requirements in Smart Cities

James Heaton and Ajith Kumar Parlikad

Abstract With the predicted world population growth of 83 million people per year (increasing 1.09% year on year) compounded with a strong trend for migration to urban centres, there is a developing interest by academics, industry and government to the digitalisation of the built environment and its potential impact on private enterprises, public services and the broader context of society. Governments around the world are aiming to guide and standardise this process by developing an array of standards to support this digitalisation, most notably on Building Information Modelling (BIM) and Smart Cities. Furthermore, the advancement of the Internet of Things (IoT) is creating a highly flexible, dynamic and accessible platform for the exchange capture and of information. There is a risk that this information on the built environment is quickly becoming unmanageable, and the value of that information is quickly becoming lost. This chapter presents a smart asset alignment framework that creates an alignment between the information captured at the infrastructure asset level and citizen requirements within a Smart City. The framework contributes to the debate on designing and developing Smart City solutions in a way that will deliver value to the citizens.

Keywords Smart cities · Asset management · Building information modelling · BIM · Smart cities framework · Citizen requirements · Smart assets

1 Introduction

The concept of using data within a city environment to inform economic, social and environmental policy decisions is not new. During the Cholera outbreak of 1854 in London Dr. John Snow theorised that the disease was being spread through contaminated water and collected data on the location of pumping stations and nearby cholera deaths [1]. John quickly realised that there were geospatial clusters of death

J. Heaton · A. K. Parlikad (✉)
Institute for Manufacturing, University of Cambridge, 17 Charles Babbage Road, Cambridge CB30FS, UK
e-mail: ajith.parlikad@eng.cam.ac.uk

A. Crespo Márquez et al. (eds.), *Value Based and Intelligent Asset Management*,
https://doi.org/10.1007/978-3-030-20704-5_2

around specific water pumps and despite the scepticism from the local authority, the pumps' handles were removed, and the deaths quickly subsided. One of the first attempts to document the Social status of citizens within a city was from Charles Booth, who mapped every street of London between 1889 and 1903 and documented the average "social class" of families on those streets [2]. Even though the maps and associated data capture techniques were considered revolutionary at the time, there is little evidence to suggest they helped inform policy and decisions regarding the city's development. During the 1940s the Los Angeles Department for Planning had developed a computer stamp-card system that they hoped could track and analyse all of the properties within the city including information on ownership, number of bedrooms and location [3]. After World War two (1939–1945) there was a growing awareness that the poorly maintained housing stock threatened the prospective health and morals of the city [4], and the planning department while alone could not address this problem. During the 1950s and '60s, the city started to investigate the integration of other data sources such as US census, police department, county assessor, aerial photos and other private and public sources [5]. This exercise was hugely successful in gaining federal funding to support the redevelopment of Los Angeles during the 1950/60/70s.

One of the initial mentions of Virtual and Digital Cities within academic literature was in 1997 by Graham and Aurigi [6], who discussed the nature and potential value of the virtual city within a social and inclusive context. The first Digital City practice was developed in Amsterdam in 1994, that gave the Internet to a large group of people for the first time and is cited as creating the first online community within a specific city and including the general public (not just computer experts, which was common at the time) [7, 8]. These examples show the first concepts of a Smart City, and the advancement of modern technology is evolving of the concept of Smart City that engages with cities' stakeholders and encompasses all of the built and natural environment.

It is accepted that the built environment including infrastructure within a city has a direct impact on the quality of life for citizens that live, work and visit the city. This relationship is generally understood at a high level but not when considering the performance of individual assets to the citizen requirements, specifically within a Smart Cities framework. This chapter addresses this gap by proposing an addition to the existing Smart Cities framework that examines the functional output of infrastructure assets and systems to create an understanding of how a city's infrastructure comes together to deliver services and meet citizen requirements. The fundamental objectives of this research are: (1) to investigate the impact of individual infrastructure asset functions and systems performance on city services and citizen requirements, (2) to investigate the relationship between citizen requirements and cities services and (3) suggest how to underpin the development of Building Information Modelling (BIM) within the concept of Smart Cities. The material presented in this chapter was first published by the authors in the Cities journal.

The chapter is structured as follows. Section 2 sets the context by reviewing Smart Cities standards and specifications alongside the current academic literature in the domain of Smart Cities which informs the smart asset alignment framework presented in Sect. 3. Finally, Sect. 4 summaries the approach and proposes future research opportunities.

2 Literature Review

2.1 *Method*

A systematic literature review allowed clear understanding of the cross-functional nature and the diversity and complexity of Smart Cities and BIM. Firstly, standards and specifications directly and indirectly related to BIM and Smart Cities were reviewed. Secondly, grey literature such as reports and organisational white papers were analysed. Specifically focused on Smart City ranking and rating reports and white papers focused on Smart Cities management services, technology platforms and implementation and integration offerings. Finally, academic literature was reviewed, utilising the research databases of Google Scholar, Direct Science and Scopus too source both peer reviewed journals and conference papers. The key search terms included Smart Cities frameworks/governance, Building Information Modelling, Engineering Asset Management, physical asset classification, Internet of Things and citizen requirements. Three discreet parts were discovered including governance (government and policy), technology (software, hardware and platforms) and people (educations and stakeholder engagement). These domains were used to structure the following two research questions (1) How can the emerging domains of BIM and infrastructure asset management aid in the development of a Smart Cities framework? (2) How does the performance of infrastructure assets impact on the city services and citizen requirements within a city?

2.2 *Smart Cities Standards, Specifications and Guidance*

Cities are either planned or evolved organically, often over a timeline of hundreds of years [9]. As an example, Saint Petersburg is a planned city with a specific date of foundation (1st of May 1703) and designed for specific function, as being the new capital of Russian political and military power. Saint Petersburg from its foundation, had a city master plan with construction rules and registrations [10]. While in contrast, Venice is a city that has evolved organically over thousands of years that has been occupied and exploited many times, with little thought to the city planning requirements [11]. While Saint Petersburg had the advantage of a well-structured top-down planning process that provided a structured approach to the city's development, it is often cited that these cities lack a sense of place, culture and community feeling due to their structured development. Because Venice had no structured approach to its development, it created a chaotic and ad hoc approach to the city's development, and history, community and culture playing a key role in the city's development [12]. This dynamic nature of cities makes it impossible to develop a "one size fits all" approach to the development of Smart Cities. The published standards, specifications and guidance have focused on the conceptual framework for how each city should develop its own Smart City objectives and strategies.

Several organisations have started developing an array of Smart Cities related standards, specification and guidance, most notably British Standards Institute (BSI), International Standards Organisation (ISO) and the International Telecommunication Union (ITU). The BSI has developed a comprehensive set of ad hoc standards that are in the form of Publicly Available Specifications (PAS) and Published Documentation (PD) that focus on developing a Smart Cities framework [13]. The ITU has primarily focused on the development of Key Performance Indicators (KPIs) to allow cities to have a credible measure of their Smart City transformation. Furthermore, based on the research developed by ITU study group the KPIs were categorised into ICT, environmental sustainability, productivity, equality and social inclusion, quality of life and physical infrastructure [14–16]. ISO, as the leading international organisation for the development of standards, have a comprehensive array of standards that directly or indirectly aid the development of a Smart City by developing specific standards for specific needs within a city including but not limited to energy, urban mobility, water, infrastructure, security and health [17–22]. The BSI specification PAS 182 (model for data interoperability within a Smart Cities framework) has been adopted as an ISO standard [23].

Even though there is a growing set of documentation around Smart Cities, there are very few enforceable standards,[1] and most of the documentation is guidance, specifications and technical reports. This is partly due to the confusion around the definition of a Smart City and the challenges in developing standards from a holistic point-of-view while still maintaining the required detail. With that being said, there are Smart Cities related standards being developed by ISO, most notably ISO 21972 developing an upper-level ontology for Smart Cities indicators and ISO 27550/1 focusing on information security within a Smart Cities framework [24]. There is no direct and official alignment between the different organisations' standards being developed, but they tend to fall under one of three categories as summarised below (see Table 1 and Fig. 1).

- **Strategic**—Aid in establishing strategies, plans and objectives for Smart Cities, providing a high-level framework for decision-making to agree and develop a holistic Smart Cities strategy with a well-defined vision and purpose, focusing on management progresses and implementation, not the technical processes.
- **Processes**—Support the development of a framework within the city that aids in the data interoperability, normalisation and classification of different datasets that can be combined to create greater informed decisions.

[1]Standard that have a measurable performance rating.

Table 1 Summary of Smart City related standards and documentation

Title	Description	Category	Reference
British Standards Institute			
Smart Cities—Vocabulary	A collection of a diverse range of terms and expressions used in discussions around Smart Cities	Strategic	PAS 180 [25]
Smart City Framework—Guide to establishing strategies for Smart Cities and communities	Proposes a Smart City Framework allowing leaders of a city to develop a Smart City strategy with a vision, objectives and success factors	Strategic	PAS 181 [26]
Smart City concept model—Guide to establishing a model for data interoperability	Guide to establishing a model for data interoperability supporting the classification of information from many data sources within a city	Technical	PAS 182 [27]
Smart Cities—Guide to establishing a decision-making framework for sharing data and information services	guide to establishing a decision-making framework for the sharing of data and information for the creation of information services to support decision-making processes	Process	PAS 183 [28]
Smart Cities—Developing project proposals for delivering Smart City solutions—guide	Guides and case studies for developing a project proposal for Smart Cities solutions.	Strategic/process	PAS 184 [29]
Smart Cities—Specification for establishing and implementing a security-minded approach	a framework for establishing Smart Cities with a security-minded approach aligns to PAS 1192-5	Technical/process	PAS 185 [30]
Smart Cities overview—Guide	Provides general guidance and approach for adoption of Smart Cities processes, focused on rapid development	Process	PD 8100 [31]
Smart Cities—Guide to the role of the planning and development process	Guide for city planning departments on how to advise and plan for the implantation of Smart Cities, including innovative technologies and approaches	Strategic	PD 8101 [32]

(continued)

Table 1 (continued)

Title	Description	Category	Reference
Automatic resource discovery for the Internet of Things—Specification	Specifies a common catalogue format that IoT sensors can be used to recognise each other	Technical	PAS 212 [33]
International Organization Standards			
Guidance on social responsibility	Provides guidance on the underlying principles of social responsibility, recognising the social responsibility and engaging stakeholders.	Strategic	ISO 26000 [34]
Sustainable cities and communities—Vocabulary	A collection of a diverse range of terms and expressions used in discussions around Smart Cities	Strategic	ISO 37100 [35]
Sustainable development in communities—Management system for sustainable development—Requirements with guidance	Establishes requirements for a management system for sustainable development in communities, including cities, using a holistic approach	Process	ISO 37101 [36]
Sustainable development of communities—indicators for city services and quality of life	Establishes definitions and methodologies for a set of city indicators to steer and measure delivery of city services and improved quality of life	Process/technical	ISO 37120 [37]
Smart community infrastructures—Review of existing activities relevant to metrics	An overview of the current metrics and processes used to measure digital infrastructure in a Smart City	Process	ISO/TR 37150 [38]
Smart community infrastructures—Principles and requirements for performance metrics	Provide principles and specifics requirements for community infrastructures performance metrics.	technical	ISO/TR 37151 [39]
Smart community infrastructures—a Common framework for development and operation	A framework for developed of smart community infrastructure, considering their characteristics	Process	ISO/TR 37152 [20]

(continued)

Table 1 (continued)

Title	Description	Category	Reference
Asset Management	Framework for establishing and adopting an asset management system for infrastructure assets	Strategic/process	ISO 55000 [40]
Master data: Quality management framework	Provides a framework for improving data quality that can be used independently or in conjunction with quality management systems	Strategic	ISO 8000 [41]
Quality management systems	Framework for a quality management system with organisations	Strategic	ISO 9000 [42]
International Telecommunication Union			
Key performance indicators for smart Internet of things and Smart Cities	Recommends KPI's for Smart Cities, which guidance on how to measure/achieve them	Process	Y.4903/L.1601/2/3 [14–16]

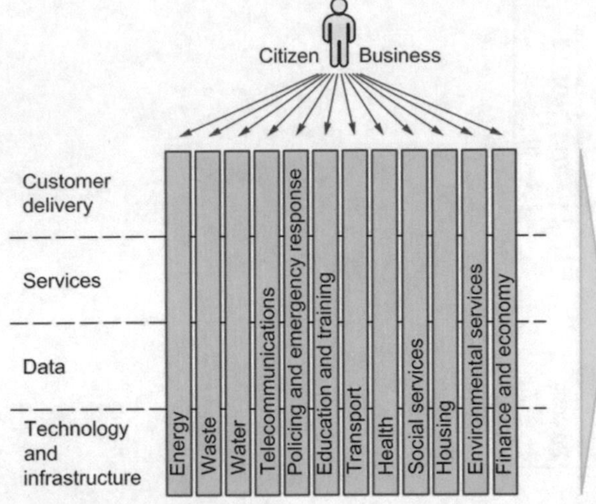

Fig. 1 Tradition cities operating model [26]

- **Technical**—Technical support and guidance on how best to develop the digital infrastructure for Smart Cities, including communication, internet protocols and sensors development.

Whilst not directly related to Smart Cities, the emergence of Building Information Modelling (BIM) is providing a catalyst for the development of Smart Cities. BIM provides a collaborative information management framework used to inform integrated decisions throughout the whole lifecycle (design, construction, operational & maintenance and disposal/renew) of built environment assets. BIM has been widely adopted in the design and construction phase, but its adoption is limited in the operational & maintenance phase [43]. BIM information management processes are governed by a set of standards and specifications that lay down the foundation for how information should be defined, collected, stored, exchanged, used and disposed of in the context of the engineered assets. The key BIM standards are summarised in Table 2 and categorised along the associated asset lifecycle.

Tables 1 and 2 provides an extensive overview of current Smart Cities and BIM related standards, specification and guidance. As can be seen within Table 1, there has been a considerable amount of work completed in developing Smart Cities specifications including strategic guidance on developing a Smart City vision and strategy, process guidance for developing an information decision framework and technical guidance for developing a city data model. Furthermore, KPIs have also been developed to validate Smart Cities' performance. While not all of the references within Table 1 are directly Smart Cities related (such as ISO 8000, 9000, 55000) they will ultimately have an impact on implementing a Smart Cities framework. While the standards within Table 1 are extensive, the Smart Cities framework proposed lacks sufficient guidance for its implementation and fails to align with current and emerging

Table 2 BIM Related Standards Summary

Title	Description	Lifecycle	Reference
Collaborative production of architectural, engineering and construction information	Provides the framework for the development of a Common Data Environment (CDE), an environment to freely share design and construction related data. The owner or principal contractor manage the CDE	Design/construction	BS 1192 [44]
Specification for information management for the capital/delivery phase of construction projects using building information modelling	Guidance in the management of BIM related data within a CDE. A strong focus on BIM management and required documentation, e.g. BIM Execution Plan	Design/Construction	PAS 1192-2 [45]
Specification for information management for the operational phase of assets using building information modelling	proposes the information management framework for the use of BIM within the operational phase, including developing organisational requirements within a BIM-enabled environment	Operational and Maintenance	PAS 1192-3 [46]
Fulfilling employer's information exchange requirements using COBie	UK government requirement for the exchange of information from project to the end user/client, in the format of organised spreadsheets	Exchange from Construction to Operational	BS 1192-4 [47]
Specification for security-minded building information modelling, digital built environments and smart asset management	Guidance on how to support BIM processes with security sensitive information and models		PAS 1192-5 [48]
Briefing for design and construction Code of practice for facilities management	Guidance on operational briefing requirements within the design and construction phase	Operational and Maintenance	BS 8536-1 [49]
Building construction—Organization of information about construction works	Defines a framework for classification of construction-related information, e.g. cost, time, models, ETC	Design	ISO 12006-2 [50]

(continued)

Table 2 (continued)

Title	Description	Lifecycle	Reference
Industry Foundation Classes (IFC) for data sharing in the construction and facility management industries	An opensource information model allowing for the exchange and transfer of 3D geometry, between different enterprise systems	All	ISO 16739 [51]
Building Information Modelling—Information Delivery Manual	A methodology to highlight the exchange of information between different actors for a specific task	All	ISO 29481 [52]
Government soft landings	Guide on how to successfully deliver built asset related information throughout the lifecycle of an asset	All	GSL [53]

processes such as BIM within the construction/operational and maintenance domain within cities.

Table 2 provides the key specifications and standards for the development of BIM information management processes throughout an engineered asset's whole-life. Furthermore, the standards provide a structured approach for the exchange of data throughout the different lifecycles and stakeholders including key milestones for when to exchange data, and the open source format this data should be in, e.g., IFC. Similar to the Smart Cities standards, the BIM standards lack any alignment with current and emerging processes within Smart cities, despite the overlaps within interoperability data models and information decision frameworks. It can be seen that both BIM and Smart City standards have been developed in parallel but in isolation to each other. Furthermore, as BIM spans the whole-life of engineered assets in the contents of information management processes, data structure and exchange protocols, it is well placed to act as an enabler to support the development of a Smart Cities framework.

2.3 *Review of Smart Frameworks*

The purpose of this review is to gain a comprehensive understanding of the current state of the concept of Smart Cities and informs the development of the Smart Asset Alignment to Citizen Requirements Framework.

Whilst there is not a single solution, there are recurring components in the literature that support the strategic development, implementation and support of a Smart City. The most recurring components can be categorised as technology (software, hardware and platforms), people (education, innovation and creativity) and institutes (government, policy and organisations). Al-Hader [54] specifies the specific components of technology within a Smart City develop as a graphic user-interface (dashboards, reports, web interface, maps), control systems (common platforms, automatic control elements) and database resources (big data, data warehouse, exchange platforms) [54]. The application of IoT has been proposed as a solution to provide a holistic platform to integrate the cities' services under one technology platform [55]. The people component is critical to the success of developing a Smart City, but it is often neglected at the expense of technology and strategic development. It is essential to understand the individual's needs within a city but also the needs of the communities, groups and neighbourhoods of the city [56]. A strong focus is required on education that will foster the knowledge and required innovation to develop and operate within a Smart City. These individuals will form smart communities that deploy ICT solutions in a consensus and agreed-upon approach to aid in meeting the requirements of the community. Institutes are essential for providing leadership, governance, guidance and lead the development of the overall vision [57]. Smart Cities and more specifically the deployment of ICT can enhance the democratic process and provide the community with a more dynamic and alternative relationship with institutes. Governance is a significant challenge for the development of a Smart

City; some traditional challenges include limited transparency, accountability, isolated city services and lack of human resources [58]. A Smart City and therefore smart governance need to address these limitations and incorporate collaboration, communication, partnership, leadership and data exchange/integration solutions.

A growing amount of research is developing (most notably, coming out of the European Commission Horizon 2020 research grants) that focus on the engagement of the stakeholders within the development of Smart, most notably in the use of digital solutions to address the city challenges. These stakeholders include citizens, businesses, city management teams and technology providers. Organicity has developed a seven-step service framework for collaboration within a city based upon Experimentation as a Service [59]. Several case studies have been developed that show how the collaborative approach of the Organicity framework enables the city communities to engage with technology providers and city management support experiments that address a specific city challenges with a digital solution [60, 61]. A more technology-focused development is the City Platform as a Service (CPaas.io). The goal of CPass.io is to provide a solution that enables Smart City innovations for all of the city stakeholders by using the platform to combine the capabilities of IoT, big data analytics, cloud services with government open data approaches and linked data approaches [62]. The platform is then made available for interested parties to engage with. CPass.io, has taken a novel approach to the management of personal data which they called citizen engagement. This uses the human-centred personnel data management processes of MyData [63] and then visualises this in a citizen privacy dashboard that allows the citizen to see when and how their data is being used [64].

The abovementioned Living Labs institutes support the development of Smart Cities frameworks by proving several approaches that aid the technical communities to develop Smart Cities frameworks in engagement with non-technical communicates. Furthermore, a core focus of the tools developed within the Living Labs is providing feedback from the non-technical communities to the technical communities to ensure that non-technical communities needs and wants are addressed within the technical solution. Living Labs is a user-centric approach to integrate current research and innovation processes often within a private-public-city partnership [65]. Several Smart Cities Living Labs have been developed over the years with the specific goals of bring together city management, city planners, sociologists, local community groups and the technical community. There are many similarities within the recent and ongoing research efforts that aim to align the wants and requirements of non-technical local community groups within the technical developments. Furthermore, the references within this section demonstrate that the technical community are testing and putting into practice several aspects of the approaches proposed.

3 Smart Assets/Cities Alignment Framework

This research integrated the industry and academic literature to generate a Smart Asset Alignment to Citizen Requirements Framework for the development of Smart Cities to incorporate the relationship and influences between the citizen's requirements within a city and the functional outputs of the cities infrastructure assets. The framework utilises the Smart Cities operational model within PAS 181 that illustrates the requirement within a Smart City to integrate all the city services through city-wide governance enabled by ICT. This is moving away from the traditional model where the citizen would have to interact with the individual service providers within the city. Figure 1 illustrates the traditional operating model within a city, where services are purely based around the service they provide and are not designed around the citizen requirements. These services are traditionally in vertical silos where organisational processes such as budget-setting, operational delivery, accountability and decision-making processes happen in isolation to the other city services and embedded within the silos of their delivery chain. This traditional approach provides two fundamental challenges in developing a Smart cities framework. Firstly, data and therefore information has typically been siloed within the individual services, both technically such as different data structures and at an organisational level, such as different data quality management processes. This limits the potential for collaboration and alignment across the city services. Secondly, individual citizens and business are required to engage with each siloed service in isolation, having to make connections themselves, rather than receiving a connected service that meets their requirements.

To support the alignment between the services and the citizen requirements it is proposed that within a Smart City framework the services have to be linked to the infrastructure assets (e.g., transport infrastructure) that support the operational requirements of that service. This is achieved by viewing the infrastructure assets within a city as a system, that when combined provide a functional output that aids to support the operational requirements of the city services. The infrastructure assets hierarchy structure follows the industry standard ISO 12006-2 for the classification of infrastructure assets [50]. Several international organisations have aimed to classify infrastructure assets functions, systems, sub-systems and products, the most comprehensive being Omniclass [60] and UNIClass [66, 67]. Figure 2 illustrates the parent-child relationship as defined within ISO 12006-2 and example definitions from UNIClass.

One key advantage for a city to classify its infrastructure assets is to understand the many different asset systems and products that support a function output and the relationship they form. For example, the functional output of heating is partly supported by the gas boiler asset system in which the thermostat is a product/component. Furthermore, as the services provided within a city as defined by PAS 182 [27] are primarily supported by infrastructure assets, it is required to create a relationship between the functional output of the assets to the city services. As an example, a Smart City must provide the service of education, which is supported by multiple functional outputs such as heating, water supply and electricity supply which are

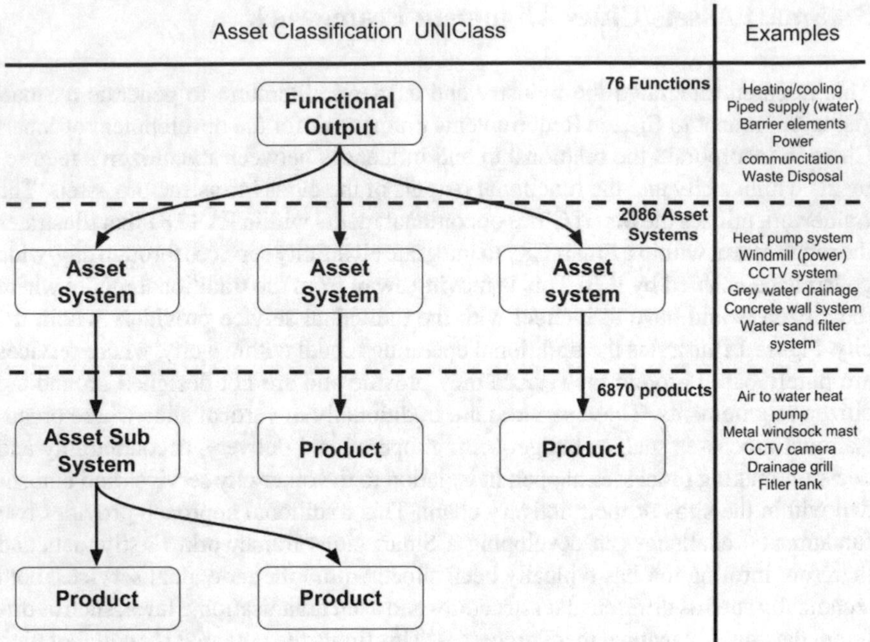

Fig. 2 Infrastructure asset classification system

themselves supported by an array of asset systems and products. Within this example, you could monitor the performance of education via the performance of the infrastructure assets that support that service. Furthermore, you could also monitor/predict the impact that the failure of infrastructure assets will have on the city services. Within this example, the failure of the water supply will have a direct impact on the performance of providing the education service and will result in lost educational hours, as you cannot operate a school without a running water supply. Classification of cities infrastructure assets makes it possible to create a tangible link between the city services and the infrastructure assets that support them.

While the classification of infrastructure assets within a part-child hierarchy relationship is not a new concept, the classification of infrastructure assets within the concept of a Smart Cities framework has not been widely explored. When classifying infrastructure assets from the point-of-view of a Smart Cities framework, the highest functional output of the infrastructure assets should be identified such as transport, communication and waste disposal that align to support the city services. By a city adopting such an approach within its framework, it enables the alignment of BIM related data and the city services, as asset classification is a key step within the BIM processes. As stated within the literature review (Sect. 2). BIM has been widely adopted within the design and construction phases, but with limited use within the operation and maintenance phases. A Smart City framework that aligns itself to BIM related classification will support the seamless transfer of BIM related data into the cities' operational services. Traditionally there is a time-lag between the completion of infrastructure assets such as a new train platform, water pump or school complex within a city and integration into the city services due to the complex nature of infrastructure asset data and information handover over from projects to the city services. A Smart Cities framework that follows BIM enabled classification processes will support a structured approach for the exchange of new infrastructure assets to the city services by providing a common structure for infrastructure asset-related data.

Creating the alignment between asset functions and city services has added benefits. Firstly, it allows the owner of the city services to have a holistic understanding of the assets that support that service and the multiple stakeholders that develop, operate and maintain them. This is especially important when cities assets have public and private owners. Secondly, it provides a scalable platform for data analysis and modelling tools that can focus on individual infrastructure assets performance impact on the cities' services and ultimately the citizen's requirements.

Figure 3 illustrates how the infrastructure assets can be amalgamated within the city services via a data integration layer. The Data Model Integration Layer (DMIL) acts as a data amalgamation platform that supports the exchange of asset related information. The arrows from the functional output to the DMIL represents the flow of asset related data into the DIML. This flow of data should be in an open-source format, ideally in one of the BIM enabled formats such asIFC or COBie, as highlighted in Sect. 2.2. If a BIM-enabled format is not possible, for example if BIM has not been widely adopted within the country, then open source formats as XML, JSON or CSV should be considered. The remaining arrow flowing from the DMIL into the city services represent the flow of data and information from the DMIL directly into their enterprise systems. Examples of such enterprise systems include reporting systems, information technology management, resource planning and fiscal management. The DMIL provides a single point of access to all infrastructure asset related data in a structured approach. As an example, the health services within a city could monitor the performance of public transport related infrastructure assets and feed this data into their appointments management and resource scheduling to respond dynamically to their performance, such as reschedule or cancel the appointment if patients are delayed due to a failed rail signal or rolling stock. Furthermore, the health department could utilise the DMIL to gain greater insight on its resilience by monitoring the performance of the water supply, energy supply, communion and environmental services within one holistic point-of-view and utilise it within their risk management processes.

The DMIL development takes concepts from BIM in the operational phase specification PAS 1192-3 that specifies the development of an Asset Information Model (AIM) [46]. The AIM acts as a single source of amalgamated information for infrastructure asset related data including graphical, non-graphical and documentation. While the DMIL does not encompass all of the concepts of the AIM, the concept of acting as a data store for infrastructure related data and exchange this with enterprise systems is a crucial concept of the DMIL.

To ensure that the Smart Cities development framework is citizen-centric and not the traditional city operational model where citizens have siloed interaction with

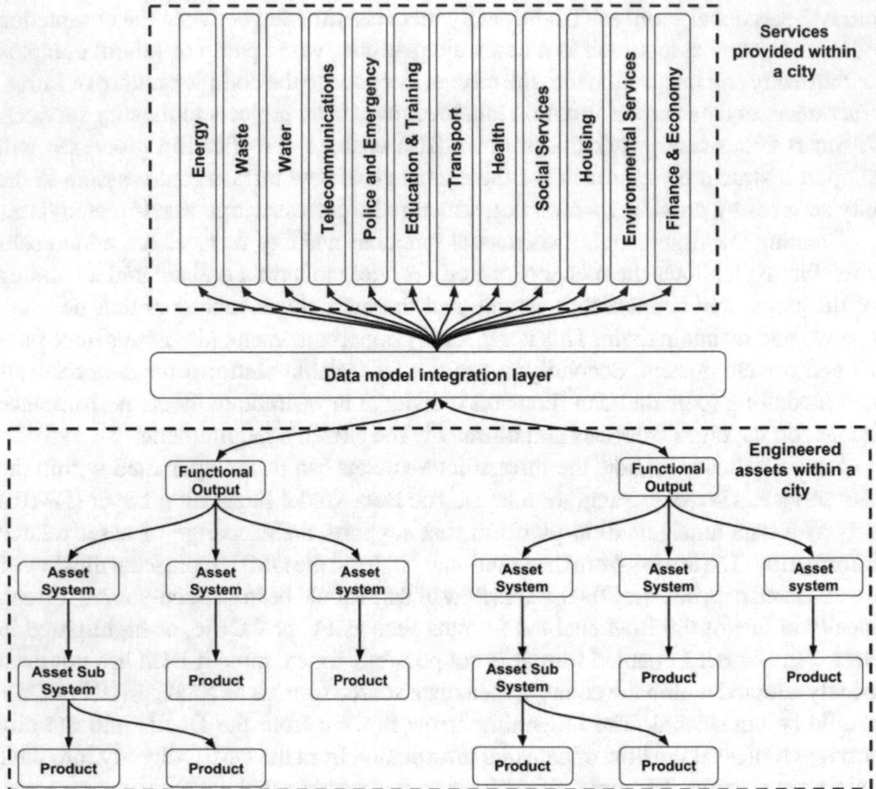

Fig. 3 Alignment of infrastructure assets to the services provided by a city

individual city services, this research has identified the need to understand the citizen requirements in the contents of a Smart Cities framework. Many of the cities citizens requirements will be supported by multiple city services, which in turn are supported by multiple infrastructure assets.

To support the development and classification of citizen requirements it is proposed to use The United Nationals Statistics high-level Classification of the Function of Government (COFOG) [68] as a reference point. The COFOG goal is to identify and classify high-level functions that a stable government should provide to its citizens. Where appropriate these functions have been adopted into citizen requirements, Table 3 summarises the high-level COFOG and associated citizen requirements were applicable.

Classifying citizen requirements in alignment to government functions aids in supporting the integration of city services, as they no longer support individual service requirements but aid to support the holistic requirements for the citizens. While there are vastly different cities around the world due to their development (organic growth or planned), culture and demographics, they must all meet a set of citizen

Table 3 Alignment of government functions with citizen requirements

Functions of government [68]	Citizen requirements
General public service	N/A
Defence	N/A
Public Order and Safety	All
Economic affairs	Work, Invest
Environmental Protection	All
Housing	Live
Health	Heal
Recreation, Culture and Religion	Socialise, play
Education	Learn
Social Protection	Grow-up, ageing (die)

requirements. Understanding citizen requirements is a complex exercise due to the diverse nature of people within cities, especially global cities such as London, New York and Beijing. The citizen requirements developed within this framework are deliberately a high-level concept that addresses all the citizen requirements, no matter the city in question. Furthermore, the high-level nature of these requirements allows for a more holistic alignment of city services to the citizen requirements.

As an example, the United Nationals within the COFOG stated that a functional government needs to provide the service of education, which as a citizen requirement is the need to learn. This citizen requirement is supported by multiple city services such as fresh drinkable water supply, transport services to get to and from the place of learning and telecommunication. The high-level nature of the requirement for learning allows this holistic point-of-view and enable city services integration. The degree to which the individual cities will value and measure the performance of each citizen requirements will depend on the current policy and objectives within the city. As an example, a newly-elected mayor who campaigned on the policy of creating a higher performing education service will result in an increase in the benchmark for performance in learning.

Figure 4 illustrates the final Smart Cities framework that incorporates the three discussed selections of infrastructure asset classification, integration of city services and categorising citizen requirements. A core advantage of implementing such a framework is providing the direct line-of-sight from the citizen requirements through the city services and alignment to the performance of the infrastructure assets. Ultimately the city could analyse the impact of poor performing or failing infrastructure assets on meeting the citizen requirements. As an example, the failure of a rail signal results in a series of cancellations of trains during the morning rush of students travelling to school. This results in students being delayed for school and impacts the level of performance that the city provides in education and ultimately impacts on the students' requirement to learn. Within the traditional Smart Cities framework, this kind of citywide impact analytics would not be possible as they don't consider the performance of infrastructure assets on providing city services.

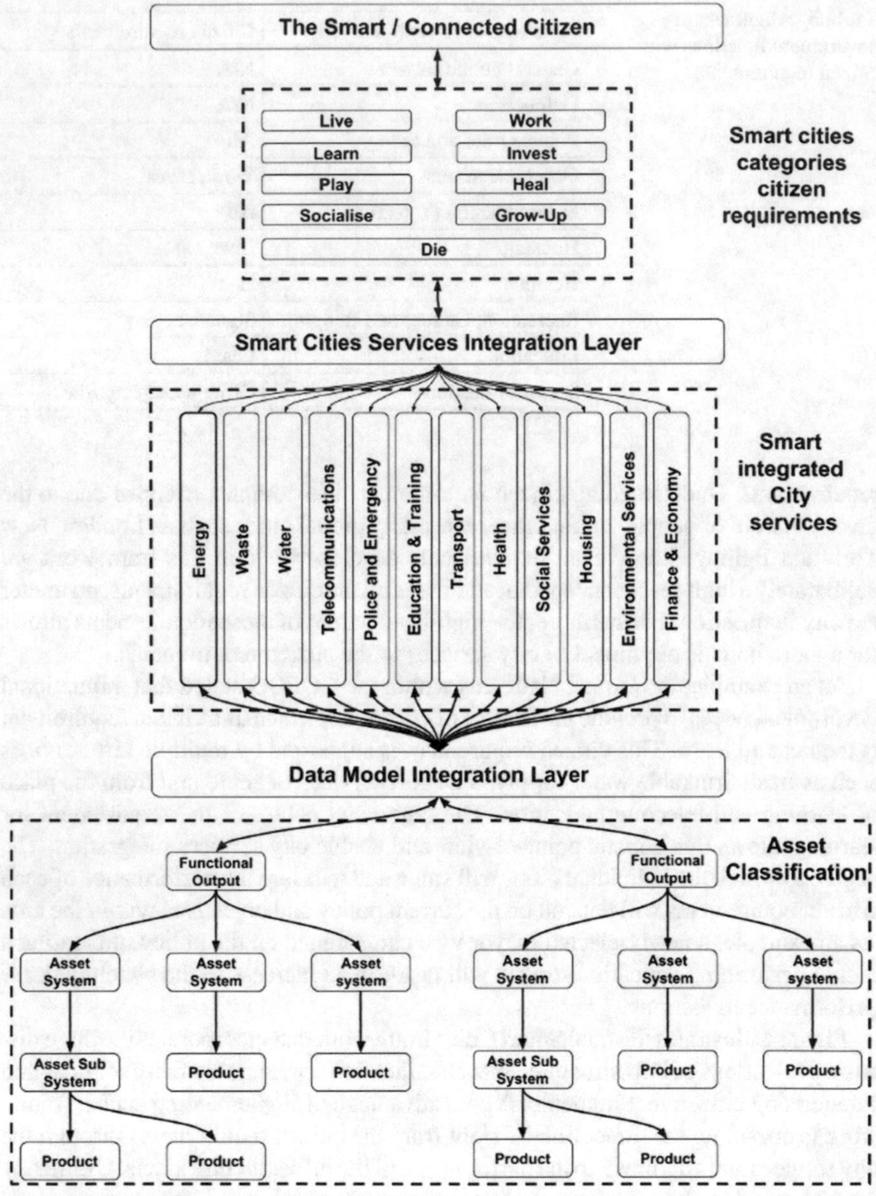

Fig. 4 Smart asset alignment to citizen requirements framework (SAACRF)

The Smart Cities Integration Layer (SCIL) acts as an amalgamation between the city services and the citizen requirements. Focusing on the arrows flowing from the city service to the SCIL, the SCIL integrates all of the city services performance data into a single standardised platform. Much like the DMIL, in a BIM-enabled Smart City this should take the form of a BIM format such as IFC or COBie as highlighted within Sect. 2.2, alternatively an open source format such as JSON, XML or CSV. While the DMIL enables the push of data from the infrastructure assets to the enterprise systems within the city services, the SCIL supports the pull of data out of the individual enterprise systems within the city services. Focusing on the arrow between the SCIL and the citizen requirements, this illustrates the integrated flow of data from the SCIL to the citizen requirements, using the data pulled in by the CSIL to validate if the performance requirements for the citizen requirements are being achieved. Ultimately the CSIL acts as the integration later and gateway to aligning the city services with the citizen requirements and the citizen themselves.

The Smart/Connected citizen is one that can seamlessly connect with the city services in an integrated and holistic solution. Instead of the traditional model where the citizen has individual interaction with the city services, the Smart/Connected Citizen can have access to multiple services through one point of interaction and the value generated from connecting the services is realised. As an example, if a new health problem impacts the mobility of a citizen, this will impact the ability for them to attend school and will need the support of a carer from social services. The proposed framework would support the seamless connection between the health service, education service and social services, without the citizen having to engage with the individual services. Ultimately, this will also allow the city leaders to validate if the citizen requirements are meant to meet by the city services.

Given the recent highly publicised events of data breaches within financial and commercial organisational and more specifically the "hijacking" of personal data for exploitation, it is important to consider the data governance and privacy within such a proposed framework. All personal data should be protected under a data privacy framework, the European Union (EU) development of the General Data Protection Regulation (GDPR) [69] is an example of a data privacy framework that organisations which store personal data within the EU must follow. Furthermore, citizen engagement is critical to ensure transparency and provide assurance that their data is being used for what they approve, Sect. 2 provides examples of Living Labs, which aims to integrate the technical community with the city citizens to provides a citizen-centric smart city solution. The data privacy framework should be implemented within the city services, as they will collect and store the bulk of personnel-related data. The infrastructure assets themselves will not collect, store or use personal data—only operational and performance data will be collected. Care should be taken with the governance of this data to ensure that security, safety and commercial dimensions of the data are highlighted and processes put in place to protect them from malicious exploitation.

The Smart Asset Alignment to Citizen Requirements Framework is a combination of a collection of research domains that have developed in isolation and aims to align key elements of those domains, most notably BIM and Smart Cities. Asset classification is derived from the domain of BIM. A BIM referenced international standard ISO 12006-2 defines the parent-child relationship for infrastructure assets structure and classification that provides the foundation for this section [50]. Furthermore, key literature from the domain of construction management, engineering asset management and information technology in construction provides examples of infrastructure assets part-child relationships and hierarchy [70, 71]. The Smart Integrated City Services is derived from Smart Cities specification PAS 181 [26], which demonstrate the often siloed services provided within a city. Finally, the citizen requirements are derived from The United Nationals Statistics high-level Classification of the Function of Government (COFOG) [68], which states the high-level functions a government must provide.

4 Discussion and Conclusion

This chapter attempts to extend the current Smart Cities frameworks. The concept of a Smart City is becoming increasingly popular, both in academic literature and in industrial applications. The review of the current standards, specification and guidance in the domain of Smart Cities revealed that various international organisations are developing Smart City-related standards within their given domains, but little focus is given to the citizen requirements within a Smart City framework. Furthermore, infrastructure assets within a city have also been neglected from Smart Cities frameworks. This is partly due to the multi-faceted concept of Smart Cities and the complexity of developing citizen requirements. The developed standards, specifications and guidance have been categories into the groups of strategic, process and technical, to describe their focus areas. The most notable and comprehensive Smart Cities standards have been developed by the BSI within the PAS 18X series, focusing on establishing strategies development, establishing an interoperability data model, establish a decision-making framework, developing project proposals and establishing a security-minded approach to Smart Cities. Furthermore, it was noted that the BIM and IoT standards are not directly related to the Smart Cities development but can act as an enabler for the development of Smart Cities throughout the infrastructure assets lifecycles.

The academic literature review discovered that there are many definitions of a Smart City, initially with a strong focus on ICT development but more recently with a focus on citizens and smart communities. Furthermore, many variations exist by replacing smart with alternatives such as digital, intelligence, knowledge and innovation. It is noted that cities are complex, unique and dynamic, led by their history, culture and citizen requirements. Due to this complex nature, it is unrealistic to assume a single framework for Smart Cities development or a one-size-fits-all solution. The most recurring themes include technology (software, hardware and platforms), people (education, innovation and creativity) and institutes (government, policy and organisations). Finally, it was noted that there are many different ways to score and rate the smartness of cities. Most reviewed indicators where ICT focused, but there was a growing need to be able to measure citizen satisfaction and wellbeing within a Smart Cities context.

The proposed smart asset alignment framework within this chapter builds on existing Smart City frameworks (notably PAS 181), it was noted that two components are missing from this framework. Firstly, it fails to identify the citizen requirements within the city. Secondly, it fails to consider the functional output of the cities' infrastructure assets and the impact of this on the citizen requirements.

The existing Smart City framework was first expanded to include infrastructure assets aligned to the city services. To support this, it is proposed to classify infrastructure assets as per the functional output they provide, this follows an industry classification standard. This supports the alignment of thousands of individual asset systems and products that support a function output and ultimately aid in support of city services.

Secondly, many of the citizen requirements are supported by multiple city services, the citizen must manually interact with individual services to meet their requirements. To support the holistic integration requirements of city services within a Smart City framework, it is needed to categorise the city citizen requirements. The high-level governance functions as defined by the United Nations was used as a framework to transform into citizen requirements.

When adding the two proposed components to the current Smart Cities framework, it will provide a direct line-of-sight from citizen requirements, the services used within the city to meet that requirements and the infrastructure assets that support the used services and ultimately validate if the citizen requirements have been fulfilled.

This chapter demonstrated that the performance of a city service is dependent on the performance of multiple different asset functions that are not traditionally considered, as an example, providing the service of health care is impacted on the performance of the public train network to get staff and patients into the hospital. Furthermore, it was noted that a single citizen requirement is often supported by multiple city services. As an example, the citizen requirements 'to learn' is support by the service of education but also by the services of transportation to support teachers and students to travel to a school. Finally, BIM was highlighted as an enabler to support the development of a Smart Cities framework by providing a structured approach for the developed, storage and transformation of built environment data throughout its whole-life cycle.

While there have been significant advancements in the Smart Cities technology solutions (such as IoT), there are still limitations in current technology and data analytics processes to support the data capture, integration and exploitation required within the proposed framework. Furthermore, the understanding of the interaction with these technologies both at the individual level and the collective level is not well understood and could limit the implementation of the framework. Often fractured national government of local policies do not provide the needed transparency and leadership required. Furthermore, the established bureaucracy in city services will be reluctant to expose their services processes and associated data to the other city services and the broader city management. A political, city services culture, technical and social transformation is required to support the development and implementation of the proposed framework. Privacy and concerns of impact on democratic governments within a Smart Cities framework need to be addressed as it becomes a growing concern for cities and society as a whole. This includes both technology-related concerns such IT safeguards of personnel data and governance concerns around the separation of power between governments, technology provides and the citizens. These concerns must be addressed for successful implementation of a Smart Cities framework.

Future research should focus on exploring the scalability of the proposed framework to incorporate the alignment to the broader regulation and government objectives and strategies. This will support line-of-sight from government policy to citizens requirements and the performance of infrastructure assets. Furthermore, due to the diverse nature of cities, the dynamic and changing aspect of citizen requirements should be investigated and inform changes in government functions. Finally, investigating the commercial business requirements might differ from individual citizen requirements and provide new insight into the relationship between business, city services and the infrastructure assets.

Acknowledgements This research was supported by the Engineering and Physical Sciences Research Council and Costain Plc through an Industrial CASE Award. The authors also thank the support of the EPSRC Innovation and Knowledge Centre for Smart Infrastructure and Construction as well as the Centre for Digital Built Britain.

References

1. Vinten-Johansen, M. R. P., Brody, H., Paneth, N., Rachman, S. (2003). *Cholera, chloroform, and the science of medicine: A life of John Snow* Oxford University Press.
2. Booth, C. (1969). *Life and labour of the people in London*. Macmillan.
3. Los Angeles Community Analysis Bureau. (1974). *State of the City II: A cluster analysis of Los Angeles*. City of Los Angeles.
4. Housing Authority of the City of Los Angeles. (1944). *A decent home: An American Right, 5th, 6th and 7th Consolidated Report*. Housing Authority of the City of Los Angeles.
5. Vallianatos, M. (2015). Uncovering the Early History of 'big data'and 'Smart city'in Los Angeles. *Boom California*. [Online]. Available: https://goo.gl/Hyjjnp. Accessed February 11, 2018.
6. Graham, S., & Aurigi, A. (1997). Urbanising cyberspace? *City, 2*(7), 18–39.
7. van den Besselaar, P. (2005). The life and death of the great Amsterdam Digital City. Digit. Cities III. *Information Technologies for Social Capital: Cross-cultural Perspectives, 3081*, 66–96.
8. Anthopoulos, L. G. (2017). Understanding Smart Cities: A tool for Smart Government or an Industrial Trick? *Public Administration and Information Technology, 22,* 5–45.
9. UN-Habitat. (2016). World cities report 2016: Urbanization and development–emerging futures. No. 8.
10. Hassell, J. (1974). The planning of St. Petersburg. *The Historian, 36*(2), 248–263.
11. Howard, D. (2002). The architectural history of Venice.
12. Lindsay, B. E., Friedmann, J., & Weaver, C. (1981). Territory and function: The evolution of regional planning. *63*(3).
13. BSI. (2014). The role of standards in smart cities. *2*(2) 1–19.

14. ITU. (2016). *Key performance indicators related to the sustainability impacts of information and communication technology in smart sustainable cities*. Switzerland.
15. ITU. (2016). *Key performance indicators for smart Internet of things and smart cities and communities*. Switzerland.
16. ITU. (2016). *Key performance indicators related to the use of information and communication technology in smart sustainable cities*. Switzerland.
17. ISO. (2015). *ISO 17752—Energy efficiency and savings calculation for countries, regions and cities*.
18. ISO. (2012). *ISO 39001—Road traffic safety (RTS) management systems—requirements with guidance for use*.
19. ISO. (2007). *ISO 24510—Activities relating to drinking water and wastewater services—guidelines for the management of drinking water utilities and for the assessment of drinking water services* (Vol. 3).
20. ISO. (2016). *ISO/TR 37152—Smart community infrastructures—Common framework for development and operation*.
21. ISO. (2014). *ISO 22313—Societal security—business continuity management systems—requirements*.
22. ISO (2016). *PD ISO IWA 18 : Framework for integrated health and care services in aged societies*.
23. ISO. (2014). *ISO/IEC 30182—Smart city concept model—guide to establishing a model for data interoperability* (pp. 1–56).
24. ISO. (2017). *ISO and smart cities*. Switzerland.
25. British Standards Institute. (2014). *PAS 180:2014 smart cities—vocabulary*. London: United Kingdom.
26. British Standards Institute. (2014). *PAS 181:2014 smart city framework—guide to establishing strategies for smart cities and communities*. London: United Kingdom.
27. British Standards Institute. (2014). *PAS 182:2014 smart city concept model—guide to establishing a model for data interoperability*. London: United Kingdom.
28. British Standards Institute. (2017). *PAS 183:2017 smart cities—guide to establishing a decision-making framework for sharing data and information services*. London: United Kingdom.
29. British Standards Institute. (2017). *PAS 184: 2017 smart cities—developing project proposals for delivering smart city solutions—guide*. London: United Kingdom.
30. British Standards Institute. (2017). *PAS 185:2017 smart cities—specification for establishing and implementing a security-minded approach*. London: United Kingdom.
31. British Standards Institute. (2015). *PD 8100:2015—smart cities overview—guide*. London: United Kingdom.
32. British Standards Institute. (2014). *PD 8101:2014 smart cities—guide to the role of the planning and development process*. London: United Kingdom.
33. British Standards Institute. (2016). *PAS 212—Automatic resource discovery for the Internet of Things—Specification*.
34. ISO. (2010). *ISO 26000 Guidance on social responsibility*.
35. ISO. (2016). *ISO 37100 Sustainable cities and communities—vocabulary*.
36. ISO. (2016). *ISO 37101—sustainable development in communities—management system for sustainable development—requirements with guidance for use* (p. 42).
37. ISO. (2014, July). *ISO 37120 sustainable development of communities: Indicators for city services and quality of life* (p. 112).
38. ISO. (2014). *ISO/TR 37150 Smart community infrastructures—Review of existing activities relevant to metrics*.
39. ISO. (2015). *ISO/TS 37151 Smart community infrastructures—Principles and requirements for performance metrics*.
40. ISO. (2014). *BS ISO 55000 series—asset management*.
41. ISO. (2005). ISO 8000—Master data: quality management framework. *Electron Bus, 01*.
42. ISO. (2015). *EN ISO 9000 : 2015 quality management systems fundamentals and vocabulary*.

43. Waterhouse, R., & Philp, D. (2016). *National BIM report* (pp. 1–28). London, UK: National BIM Library.
44. British Standards Institute. (2007). *BS 1192-2007 + A22016: Collaborative production of architectural, engineering and construction information.*
45. British Standards Institute. (2013). *PAS 1192-2:2013 Specification for information management for the capital/delivery phase of construction projects using building information modelling* (No. 1, pp. 1–68).
46. British Standards Institute. (2014). *PAS 1192-3:2014 specification for information management for the operational phase of assets using building information modelling* (No. 1, pp. 1–44). British Standards Industries (BSI).
47. British Standards Institute. (2014). *BS 1192-4:2014 collaborative production of information part 4 : Fulfilling employer' s information exchange requirements using COBie—code of practice* (p. 58). British Standards Industries (BSI).
48. British Standards Institute. (2015). *PAS 1192-5-2015 specification for security-minded building information modelling, digital built environments and smart asset management*. British Standards Industries (BSI).
49. British Standards Institution. (2015). *BS 8536-1-2015_Briefing for design and construction—-part 1 : Code of practice for facilities management (Buildings infrastructure).*
50. ISO. (2015). *BS ISO 12006-2:2015 building construction organization of information about construction works Part 2: Framework for classification.*
51. ISO. (2013). *ISO 16739:2013—Industry Foundation Classes (IFC) for data sharing in the construction and facility management industries.*
52. ISO. (2016). *BS ISO 29481-2:2016—Building Information Modelling—Information Delivery Manual.*
53. C. Office, "Section 2 GSL Lead and GSL Champion," p. 10, 2013.
54. Al-Hader, M., Rodzi, A., Sharif, A. R., & Ahmad, N. (2009, November). SOA of smart city geospatial management. In *2009 Third UKSim European Symposium on Computer Modeling and Simulation* (pp. 6–10). IEEE.
55. Zanella, A., Bui, N., Castellani, A., Vangelista, L., & Zorzi, M. (2014). Internet of Things for smart cities. *IEEE Internet Things J., 1*(1), 22–32.
56. Chourabi, H., Nam, T., Walker, S., Gil-Garcia, J. R., Mellouli, S., Nahon, K., … & Scholl, H. J. (2012, January). Understanding smart cities: An integrative framework. In *2012 45th Hawaii international conference on system sciences* (pp. 2289–2297). IEEE.
57. Meijer, A., & Bolívar, M. P. R. (2016). Governing the smart city: A review of the literature on smart urban governance. *International Review of Administrative Sciences, 82*(2), 392–408.
58. Joshi, S., Saxena, S., & Godbole, T. (2016). Developing smart cities: An integrated framework. *Procedia Computer Science*, *93*, 902–909.
59. Pye, L., & Schaaf, K. (2018). Organicity playbook How to launch experimentation as a Service in your city.
60. Gutiérrez, V., Amaxilatis, D., Mylonas, G., & Muñoz, L. (2018). Empowering citizens toward the co-creation of sustainable cities. *IEEE Internet Things Journal, 5*(2), 668–676.
61. Amaxilatis, D., Boldt, D., Choque, J., Diez, L., Gandrille, E., Kartakis, S., et al. (2018). Advancing experimentation-as-a-service through urban iot experiments. *IEEE Internet of Things Journal*. 1.
62. Haller, S., Neuroni, A. C., Fraefel, M., & Sakamura, K. (2018, May). Perspectives on smart cities strategies: sketching a framework and testing first uses. In *Proceedings of the 19th Annual International Conference on Digital Government Research: Governance in the Data Age* (p. 42). ACM.
63. Kuikkaniemi, K., Poikola, A., & Honko, H. (2015). *MyData—A Nordic Model for human-centered personal data management and processing* (p. 12). ISBN: 978-952-243-455-5.
64. CPaaS.io. (2018). City platform as a service. [Online]. Available: https://cpaas.bfh.ch/. Accessed September 02, 2018.
65. Eriksson, M., Niitamo, V. P., & Kulkki, S. (2005). State-of-the-art in utilizing living labs approach to user-centric ICT innovation—a European approach *1*(13), 131.

66. OCCS. (2017). Omniclass.. [Online]. Available: http://www.omniclass.org/. Accessed Februarys 22, 2018.
67. Delany, S. (2016). UNICLASS calssification. *NBS*. [Online]. Available: https://toolkit.thenbs.com/articles/classification. Accessed November 15, 2016.
68. United Nation Statistics Divison. (2018). *Classification of the Functions of Government*. [Online]. Available: https://unstats.un.org/unsd/cr/registry/regcst.asp?Cl=4&Lg=1&Top=1. Accessed February 20, 2018.
69. European Union. (2016). Regulation 2016/679. Official Journal of Europe Communities, *2014*, 1–88.
70. Oxenford, J. L. et al. (2012, April). Key asset data for drinking water and wastewater utilities.
71. Becerik-Gerber, B., Jazizadeh, F., & Li, N. (2011). Application areas and data requirements for BIM-enabled facilities management. *Journal of Construction Engineering and Management, 138*(March), 431–442.

Chapter 3
Application of a Performance-Driven Total Cost of Ownership (TCO) Evaluation Model for Physical Asset Management

Irene Roda and Marco Garetti

Abstract The core concept of this chapter is the total cost of ownership (TCO) of industrial asset and its relevance in supporting decision making, if properly evaluated through the analysis of the technical performances of the asset. The chapter is based on a framework that systematizes benefits and potential applications of TCO for different kind of stakeholders at different stages of the life cycle of the asset supporting different kind of decisions. The aim is to present an experimental case study that has been implemented in order to show the empirical evidence of what is in the framework by focusing on one of the primary companies in the chemical industry in Italy. The application proposes a modeling approach for trying to overcome one main gap that still subsists when referring to TCO models that is that most of the existing ones lack of the integration of technical performances evaluations into the cost models or are based on very limiting hypothesis. In this chapter a comprehensive methodology for the evaluation of Total Cost of Ownership of industrial assets that has being developed within a research activity carried out at the Department of Management, Economics and Industrial Engineering of Politecnico di Milano is presented.

Keywords Asset management performance · Total cost of ownership · Asset performance measurement

1 Introduction

In order to meet the challenges of global competition and changing market conditions, production companies need to adopt an asset management strategy and system to sustain or improve the life cycle profits of the original investment [1, 2]. With this regard, one of the challenges in the physical asset management field is to keep a life cycle perspective whenever a decision is taken both for acquisition or configuration

I. Roda (✉) · M. Garetti
Department of Management, Economics and Industrial Engineering, Politecnico di Milano, Piazza Leonardo da Vinci, 32, 20133 Milan, Italy
e-mail: irene.roda@polimi.it

A. Crespo Márquez et al. (eds.), *Value Based and Intelligent Asset Management*,
https://doi.org/10.1007/978-3-030-20704-5_3

and management actions on any asset. Through this perspective, it is essential to improve the quantification process of costs, in order to be able to evaluate the total cost of operating a production system throughout its life cycle [i.e. the so called Total Cost of Ownership (TCO)] as a supporting evidence that allows informed decision-making [3].

More in detail, this work refers to the concept of TCO intended as the actual value of the sum of all significant costs involved for acquiring, owning and operating physical assets over their useful lives [4]. TCO is strictly related to the concept of Life Cycle Cost (LCC) and they are often used without distinction in literature. The widely shared idea is that TCO provides a selected perspective on LCC. In contrast to LCC, it focuses on the ownership perspective of the considered object and all the costs that occur during the course of ownership [5]. Moreover Clarke [6] and other authors later on, used it with a more strategic connotation, giving to TCO the meaning of a supporting information for strategic choices regarding both investment decisions and operational strategies.

It is widely accepted in the academic literature [7] that TCO should be an integral part of an asset management strategy and the same is assessed by the ISO 55,000 series of standards for asset management. In particular, the latter puts into evidence the relevance of being able to quantify the TCO of an asset, being it an industrial system or a single equipment, and it is indicated that: "[…] Life cycle cost, which may include capital expenditure, financing and operational costs, should be considered in the decision-making process. […] When making asset management decisions, the organization should use a methodology that evaluates options of investing in new or existing assets, or operational alternatives" [8]. On the industry side, companies are more and more acknowledging that a TCO model can represent a reliable economic-sound support for taking decisions and conveying the information it represents to both internal and external (costumers/suppliers) stakeholders [9].

This paper refers to the framework (Table 1) that the authors developed based on an extensive literature review aiming at highlighting the benefits of the adoption of a TCO model in decision making support for asset management [10]. Developing the framework, three main dimensions have been identified:

a. type of stakeholder: given the meaning itself of TCO, it is evident that asset users (asset owners/managers) are the main stakeholders; nevertheless, asset providers (asset builders/manufacturers) have also interest in evaluating the TCO of assets they build/sell.
b. type of supported decision: a TCO model has potentiality to support different kinds of decisions and in the framework two main categories have been identified: (i) configuration decisions and (ii) management decisions.
c. phase of the life cycle: TCO analysis can be carried out in any and all phases of an asset's life cycle [Beginning of Life (BOL), Middle of Life (MOL) and End of Life (EOL)] to provide input to decision makers.

The framework shows which benefits a TCO model can bring to each of the two types of stakeholder at each lifecycle phase by supporting different kinds of decisions (configuration or management decisions).

Table 1 Framework of benefits of TCO adoption in decision making for asset providers/users

	Asset provider		Asset user	
	Configuration	Management	Configuration	Management
BOL	– Evaluation of project alternatives – Comparison and optimization of design alternatives – Components/equipment procurement and construction alternatives evaluation – Spare parts requirements estimation	– Communicating value to the customer and selling support – Propose to the clients specific design solutions – Pricing – Contracting maintenance services provision	– Evaluation of design alternatives offered by a provider	– Suppliers and tenders evaluation and selection – Maintenance service contract evaluation – Investment, budget planning, cost control
MOL	– Proposal of re-configuration solutions	– Maintenance service provision offering – Spare parts provision offering	– Reconfiguration decisions – WIP sizing	– Maintenance scheduling and management – Repair level analysis – Asset utilization and production strategies
EOL	– Proposal of reconfiguration for EoL optimization	– Evaluation and proposal of rehabilitation strategies	– Reuse strategies for components/machines	– Evaluation of rehabilitation strategies

2 Problem Statement and Objective

Even if it clearly emerges from the literature that TCO has got positive effect in supporting decision making for asset management; however, many limitations exist up to day. The main issue is that most of the TCO methods developed so far only consider the cost but neglect the performance of the system, which has significant limitations [11]. A crucial point in order to understand the applicability of a TCO model for supporting physical asset management is that the evaluation criteria for the costs elements definition should encompass not only all incurring cost elements along the asset life cycle but also system performance characteristics, like system availability, in upfront decisions for achieving the lowest long-term cost of ownership [4, 6, 12].

Indeed, some main issues should be considered when approaching the TCO evaluation of a production system as a support for decision making:

i. a large number of variables directly and indirectly affect the real cost items and are affected by uncertainty in their future evolution (e.g. inflation, rise/decrease of cost of energy, cost of raw material, cost of labor, budget limitations, etc.) [13, 14];
ii. the evolution of asset behavior in the future is difficult to predict (e.g. aging of assets, failures occurrence, performance decay) and 'infinitely reliable' components or systems do not exist [15];
iii. complex relationships in the assets intensive system dynamics, due to the presence of many coupled degrees of freedom, make it not easy to understand the effects of local causes on the global scale [16];
iv. conventional cost accounting fails to provide manufacturers with reliable cost information due to the inability of counting the so-called invisible and, in particular, intangible costs, and thus there is inaccuracy in calculating total costs [17].

It is evident that additional research is required to develop better TCO models to quantify the risks, costs, and benefits associated with physical assets including uncertainties and system state and performance evaluations to generate informed decisions [18]. The objective of this chapter is to present a comprehensive methodology for the evaluation of the TCO of industrial assets that has being developed within a research activity carried out at the Department of Management, Economics and Industrial Engineering of Politecnico di Milano. The methodology is based on an integrated modelling approach putting together a technical model for the evaluation of the technical performances of the asset over its lifecycle (by accordingly generating the asset failure, repair and operation events) and a cost model for evaluating the final cost breakdown and the corresponding TCO calculation (Fig. 1). An industrial application case study has been implemented and first experimental findings of developed methodology are presented showing the relevance and potentialities of such approach for companies.

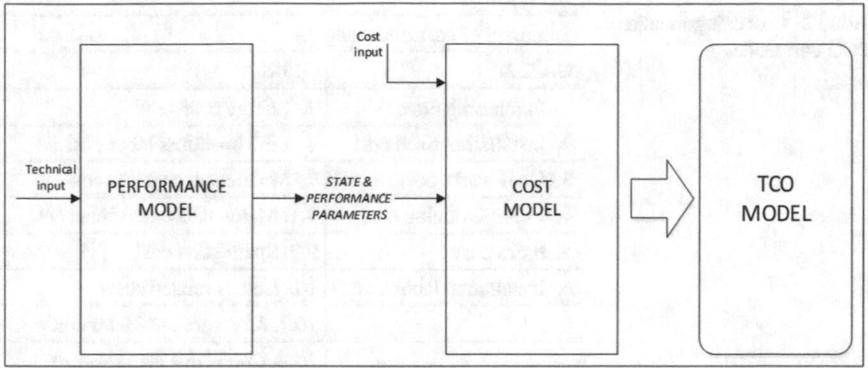

Fig. 1 Concept of integrated TCO evaluation model

3 Performance-Driven TCO Evaluation Methodology

The TCO methodology that is presented in this chapter, is based on the idea that only by the integration of a performance model and a cost model is it possible to develop a reliable TCO model to support strategic decision-making (Fig. 1).

The underlying assumption is that proper system modeling has to be introduced for availability, maintainability and operation and that it must be integrated with a cost model for economic evaluations.

Cost model: Whilst there is general agreement that all costs occurring along the life cycle of an asset should be included in the related TCO model, opinion varies as to their precise identification [4]. Several cost models have been proposed in literature and different ways to categorize the main cost items can be found. Some models group cost items depending on the life cycle phases of the asset, others refer to the two main categories CAPEX and OPEX. In spite of these different categorization approaches, in the end the detailed costs items list will depend upon the particular system under consideration and a cost break down structure (CBS) approach is commonly adopted [12, 19]. The important point is that the cost structure must be designed so that the analyst can perform the necessary TCO analysis and trade-offs to suit the objectives of the project and the company concerned [4]. Table 2 shows which is the CBS that has been defined for the specific case study that is presented in Sect. 4. The specific cost model includes the main cost items that are usually considered by a manufacturing company for evaluating different design solutions for production systems.

A relevant issue that must be taken into account and that is mostly undervalued in practice, is the need to include those cost elements that depend on the performance of the system within the cost model. For example, it has to be considered that when an asset fails, the cost is not limited to the cost of repair or replacement (in terms of manpower and material), but the money lost because the asset is out of service must be included as well [20]. The same is valid for other performance losses consequences (ex. quality losses, speed losses etc.). All these aspects must be considered within a

Table 2 Cost categories in TCO cost model

Summary of costs per category	
CAPEX	OPEX
1. Purchasing price	7. Energy cost
2. Installation fixed cost	8. Line operators labor cost
3. Civil works cost	9. Maintenance visible cost
4. Commissioning cost	9.1. Maintenance personnel cost
5. Extra cost	9.2. Spare parts cost
6. Installation labor cost	10. Losses related costs
	10.1. Management losses costs
	10.2. Corrective maintenance downtime losses costs
	10.3. Speed losses costs
	10.4. Non-quality costs
	10.5. Labor Savings
	End of life costs and savings
	11. Decommissioning costs

complete TCO model, hence it is necessary to evaluate and to quantify factors that allow predicting the form in which the production processes can lose their operational continuity due to events of accidental failures and to evaluate the impact in the costs that the failures cause in security, environment, operations and production [14, 21].

To this regard, a widely used performance measure in the manufacturing industry is overall equipment effectiveness (OEE) originally introduced by Nakajima [22] and Jönsson et al. [23]. It is clear that for making asset management decisions it is important to have a thorough insight into all involved costs and their impacts on the profit and competitiveness. Managers need to consider the trade-offs between the amount of investment and its impact on the OEE and TCO becomes an indicator required for competitiveness analysis [24]. The following Fig. 2 shows which are the losses that have been considered into the cost model in the methodology herein proposed, by referring to OEE. The identified losses (availability, performance and quality losses) lead to specific cost items in the OPEX category of the cost model (Table 2, cost items under category 10) and they must be evaluated through a performance model as it is detailed in next section.

Performance Model: As assessed above, system state and performance evaluation is an essential step that needs to be developed to feed with the proper inputs the cost model, hence to evaluate the real TCO referring to an asset.

Obviously in complex systems, OEE should be calculated at system level, by correctly considering the result of dynamic interactions among various system components (i.e. individual assets). This issue has been identified by Jonsson and Lesshammar [25]; Muchiri and Pintelon [26]; Muthiah and Huang [27]; and the latters introduced the term overall throughput effectiveness (OTE) as a factory-level version of OEE that takes the dependability of equipment into account. Some approaches have been proposed in literature in order to try and face the quantification of costs related to system unavailability. On one hand, the most traditional approach is to use ex-post analysis as a calculation based on historical or actual data; applying the traditional RAM analysis based on statistical calculations or probabilistic fittings. On the other hand, great potentialities are added by applying ex-ante estimation aiming at a static or dynamic prediction of total costs through estimated behavior over the life cycle [28]. Within this second perspective, some works have been proposed in literature suggesting the use of stochastic point process [5, 14, 29] and some others propose the use of simulation based on the Monte Carlo technique [18, 30, 31]. In this work, the stochastic simulation is proposed for modeling the casual nature of stochastic phenomena and the Reliability Block Diagram (RBD) logic is used to express interdependencies among events thus evaluating how individual events impact over the whole system (Fig. 3).

The Monte-Carlo method is used for generating random events relying on the statistical distribution functions of the time before failure (TBF) and time to repair (TTR) variables given as input values at component level. Both failures modes and stops of the system related to other reasons (such as operations problems) can be considered.

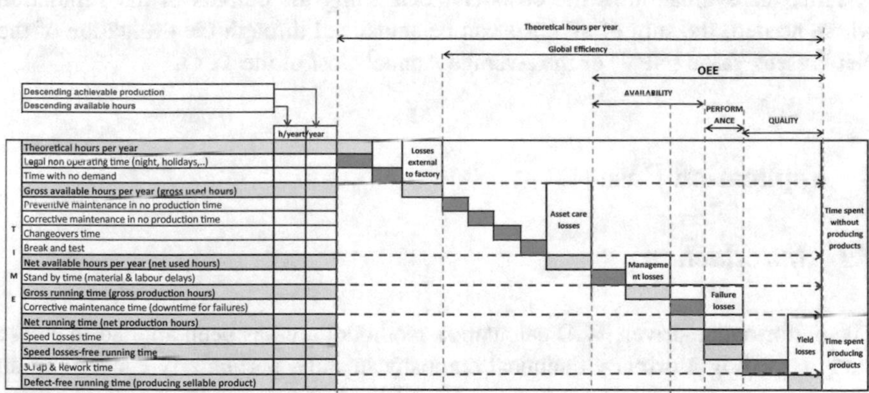

Fig. 2 Outline of losses and OEE calculation scheme

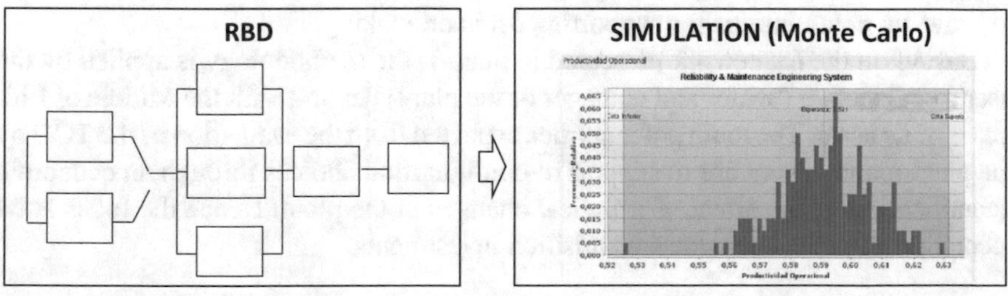

Fig. 3 Tools for the performance evaluation model

Using the simulation technique, the system behavior can be generated in a series of random iterations by calculating as a final result, a statistical estimate value of operational availability and OTE for the complete system. One of the main disadvantages of the use of simulation is the high effort that it requires for making the system model and data preparation [14]. To this regard, new approaches are introducing the use of some conventional modeling techniques such RBD for simulation purposes [32–34]. In fact the RBD logic has the advantages of giving a systemic, integrated and very compact view of the system with a bottom-up perspective while keeping an easy implementation approach. In order to ease application, these concepts have been embodied recently in several software based tools for asset management which use simulation (such as for example Availability Workbench™ by ARMS reliability; Relex or R-MES Project©). Within this approach, aspects that go beyond the pure unavailability evaluation determined by asset failures can be considered such as production losses due to system performance or quality reduction.

This approach has been adopted in the proposed model for the evaluation of technical performances. The performance model allows evaluating the OTE of a system by taking into account assets behavior and dependability during equipment lifecycle. Such information is a relevant input for the evaluation of the OPEX cost components within the cost model (Table 2).

After the evaluation of the costs elements using the outputs of the simulation where needed, the sum of all costs can be actualized through the evaluation of the Net Present Value (NPV) or the Average Annual Cost of the TCO.

4 Application Case

4.1 Introduction

The performance-driven TCO calculation methodology has been applied in a case study regarding a primary chemical company in Italy, particularly concerning an industrial line for rubber production. Next Fig. 4 shows the basic process flowsheet and the main equipment composing the plant section under analysis. The main objective of the case study is to apply the developed TCO evaluation methodology to prove its potentialities for supporting decision making.

Basing on the framework presented in Sect. 1, the methodology is applied by the user's perspective (owner and manager of the plant) dealing with the Middle of Life phase of its asset. The main potentialities expected from the evaluation of the TCO by the plant management are to support re-configuration choices through an economic quantification of the effect of technical changes in the plant. Hence the focus is on reconfiguration decisions/new acquisition investments.

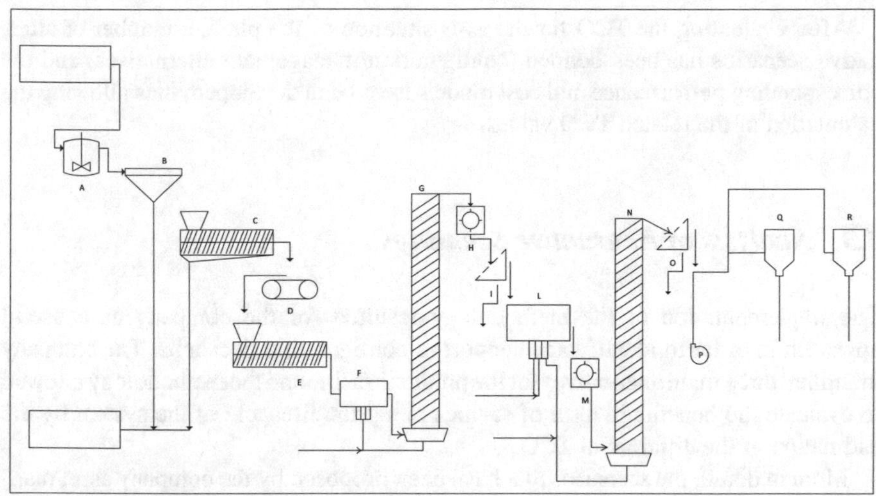

Fig. 4 The case-study production line

4.2 *TCO Evaluation Procedure*

The case is based on the use of the TCO evaluation methodology that has been presented above based on the cost model and performance model. In particular, the following steps have been developed for the application case.

Performance evaluation:

STEP 1. Process understanding and system's components identification.
STEP 2. Identification of failures modes or stop causes of each component.
STEP 3. Reliability, maintainability and operation data acquisition (TBF and TTR)
STEP 4. Modeling of the as-is system through RBD logic
STEP 5. Simulation (Monte Carlo)
STEP 6. Technical performance calculation of the system

On the basis of the given situation, 156 equipment have been put in the model and simulation runs (200 runs) were conducted in order to calculate the operational availability and OTE of the as-is situation over a time span of 5 years.[1] Such data was used as one of the inputs for the following cost modelling phase.

Cost evaluation:

STEP 7. Cost model setting (Table 2)
STEP 8–9. Cost data acquisition and Calculation of TCO

[1]The reliability oriented engineering software R-MES Project© (Reliability Maintenance Engineering System Project) is used for performing the above mentioned modelling and calculation steps from 4 to 6.

After evaluating the TCO for the as-is situation of the plant, a number of alternative scenarios has been defined (configuration/management alternatives) and the corresponding performance and cost models have been developed, thus allowing the calculation of the related TCO values.

4.3 Analysis of Alternative Scenarios

The implementation of the methodology resulted for the company as a useful approach in order to identify and support re-configuration decisions. The company identified three main alternatives for the production line and the methodology allowed to evaluate the benefits in term of savings along the lifecycle of the system by the estimation of the differential TCO.

More in detail, the scenarios that have been proposed by the company asset managers for comparative evaluations are the following:

- Scenario A: the installation of a second machine of type E to be kept in stand-by with the already existing one;
- Scenario B: the disposal of the mechanical transport machine N and its substitution with a pneumatic transport system;
- Scenario C: The installation of three more screens in stand-by to the existing ones.

After implementing the methodology for the as-is situation and the three alterative ones, the technical outputs in terms of OEE (that are showed in Table 3) have been used to make the economic evaluation by combining them with the related cost inputs.

In particular, for each scenario, the differential costs and savings with respect to the as-is situation have been considered (such as, energy consumption, acquisition and installation costs, end of life disposal cost for the new equipment etc.), as well as the additional margin resulting from the increase in production volume.

After establishing a lifetime period for the evaluation of the various scenarios, the TCO cost calculation model allowed the company estimating the money cash-flow over the asset lifecycle and the payback time related to the investment required by each scenario. These data are not presented due to confidentiality reasons, however the results were very promising and attracted the attention of the company management asking for a more detailed estimation work.

Table 3 Results of OEE improvements in the investigated scenarios

	Scenario A (%)	Scenario B (%)	Scenario C (%)
Delta OEE	+4.52	+0.73	+2.58

4.4 *Benefits and Limitations*

After the case was developed and results generated, the plant management confirmed the usefulness of the model as a tool for supporting investment decisions by proving the return of an investment taking into account the life of the asset and its performance along it, going beyond the pure acquisition cost. The use of RAMS modeling techniques combined with Monte Carlo simulation engine provided a fast way to evaluate trade-offs among availability and redundancy. It resulted that performance analysis and reliability engineering are fundamental for financial and economic evaluations referring to capital-intensive asset systems. During the development of the case some criticalities emerged that need to be overcome in the future. In particular, the main limit was found at the data acquisition step. In fact, data regarding the past failures and repair events where spread among different sources and not complete to be used. This limit was overcome through the use of estimations asked to the plant experts of TBF and TTR values. The estimations allowed building triangular distributions for the two variables for each component to be used for the simulation. Anyways, it is evident that a reliable and complete historical data base would have made the calculations more precise through a fitting of the distributions over the real data.

5 Conclusions

TCO is seen a useful indication for guiding asset managers in the decision making process by companies and the main value is that it is a synthetic economic value including in itself a lifecycle vision and technical evaluations. TCO can be used as a management decision tool for harmonizing the never ending conflicts by focusing on facts, money, and time [35] and, if properly estimated it does represent a competitive advantage for companies.

Up to day, there are still a number of difficulties that limit a TCO model widespread adoption by industry and there is no single model that has been accepted as a standard. As it is pointed out by Al-Hajj and Aouad [9] the desire to implement life cycle costing was much talked about but little practiced. This can be attributed to several major obstacles which also emerged through the application case: (i) absence of a database and systematic approach to collect and analyze the significant amount of information generated over the life of projects [4], (ii) general lack inside the organizations of the adequate consideration of the entire asset life cycle that requires inter-functional cooperation and alignment [36–38], (iii) establishment of the more appropriate modelling approach for evaluating the technical performances of the asset over its lifecycle by accordingly generating the asset failure, repair and operation events.

The research work presented in this chapter is following these issues moving in the direction of integrating technical performance and cost models so to be able to develop a realistic evaluation of the TCO of an asset over its estimated lifecycle. By using simulation together with RBD modeling of the system under study, allows to easily evaluate the technical performances of production systems in a computer environment. On the other hand, the use of an appropriate cost model can support management in a decision making process which is oriented to the whole asset life cycle. This approach allows combining the reliability engineering concept to the economic and financial evaluation of investments translating them into the money-language which is essential to make the connection between asset management and profitability. Future research must include in the models also intangibles problems that are not necessarily related to production losses, but that lead to costs for the company. Moreover, more case studies may be developed to make the methodology generalizable.

Acknowledgements This research work has taken advantage from the application of TCO modelling methodology in an industrial case-study, hosted in the production premises of ENI Versalis in Ravenna (Italy). Many thanks are due to Mr. Saverio Albanese for his trusting in this research topic and to the Versalis Ravenna crew for their kind and passionate support.

References

1. ISO 55000:2014(E). (2014). Asset management—Overview, principles and terminology.
2. Komonen, K., Kortelainen, H., & Räikkonen, M. (2006). An asset management framework to improve longer term returns on investments in the capital intensive industries. In J. Amadi-Echendu (Ed.), *Engineering asset management* (pp. 418–432). London: Springer.
3. Parra, C., Crespo, A., & Moreu, P. (2009). Non-homogeneous Poisson process (NHPP), stochastic model applied to evaluate the economic impact of the failure in the life cycle cost analysis (LCCA). *Safety, Reliability and Risk Analysis: Theory, Methods and Applications*, 929–939.
4. Woodward, D. (1997). Life cycle costing—Theory, information acquisition and application. *International Journal of Project Management, 15*(6), 335–344.
5. Lad, B. K., & Kulkarni, M. S. (2008). Integrated reliability and optimal maintenance schedule design: A life cycle cost based approach. *International Journal of Product Lifecycle Management, 3*(1), 78.
6. Clarke, J. (1990). *Life cycle cost: An examination of its application in the United States, and potential for use in the Australian Defense Forces*. Monterey, California: Naval Postgraduate School.
7. Schuman, C. A., & Brent, A. C. (2005). Asset life cycle management: Towards improving physical asset performance in the process industry. *International Journal of Operations & Production Management, 25*(6), 566–579.
8. ISO 55001:2014(E). (2014) Asset management-management systems-requirements, Section 6.2.2.4.
9. Al-Hajj, A., & Aouad, G. (1999). The development of an integrated life cycle costing model using object oriented and VR technologies. In *Proceedings of the Information Technology in Construction: CIB W78 Workshop Durability of Building Materials and Components*.
10. Roda, I., & Garetti, M. (2014). TCO evaluation in physical asset management: Benefits and limitations for industrial adoption. In *APMS Proceedings 2014, to be published* (Vol. 2014).
11. Chen, G., Zheng, S., Feng, Y., & Li, J. (2013). Comprehensive analysis of system reliability and maintenance strategy based on optimal lifecycle cost. In *2013 International Conference on Quality, Reliability, Risk, Maintenance, and Safety Engineering (QR2MSE)* (pp. 654–658).

12. Kawauchi, Y., & Rausand, M. (1999). *Life cycle cost (LCC) analysis in oil and chemical process industries*. Chiba: Toyo Engineering Corporation.
13. Durairaj, S., Ong, S., Nee, A., & Tan, R. (2002). Evaluation of life cycle cost analysis methodologies. *Corporate Environmental Strategy, 9*(1), 30–39.
14. Parra, C., & Crespo, A. (2012). Stochastic model of reliability for use in the evaluation of the economic impact of a failure using life cycle cost analysis. Case studies on the rail freight and oil. *Journal of Risk and Reliability, 226*(4), 392–405.
15. Saleh, J. H., & Marais, K. (2006). Reliability: How much is it worth? Beyond its estimation or prediction, the (net) present value of reliability. *Reliability Engineering & System Safety, 91*(6), 665–673.
16. Xu, Y., Elgh, F., & Erkoyuncu, J. (2012). Cost Engineering for manufacturing: Current and future research. *International Journal of Computer Integrated Manufacturing*, 37–41 (May 2013).
17. Chiadamrong, N. (2003). The development of an economic quality cost model. *Total Quality Management and Business Excellence, 14,* 999–1014.
18. Shahata, K., & Zayed, T. (2008). Simulation as a tool for life cycle cost analysis. In *Proceedings of the 40th Conference on Winter Simulation* (pp. 2497–2503).
19. Asiedu, Y., & Gu, P. (1998). Product life cycle cost analysis: State of the art review. *International Journal of Production Research, 36*(4), 883–908.
20. Waghmode, L. Y., & Sahasrabudhe, A. D. (2012). Modelling maintenance and repair costs using stochastic point processes for life cycle costing of repairable systems. *International Journal of Computer Integrated Manufacturing, 25*(4–5), 353–367.
21. Woodhouse, J. (1991). Turning engineers into businessmen. In *14th National Maintenance Conference*. London.
22. Nakajima, S. (1988). *Introduction to TPM*. Cambridge: Productivity Press.
23. Jönsson, M., Andersson, C., & Ståhl, J.-E. (2013). Conditions for and development of an information technology-support tool for manufacturing cost calculations and analyses. *International Journal of Computer Integrated Manufacturing, 26*(4), 303–315.
24. Jabiri, N., & Jaafari, A. (2005). Promoting asset management policies by considering OEE in products' TLCC estimation. In *IEEE International Engineering Management Conference Proceedings* (pp. 480–484).
25. Jonsson, P., & Lesshammar, M. (1999). Evaluation and improvement of manufacturing performance measurement systems—The role of OEE. *International Journal of Operations & Production Management, 19*(1), 55–78.
26. Muchiri, P., & Pintelon, L. (2008). Performance measurement using overall equipment effectiveness (OEE): Literature review and practical application discussion. *International Journal of Production Research, 46*(13), 3517–3535.
27. Muthiah, K. M. N., & Huang, S. H. (2007). Overall throughput effectiveness (OTE) metric for factory-level performance monitoring and bottleneck detection. *International Journal of Production Research, 45*(20), 4753–4769.
28. Thiede, S., Spiering, T., & Kohlitz, S. (2012). Dynamic total cost of ownership (TCO) calculation of injection moulding machines. In *Leveraging technology for a sustainable world* (pp. 275–280). Berlin: Springer.
29. Karyagina, M., Wong, W., & Vlacic, L. (1998). Life cycle cost modelling using marked point processes. *Reliability Engineering & System Safety, 59*(3), 291–298.
30. Heilala, J., Helin, K., & Montonen, J. (2006). Total cost of ownership analysis for modular final assembly systems. *International Journal of Production Research, 44*(18–19), 3967–3988.
31. Rühl, J., & Fleischer, J. (2007). Life cycle performance for manufactures of production facilities. In *14th CIRP International Conference on Life Cycle Engineering* (pp. 11–13). Tokyo, Japan.
32. Macchi, M., Kristjanpoller, F., Garetti, M., Arata, A., & Fumagalli, L. (2012). Introducing buffer inventories in the RBD analysis of process production systems. *Reliability Engineering & System Safety, 104,* 84–95.
33. Manno, G., Chiacchio, F., Compagno, L., D'Urso, D., & Trapani, N. (2012). MatCarloRe: An integrated FT and Monte Carlo Simulink tool for the reliability assessment of dynamic fault tree. *Expert Systems with Applications, 39*(12), 10334–10342.

34. Roda, I., Garetti, M., Arata, A., & Heidke, E. (2013). Model-based evaluation of asset operational availability. In *XVIII Summer School "F. Turco". 11/9/2012-13/9/2012, Senigallia.*
35. Barringer, H. (2003). A life cycle cost summary. In *Conference of Maintenance Societies (ICOMS®-2003)* (pp. 1–10).
36. Amadi-Echendu, J. (2004). Managing physical assets is a paradigm shift from maintenance. In *Proceedings of the 2004 IEEE International Engineering Management Conference, 2004* (pp. 1156–1160).
37. Amadi-Echendu, J., Willett, R., & Brown, K. (2010). What is engineering asset management? In *Definitions, concepts and scope of engineering asset management* (pp. 3–16). London: Springer.
38. Markus, G., & Werner, S. (2012). Evaluating the life cycle costs of plant assets: A multidimensional view. *Serbian Journal of Management, 7*(2), 287–298.

Chapter 4
Defining Asset Health Indicators (AHI) to Support Complex Assets Maintenance and Replacement Strategies. A Generic Procedure to Assess Assets Deterioration

Adolfo Crespo Márquez, Antonio de la Fuente Carmona, Antonio J. Guillén López, Antonio Sola Rosique, Javier Serra Parajes, Pablo Martínez-Galán Fernández and Juan Izquierdo

Abstract An Asset Health Index (AHI) is a tool that processes data about asset's condition. That index is intended to explore if alterations can be generated in the health of the asset along its life cycle. These data can be obtained during the asset's operation, but they can also come from other information sources such as geographical information systems, supplier's reliability records, relevant external agent's records, etc. The tool (AHI) provides an objective point of view in order to justify, for instance, the extension of an asset useful life, or in order to identify which assets from a fleet are candidates for an early replacement as a consequence of a premature aging. The purpose of this Chapter is to develop a generic procedure to easy obtaining an AHI that can reasonably measure the current degradation of the assets, offering the pos-

---

A. Crespo Márquez (✉) · A. de la Fuente Carmona · A. J. Guillén López · J. Serra Parajes · P. Martínez-Galán Fernández
Intelligent Maintenance System Research Group (SIM), Department of Industrial Management, School of Engineering, University of Seville, Camino de los Descubrimientos s/n, 41092 Seville, Spain
e-mail: adolfo@us.es

A. de la Fuente Carmona
e-mail: antoniiodela84@gmail.com

A. J. Guillén López
e-mail: ajguillen@us.es

J. Serra Parajes
e-mail: jserra@enagas.es

P. Martínez-Galán Fernández
e-mail: pablomgf93@gmail.com

A. S. Rosique
INGEMAN. Association for the Development of Maintenance Engineering, School of Engineering, Camino de los Descubrimientos, 41092 Seville, Spain
e-mail: asrasrasr@telefonica.es

J. Izquierdo
Ik4-Ikerlan Technology Centre, Operations and Maintenance Technologies Area, Gipuzkoa, Spain
e-mail: jizquierdo@ikerlan.es

A. Crespo Márquez et al. (eds.), *Value Based and Intelligent Asset Management*,
https://doi.org/10.1007/978-3-030-20704-5_4

sibility to compare their status to the one expected for the at their age and functional location. As a result of the procedure, an organization will have an objective tool to prioritize interventions, attention and the renewal of significant equipment. In this Chapter we first review the concept and the most relevant models in the literature, then we introduce the methodology and different examples related to the different steps. Finally conclusions of the chapter are presented.

Keywords Asset management · Capital investment · Operation and maintenance decision-making · Life cycle analysis · Assets health

1 Introduction and Scope

Asset Health is a measure of the condition of an asset and the proximity to the end of its useful life, as a consequence of its deterioration. As the health of an asset deteriorates (i.e. its condition worsens), the likelihood that it will fail due to its condition increases. The purpose of an asset health assessment is to detect and quantify its long-term degradation and to provide a means of quantifying remaining asset life. This includes identifying whether the asset is at or near end of life and assets that are at high risk of generalized failure that will require major capital expenditures to either refurbish or replace the assets.

It is important to notice that in this chapter we concentrate on long-term asset degradation and asset replacement, not in the study of assets functional failures and corrective maintenance. The difference of this two perspectives are clear [1]:

- *Functional failures* are associated with failure modes in the ancillary systems that affect operation and reliability of the asset well before its end of life. These failures do not normally affect the life of the asset itself, if detected early and corrected. Defects are routinely identified during inspection and dealt with by corrective maintenance activities to ensure continued operation of the asset.
- *Long-term degradation* is generally less well defined and it is not easily determined by routine inspections.

In any case, there are also AHI models that are used as tools in the field of Prognostics and Health Management (P.H.M.) [2–4].

2 The AHI Concept and the Complexity of the Asset

Asset Health indices (HI) represent a practical method to quantify the general health of a complex asset. Most of these assets are composed of multiple subsystems, and each subsystem can be characterized by multiple modes of degradation and failure.

In some cases it can be considered that an asset has reached the end of its useful life, when several subsystems have reached a state of deterioration that prevents the

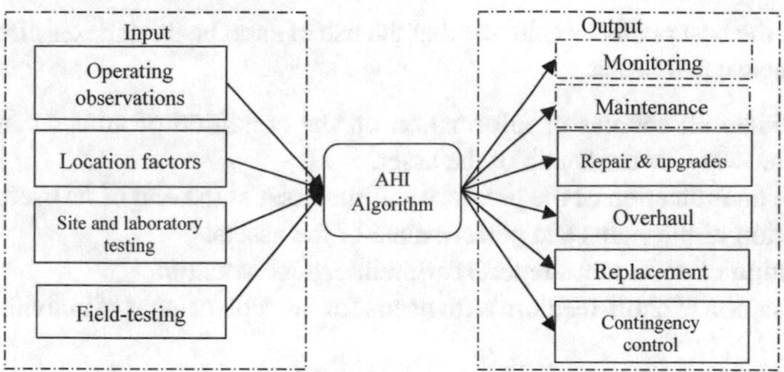

Fig. 1 The AHI concept

continuity of the service required by the business [5]. For this reason, the health index, based on the results of operational observations, field inspections and laboratory tests, produces a single objective and quantitative indicator. Health index can be used as a tool to manage assets and identify capital investment needs and potential improvements of maintenance programs [6–8], see Fig. 1.

For instance Asset Health Indicators are widely used in supporting maintenance and replacement strategies based on asset condition and performance in UK electricity companies, to justify asset replacement schemes to the regulator. They are normally a composite metric intended to provide an indication of asset capability [see 9].

In addition to the factors related to the asset condition and operation, the health index also has to contain static factors linked to its location. That is, independent of the asset itself and that do not change over time unless the environmental conditions are modified [10].

Taking into account all previous considerations, a proper design of a health index should meet the following requirements [5]:

- The index must be indicative of the suitability of the asset to provide continuity to the service and representative of the general health of the asset.
- The index should contain objective and verifiable measures of the condition of the asset, instead of subjective observations.
- The index must be understandable and easily interpreted.

3 New Capabilities and Challenges for the AHI Implementation

Currently, in order to respond to the increasingly demanding requirements in terms of asset management, the application of the AHI models offer the possibility of improving the decision making process of the asset manager. After a review of the literature, the best practices coincide that the use of asset health indexes offers a wide range of new capabilities:

- Consolidate all sources of information on the condition of an asset in a single integrated view of the health of the asset.
- Provide an indication of the proximity of the asset at the end of its useful life.
- Evaluation of the status and performance of the assets.
- Generation of continuous reports for maintenance attention.
- Identification of short-medium term needs for the replacement of individual equipment.
- Predicting the long-term needs of replacing a large volume of assets, identifying potential peaks with investment requirements.
- Identify problems, risks and opportunities for maintenance management.
- Provide information on trends of rapid deterioration of assets that do not correspond with the rates of natural processes of aging, which can be useful to plan appropriate maintenance strategies.
- Comparison of the condition of assets between classes and locations, which allows taking actions in the operation and maintenance strategy of the organization.

But besides these interesting new capabilities provided, organizations may face new challenges for the introduction of AHIs to support their strategic processes. Deloitte & the Canadian Electricity Association [11] found that in order to improve asset management through the use of AHIs, the following considerations must be taken into account:

- Data collection has a cost. The capture of certain information requires a person in the field to carry out the inspection and record the data.
- The location and scale of the assets influences the analytical priorities. Organizations with large asset bases will have to expand data collection efforts, while smaller ones will have to concentrate their efforts.
- Uncertainty in evaluating asset conditions can create inconsistencies in data collection.
- The inability to judge what information should be collected to assess the condition of the asset can affect the accuracy of the ratings.
- The service levels expected by the assets are not satisfactory.
- Uncertainty about the return on investment, the valuation of costs and the financing of replacement/renewal of assets, can make it difficult to determine the information.
- The aging of the infrastructure and the work of the staff can affect the ability of the organization to implement the processes of asset health indexes.
- The lack of consistent and compatible methods to record, store and reference information can cause errors in the analytical phase.

4 Previous Relevant AHI Models to This Research

All models for AHI in the literature have as inputs data of the condition, operation of the equipment and, in some cases, the information of the geographic location and the reliability of the spare parts used in the maintenance. As for the algorithms for the calculation of the indicator, it will be seen how in the different models all the sources of information are integrated through the weighting of factors and that it will depend on the level of maturity that the implementation of the model has in the sector. Likewise, the study of the output of the models and the recommendations that each one makes is carried out.

M1. Kinetrics Model. This model developed by Kinetrics, Canada, proposes the evaluation of the general condition of the condition of transformers.
M1.1. Inputs. The inputs to the model are service and diagnosis data, historical data of the operation variables, such as the load to which the transformer has been subjected, the number of operations in the Tap Changer, the results of the analysis of the laboratories of oil samples of the transformer (Oil quality, dissolved gas content, acidity, etc.), on-site tests carried out by technicians such as insulation tests, thermographs, checking the corrosion status, etc.
M1.2. Methodology for calculating the index. Using input variables data, the methodology assigns scores to the different subsystems of the asset (transformer) through personalized evaluation algorithms [6]. For each of one of the variables, the author proposes a weighting between 1 and 10, values close to 1, are assigned to those variables whose relationship with the aging of the equipment is very small or almost nil. On the contrary, for those variables that take higher values (greater than 5 points), they are condition variables that more accurately reflect the aging of the equipment. The variables of operation, such as the load factor of the transformers and the power factor, are also related to the speed of aging of the equipment, because they are good indicators that influence when the equipment operates outside the design conditions of the same.
M1.3. Outputs from the model. As an output of the model, the author proposes the composition of all normalized variables, in a single indicator that varies between 0 and 100. The value of 100 corresponds to a value of new equipment and the value of 0 corresponds to an item that has reached the end of its useful life and needs to be replaced because it has been left out of service. To do this, it has divided the variables into two large groups, on the one hand, the variables related to the Tap Changer of the transformer and on the other hand the others. For the variables of the first group, it has identified that most of the times that a transformer is out of service, it is a consequence of aging and failure modes related to parts of the equipment belonging to the Tap Changer, so they have a greater weight in the Health indicator HI. The value of the index, in turn, is related to a probability of failure of the equipment and also can be divided into different ranges with their respective interpretations and recommendations. In Fig. 2, for each value of the index, it is appreciated that as it decreases, the probability of failure increases [7].

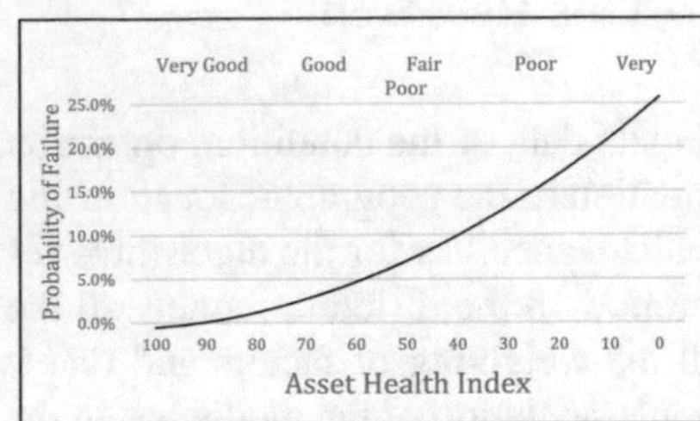

Health index and transformer expected lifetime

HI%	Condition	Expected Lifetime	Requirements
85 – 100	Very good	More than 15 years	Normal maintenance
70 – 85	Good	More than 10 years	Normal maintenance
50 – 70	Fair	From 3 – 10 years	Increase diagnostic testing, possible replacement depending on criticality
30 – 50	Poor	Lessthan 3 years	Start planning process to replace
0 - 30	Very poor	Near to the end of life	Immediately assess risk; replace or rebuild based on assessment

Fig. 2 PoF versus AHI and AHI implications in the Kinetrics model

In the case of the different health ranges, for each one of them, the author proposes a series of recommendations and measures to take into account for decision making in maintenance management. These can be very useful when planning future equipment substitutions based on the general condition of the assets and not based on age. The following table shows in a very general way the recommendations that the author proposes for each one of the ranges of the health index.

M2. Asset Health Index calculation model by DNV GL. This model developed by DNV GL, Arnhem, the Netherlands, proposes a methodology for the calculation of a health index, that together with the criticality of the asset and the business reliability constraints, helps in prioritizing maintenance, overhaul and substitutions of equipment [12].

M2.1. Inputs. The model uses statistical data of the useful life of the type of equipment that is to be studied; this allows a first estimate of the useful life that will later be corrected with the information of the asset condition (age, load, on-site condition analysis, number of PMs, etc.) and reliability (failure modes and frequency) throughout its life cycle of the asset.

M2.2. Methodology for calculating the index. The process to obtain the index is separated into three large blocks: degradation function, static function and condition function. The application of the model requires a previous estimate of the average age of the asset based on historical data (Statistic function). This average life becomes an average technical life that is subsequently corrected with the specific condition data of the asset. In a simple way, the model increases or decreases the end of the asset's technical life using this data (Condition function). At the same time, in parallel, the model assumes that the degradation of the asset depends on its use. The load influences the acceleration of equipment aging (number of operations, load, etc.), and equipment with a history of operation with more load will have shorter life than another with less historical load for the same age of the team (degradation function).

M2.3. Index outputs. The output of the model is the remaining useful life of the equipment in years. Once all the proposed methodological functions have been calculated, they are combined to give the result of the remaining life of the asset. To do this, the degradation functions are executed in parallel with the statistic and condition

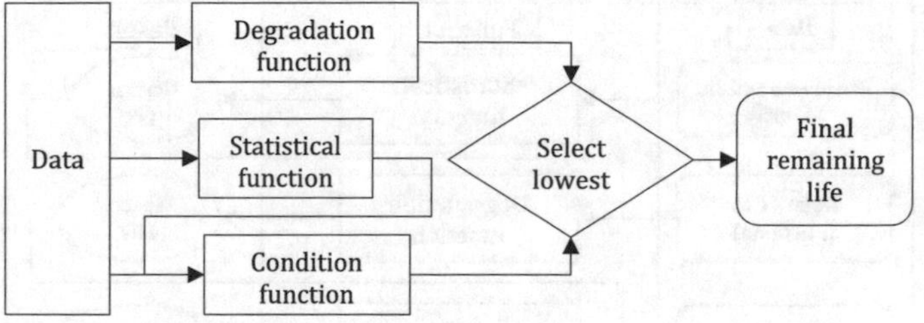

Fig. 3 DNV model process for remaining useful life calculation

function. The result takes into account the lowest estimate of the two branches [12], as in Fig. 3.

M3. UK DNO common network asset indices methodology [13]. This model is presented within a so-called framework of common reference, presenting the principles and calculation methodology adopted by all British power network operators for the regulatory assessment, prediction and report of the risk of assets. This framework is in compliance with the requirements of the standard condition 51 (SLC 51) of the electricity distribution license for RIIO-ED1 (1 April 2015 to 31 March 2023).

M3.1. Inputs. The model also uses statistical data for the useful life of the type of equipment that is to be studied; this allows a first estimate of the useful life that will later be corrected with the information of the functional location and duty of the asset, as well as with relevant condition and reliability facts (health and reliability modifiers) that may occur throughout the life cycle of the asset.
M3.2. Methodology for calculating the index. The methodology proposed is a sequential calculation of the index following three steps/blocks. The blocks can be considered similar to the previous model statistic, degradation and condition function. The degradation takes into account, besides the expected life cycle, the functional location features that could impact on degradation (duty, environment, geographical location, etc.), finally the degradation function is modified according to observed condition in so-called health and reliability modifiers. The model proposed in this chapter has many points in common with this model, which we will later describe in details (Fig. 4).
M3.3. Index outputs. The output of the model is the AHI, a value within the interval (0.5–10), offering information about the remaining useful life of the asset. Once all the proposed methodological functions have been calculated, they are sequentially combined to give the result of the AHI of the asset, inked to its probability of failure.

M4. Asset Health Index calculation model by TERNA. This model developed by Terna Rete, Italy, proposes the calculation of the equipment health index based on static and dynamic parameters. Static parameters are associated with the location where the equipment is located, which are invariable in time and independent of the asset, for example, the recurrence of catastrophic phenomena, the probability of electrical storm, etc. Dynamic parameters are associated to the equipment and can

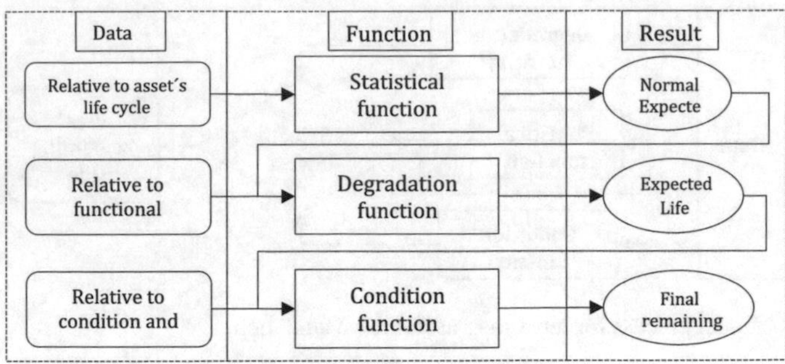

Fig. 4 DNO common network assets indices methodology

be measured in situ by functional and visual tests, as well as in laboratory tests by analysis of oil samples, lubricants, etc. The output of the model is an index between 0 and 0.5 which refer to the state as new and critical respectively. That is intended to justify technically and economically, making decisions for the investment of capital in replacement of equipment [10, 14].

M4.1. Inputs. The model author proposes static and dynamic variables for the model's inputs. Static variables do not depend on the asset itself but depend on the location (lightning frequency, catastrophic events, etc.). The dynamic variables proposed depend on the asset, and their value changes with the asset ages. Therefore, by capturing the change over time and comparing the maximum and minimum admissible for each kind of equipment, the condition status can be estimated at each moment of the asset life. The following condition parameters are those that are taken into account as inputs to the model [15], each parameter is known as health index (HI).

HI dielectric: parameters related to dielectric and thermal condition, as it may be obtained from dissolved gas analysis. These parameters are able to provide information on electrical (partial discharges, low energy discharges, arcing) and thermal problems (hot spots, overloads);
HI thermal: parameters related to pure thermal condition of the insulating paper, as they may be obtained from the CO_2, CO and further periodical determinations;
HI mechanical: parameters related to mechanical condition of the transformer, as they may be obtained from on-site electrical tests (inductance measurements, Sweep Frequency Response Analysis or SFRA, Frequency Domain Spectroscopy or PDC/FDS);
HI oil: parameters related to insulating oil condition, as they may be obtained by water content, acidity, 50–60 Hz breakdown voltage (BDV) and dielectric dissipation factor (DDF) determinations.

M4.2. Index calculation methodology. Due to the fact that different condition factors are very different from each other, they must first be standardized with their corresponding weights. In order to transform the value into a non-dimensional number, international guidelines and regulations (IEC, IEEE, CIGRE, etc.) are used.

Table 1 Health Index (HI) evaluation

Health index (HI)	Condition
0–0.10	Very good
0.10–0.20	Good
0.20–0.30	Fair
>0.30	Doubtful

Once the parameters have been standardized, from the following Eq. 1, the HI is calculated:

$$HI = \frac{HI_{dielectric} + HI_{thermal} + HI_{mechanical} + HI_{oil}}{HI_{MAX}} \quad (1)$$

where HIMAX is a prefixed number and, as a consequence, the HI of each asset may be expressed per units (p.u.).

M4.3. Model outputs. The model output is a HI value between 0 and 0.5. Higher and lower HI values are associated, respectively, to lower or higher levels of asset reliability. Depending on their HI, assets are classified in four classes. In Table 1, assets classified in "very good" and "good" condition may be managed following the common and standard maintenance practices, assets classified as "fair" or "doubtful" need an increase of analysis frequency or a deeper investigation [13].

The models that will be introduced in the Chapter are relevant, among other things, because:

- Asset managers need models to study options that maximise the value of an asset as it approaches the end of its useful life. Options may include (for example) changing the operating regime, partial asset replacement/refurbishment to extend useful life, or an indefinite ongoing 'patch-and-continue' programme, perhaps involving suppliers to provide necessary parts or services.
- Predicted performance supported by knowledge and asset information is available in many companies normally based on good understanding of how assets degrade but not incorporated in formal processes for capital investment. These models contribute to the decision process that seeks the optimal life cycle value.

5 A Practical Method for Asset Health Indexing

5.1 Background and Basic Definitions

The method to be introduced now takes into account several features of the previous models, with the intention to reach a practical procedure that can be applicable to any complex asset, regardless its technology, industrial sector, or location where it is used.

This procedure takes as a basic reference for its elaboration the last model reviewed in the previous section [13]. For instance, the definition of the following features are similar:

The asset health index (AHI) is considered as a dimensionless number between 0.5 (which corresponds to its status or condition as new equipment) and the value of 10 (corresponding to the condition of the equipment at the end of its useful life). The behavior pattern of the AHI, is supposed to be exponential along the age of the asset. The following figure shows the different 5 sections into which the health index of the asset is divided (Fig. 5).

The HI1 range comprised $0.5 \leq \text{AHI} \leq 4$ values; for which the behavior of the equipment is assumed to resemble as new equipment. The HI2 range considers AHI values within the interval $4 < \text{AHI} \leq 6$ and corresponds to the period of time when the first signs of deterioration begin to appear in the equipment. In this range, the value corresponding to $\text{AHI} = 5.5$, is assumed to be the health index value equivalent to the normal life expected for the equipment category. From this point, three intervals are considered in the methodology: HI3, HI4 and HI5 as the AHI exceeds the values of 6, 7 and 8 respectively. The methodology assumes that exceeded the value of AHI $= 8$, the equipment is at the end of its useful life.

The asset location factor. This is a factor is considered to be inherent to the functional location of the equipment, in the facility and the geographical area where it is located. Exposure to environmental agents, whether or not they are protected from the outside, distance from the coast or working at a certain altitude will have a different effect on the health of the equipment.

Some other features that are considered differ in their definition from previous existing models:

The asset load factor. The load factor of each equipment is defined by the relationship between the load of the equipment at its expected operating point (many times named *warranty point*) due to the fact that it is placed in a functional location of the installation, and the maximum admissible load that the equipment could support.

Fig. 5 AHI intervals

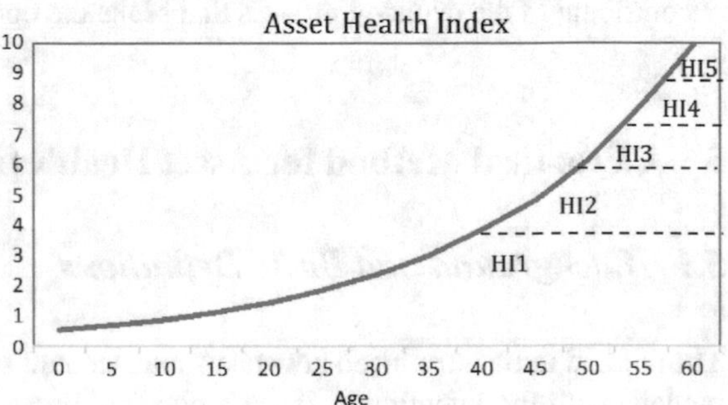

The load factor is therefore a consequence of the operating conditions for which the installation has been designed, and will take values within the interval [0, 1].
Health and reliability modifiers. Health modifiers are parameters that contain additional information about current health of the asset. The health modifiers are related to current load, condition of the asset (result of inspections and checks performed on it), and operation (result of captures made of operational variables of the asset, for example those existing in the plant information systems).
The load modifier ($M_L(c)$) will be introduced in the model when operating states of the equipment differ from the warranty point expected for its technical location. In general, this modifier will take values within the interval [0, 1]. When the load of the equipment is close to the warranty point of the technical location, it will be assigned the higher values within that interval, close to 1.

The condition and operation modifiers of the asset will take values within the range [1, 1.5]. Superior values will be assigned to modifiers or factors penalizing the health of the assets more, close to 1.5. As examples of modifiers that may appear, always depending on the equipment category, the following are cited below:

- Results of analyses and tests performed on the equipment, such as oil analyses, isolation tests, endoscopies, etc.
- Measures of continuous physical variables, such as the measurement of vibrations, etc.
- Analyses of operating data for a selected period, such as flow rate, rotation speed, operating temperature, equipment load, etc.

Reliability modifiers apply to assets for which reliability can significantly differ within the same category. In our model, these modifiers will have a value within the interval [0.9, 1.2]. The different reliability modifiers can be associated with:

- The manufacturer brand and the type or model of asset.
- The construction and integrity of the asset, the material, etc.
- Surface applied treatments.
- Number of overhauls and large maintenance performed on the equipment.
- Hours of inactivity of the equipment.
- The use of original spare parts.
- Etc.

5.2 Data Requirements

Data requirements for the procedure to be implemented will be as follows:

The identification of the asset. The category of the asset, the current age, the expected lifetime, name of the manufacturer, the model and functional location in the company.

- *Operation and Maintenance data.* Recorded during a certain period of time, added conveniently to know the time in which the equipment has been subjected to stress, number of starts and stops, energy consumption, number of overhauls, etc.
- *The condition of the equipment*. The results of the analyzes performed on the equipment in situ, results of readings of physical variables such as temperature and vibrations, results of visual inspections, insulation, thermography, etc.

5.3 Procedure

The methodology in this chapter is based on a procedure consisting in 5 consecutive steps. Starting from an estimated normal life associated with a category of equipment, the current health index of an asset is reached. With this purpose, a great variety of factors related to the functional location, and with its specific operation and condition are taken into account. This procedure is presented in Fig. 6.

Step 1. Asset selection and category definition. Capture of functional location data, physical asset data and obtaining the estimated normal life of the asset.

In this first step, the identification of the asset and all the information related to the functional location, such as:

- Functional location of the location in the plant management system where the equipment is located.

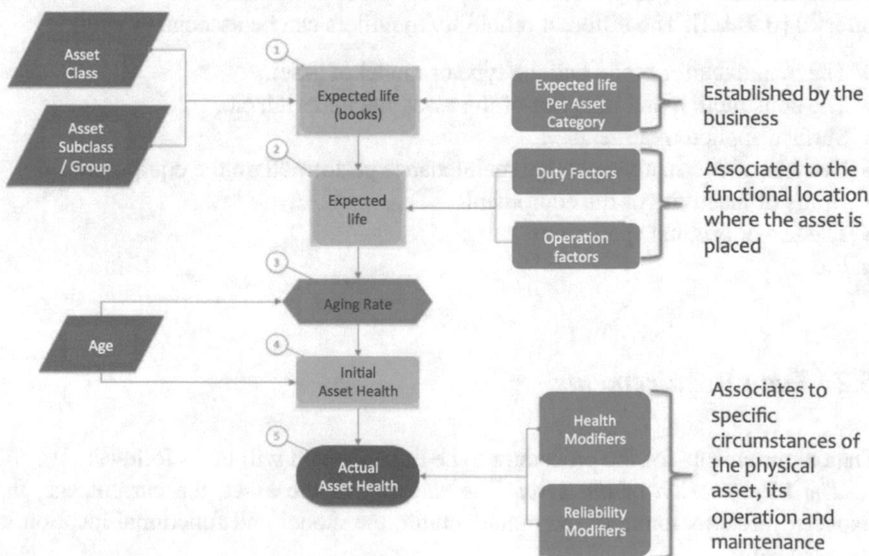

Fig. 6 Procedure to calculate the AHI

- Inside/outside situation of the asset. This parameter will be considered to determine the exposure to external agents that may affect the health of the asset.
- The distance to the coast. This parameter will be taken into account, like the previous one, to know the possibility that the humidity and the corrosive environment due to being close to the coast, may cause damage to the health of the asset.
- Average outside temperature. It is understood by the average of the exterior temperature, at the average annual temperature registered in the site and that can directly affect the performance of the equipment located there.
- Exposure to agents such as: dust in suspension and corrosive atmosphere. The proximity of the facilities to industrial emission sources of dust in suspension and corrosive agents, can cause the acceleration of equipment deterioration in the long-term.

Regarding the equipment information, we identify manufacturing data, model and technical design specifications for the preparation of tables that will be used later in the definition of health modifying parameters.

The estimated normal life of the asset is a data that, in general, comes from the technical direction of the company, considering the experience accumulated so far and the information provided by the different manufacturers.

The value of the estimated normal life is used as a starting point for the realization of all the calculations that will be seen below. Keep in mind that its value is approximate and only depends on the asset category. As we will see below, it will be modified by the characteristics of the location and loading.

Step 2. Impact's evaluation of load and location factors by type of asset, technical location and estimated life expectancy.

Once all the information in the previous point has been compiled, the location and loading factors are evaluated (unambiguously associated with the asset's technical location, as previously mentioned). Since there is more than one variable that affects each one of the factors, for each of them a single combined factor must be calculated. Below is its collection.

The location factor (F_L) is calculated from the information gathered from the asset's technical location. Tables 2, 3 and 4, serve as an example of how to translate possible location factors into values to be used in the calculation model.

Table 2 Sample conversion table for the "distance to coast" factor

Distance to the coast	
Distance	Value of the factor (F_{DC})
0 km–1 km	1.2
1 km–5 km	1.15
5 km–10 km	1.1
10 km–20 km	1.05
Superior a 20 km	1

Table 3 Sample conversion table for the "average annual temperature" factor

Average annual temperature	
Temperature (°C)	Value of the factor (F_T)
0–10	1
10–20	1.1
20–30	1.2
>30	1.3

Table 4 Exposition to external agents (corrosive atmosphere)

Exposition to external agent i (for instance: corrosive atmosphere)	
Assets location	Value of the factor (F_{EA})
No exposition	1
Low exposition	1.1
Average exposition	1.2
High exposition	1.3

For instance, depending on whether the asset is indoor or outdoor, there are two ways to estimate this factor. If the technical location of the asset is outdoor, the calculation of the combined location factor is calculated following Eq. 2, assuming that the factors listed above would be the relevant factor for the example. For situations in which the location of the asset is indoor, the location factor is equal to 1 ($F_E = 1$), except for assets in which other factors must be taken into account (for instance outside temperature).

$$F_{TL} = \max(F_{DC}, F_T, F_{EA}) \tag{2}$$

where:

F_{TC}: Combined technical location factor.
F_{DC}: Factor distance to the coast.
F_T: Factor average annual temperature.
F_{EA}: Factor exposure to external agents.

Notice that the load factor (F_L), as well as the location factor (F_{TL}), is inherent in the technical location. This factor measures the load request that is made on the equipment in that location, in front of the maximum admissible load. Normally, this data is obtained from the start-up and delivery of the equipment by the manufacturer, being recorded in the technical specifications of the equipment. In general, the following Eq. 3 is used:

$$\text{Load factor} = F_L = \frac{\text{Load under normal conditions or at warranty point}}{Maximumadmissibleload} \tag{3}$$

The Estimated Life of the asset is then defined as in Eq. 4.

$$Estimatedlife = \frac{Estimatednormallife}{F_{TL} \cdot F_L} \tag{4}$$

Step 3. Calculation of the aging rate

A fundamental hypothesis of the chosen methodology is that the aging of an asset has an exponential behavior with respect to its age. The aging rate is the parameter of the model that allows us to express mathematically this mode of behavior, and that incorporates the different phenomena that the asset can suffer throughout its useful life, such as corrosive phenomena, wear, oxidation of oils, breakage of insulation, etc.

The aging rate (β) it is determined by the relationship between the natural logarithm of the quotient between the health corresponding to the new asset and the health the asset will have when reaching the estimated life, and the calculated estimated life (see Eq. 5).

$$\beta = \frac{\ln \frac{HI_{new}}{HI_{estimatedlife}}}{\text{Estimated Life}} \tag{5}$$

where:

B: Aging rate.
$HI_{new} = 0.5$ Health index corresponding to a new asset;
$HI_{estimated\ life} = 5.5$ Health index corresponding to an asset arriving to its estimated life;

Step 4. Obtaining the Initial Health Index at the age t

The health index, as mentioned before is a dimensionless number between 0.5 and 10, with an exponential behavior with respect to the age of the asset (*t*), which is characterized by the aging rate of the equipment. For the calculation of the initial health index (HIi) of an asset, Eq. 5 is used, where *t* is the current age of the asset (in units of time) and the aging rate β is calculated in step Eq. 6.

$$HI_i(t) = HI_{new}e^{\beta t} \tag{6}$$

Figure 7 represents the evolution of the initial health index of an asset with an estimated normal life of 50 years, with a factor of location and load inherent to a particular technical location.

Step 5. Evaluation of the impact of load, health and reliability modifiers, and calculation of the Current Health Index

The health index (*HI*) adjusts the initial health index, as obtained in step 4, through the use of several load, health and reliability modifiers. In a first step, the initial health index of an asset (*HIi*) is modified to obtain what we call the modified initial health index (*HIMi*), taking into account the load modifier registered over the age of the asset, $M_L(t)$, using Eq. 7. The load modifier $M_L(t)$ will take values within the interval

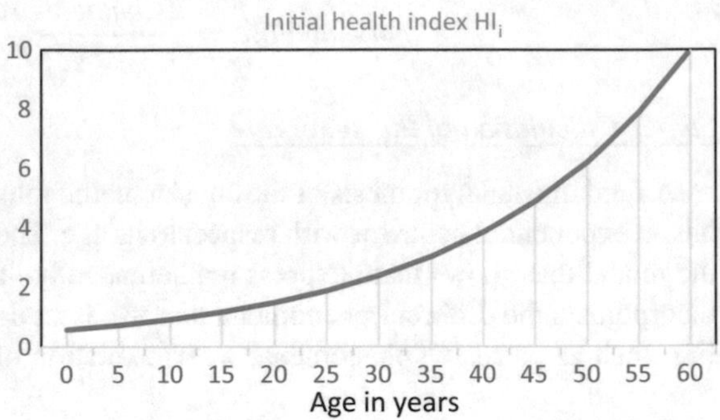

Fig. 7 Graph of the initial health index of an asset (HI_i)

[0, 1]. When the load of the equipment is close to the warranty point of the technical location, it will be assigned the higher values within that interval, close to 1. HIM_i captures the asset deterioration as compared to initially expected for the load factor (F_L) of the functional location.

$$HIM_i(t) = HI_{new} e^{\frac{\beta t}{M_L(t)}} \tag{7}$$

In a second step, the health index will be finally determined by the asset status, operating conditions and reliability conditions, at the time of evaluation, as in Eq. 8.

$$HI(t) = HIM_i(t) \cdot M_H(t) \cdot M_R(t) \tag{8}$$

where:

t: Age of the asset
HIM_i: Is the modified initial health index.
M_H: Is the asset's health modifier (condition and operation).
M_R: Is the asset's reliability modifier

The health modifier (M_H) in Eq. 8 can be evaluated through the assessment of condition and operation variables measuring potential asset degradation. As an example of these variables, Table 5 presents sets of health modifiers affecting different category of assets for a certain company. The weights, which are measuring the extent of the impact of each one of these variables on asset health, must also be established.

For the reliability modifier (M_R), and depending on the category of asset, model and manufacturer, tables can be drawn with the value of this parameter. Table 6 presents sets of reliability modifier.

In general, equations to calculate the values of the health modifier (M_H) and the reliability modifier (M_R), over time, could be similar to Eqs. 9 and 10 respectively.

$$M_H(t) = \prod_{j=1}^{j=n} M_{Hj}(t) \tag{9}$$

Table 5 Example of business available health modifiers per asset category

Measurable variable	Cryogenic pumps	Gas compressors	Turbo-compressor train	Alternative compressor
Operating flow	x	x		x
Fluid intake temperature	x	x		
Thermal jump between stages	x	x		x
No. startups/shutdowns	x	x	x	x
No. of alarms and shots due to low intake pressure	x			
Speed (rpm)			x	x
Borescope inspection			x	x
Exhaust gases analysis			x	x
Oil analysis		x	x	x
Vibration analysis	x	x	x	x
Motor isolation analysis		x		x

Table 6 Example of business available reliability modifiers per asset category

Reliability modifier	Cryogenic pumps	Gas compressors	Turbo-compressor train	Alternative compressor
Asset inactivity	x	x	x	x
Manufacturer's reliability	x	x	x	x
Overhaul number	x	x	x	x

where,

$j = 1 \ldots n$ number of health modifiers
$M_{Hi}(t)$: Is the health modifier *i* at current *age*.

And

$$M_R(t) = \prod_{k=1}^{k=m} M_{Rk}(t) \tag{10}$$

where,

$k = 1 \ldots m$ number of reliability modifiers
$M_{Rk}(t)$: Reliability modifier k at current *age*.

Both equations define the modifiers as products of factors taking values within the interval [1, 1.x]. Where x depends on the importance of the variable as a health modifier. To simplify, we will assume 1.5 (x = 5) to be the maximum potential impact of a modifier, and that their effect is a multiplicative effect. In addition, the result of multiplication will then have to be conveniently corrected, so that the value of HI can never be greater than 10, as initially stipulated.

In this way, a graphic representation of the evolution of the health index can be obtained for each of the subsystems of an asset, where the degradation speed of each of them is different, which leads to the planning of large maintenance to throughout the life cycle of the asset (Fig. 8).

Step 6. Presenting AHI results

Presenting results is a key point. There should be an easy way to appreciate the value of HIi (t) compared to HI (t) for an asset, to see extra deterioration. It is also important to show age compared to operating hours (an old item could not be very much utilized and vice versa) of the asset. Finally, it is very important to see relative health of a selected set of assets at a glance, to compare and concentrate attention to specific assets. Young assets with early deterioration are the main point to focus concerning their current maintenance assessment. Comparing old critical assets with relevant deterioration will be a must to determine assets replacement/restoration schedules.

This method proposes an XY graph where X is the age of the asset and Y is the asset health index (current vs. initial). Each asset is represented with a ball, current asset health (or deterioration), compared to expected, will be represented by the diameter of the ball (HI(t)–HIi(t)).

An example of application of the methodology is the case of a fleet of process pumps that are part of the critical systems of a plant. The functional loss of any of these pumps can result in unacceptable situations for the business, such as plant shutdowns, production losses or some type of impact with associated industrial or environmental damage.

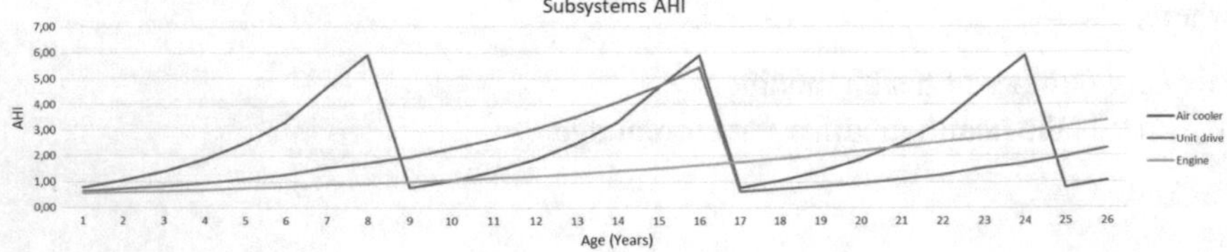

Fig. 8 Health index for the different subsystems of an asset

In order to establish the appropriate operation scenarios, the plans for major maintenance and equipment substitutions are made based on the health index and not depending on the hours of operation as usual.

The score obtained for the critical process pumps fleet (Fig. 9), appears ordered from lowest to highest AHI. Pumps with a lower AHI (H < 4), which are in the left area of the graph, have a condition and a failure rate that can be similar to that of a new equipment. On the other hand, pumps with a higher health index (H > 4) begin to behave like more aging equipment in which their failure rates begin to increase and consequently, the associated risk increases from the point of criticality.

The size of each circumference indicates the difference between the initial health index (HIi) and the health index (AHI). Circumferences with larger size indicate that the aging of the equipment has been greater than expected, so it will be necessary to pay more attention to these equipment and increase surveillance. The number that is inside each of the circles indicates the number of overhauls that have been made to the equipment.

For those pumps in which the AHI has exceeded 5.5 points, it is necessary to consider major repair or replacement in relatively close periods, including stopping operations until a deep diagnosis has been made to rule out possible problems during its functioning. In the example, pumps 4 and 10 should stop operating and raise the overhaul or replacement. Pump 11 with more hours of operation than pump 6 (17,000 vs. 15,000) has a lower aging (AHI = 3.5 vs. AHI = 4.1). This is because the load to which the pump is subjected and the results of the variables of the health modifiers are less aggressive in 11 than in 6. Probably, if the operating and contour conditions do not change, the overhaul of pump 11 will be carried out later (greater than 20,000 h of operation), however, pump 6 will have to be done before the manufacturer's recommendation.

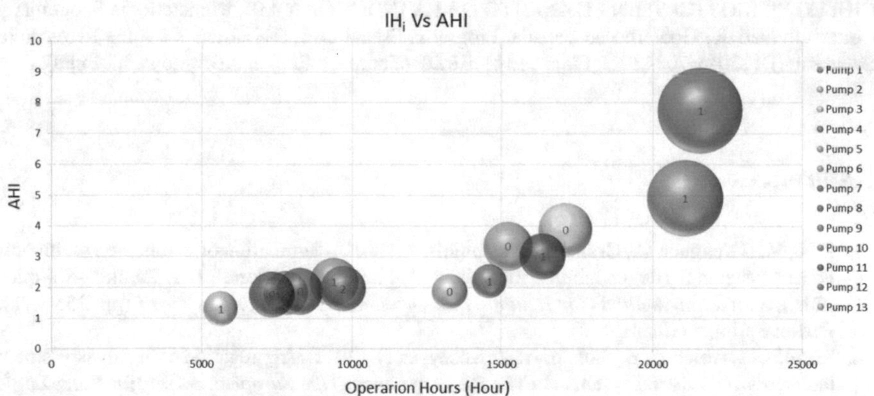

Fig. 9 Selected graph for asset health index (AHI) representation

6 Conclusions

Today, the use of tools for decision making about long-term renewal and replacement of equipment for organizations is quite extended. Thanks to life-cycle cost analysis (LCC), it is possible to know from an economic point of view, the cost of an asset over its useful life and to estimate the time for replacement if needed. The disadvantage in many cases, is the large amount of variables that must be handled when estimating the real cost of an asset over its useful life, generating a scenario of high uncertainty. At that moment, it is where the AHI comes into play as a support tool, having a completely different calculation methodology, estimated from lab tests in order to know the asset condition, visual inspections, operation and maintenance history and the age of the equipment and its components.

Simple variations of the procedure in this chapter gives the readers the possibility to adapt the methodology to their assets and sectors, also they will have the flexibility to fit the method to the existing information for the assets in the organization. Example of different aspects of each step of the procedure are provided.

Using asset health index offers a lot of advantages, such as, provide an approaching indication of the asset at the end of its useful life, prediction of long-term needs replacement in large volumes of assets, identifying potential peaks with investment requirements, identify problems, risks and opportunities for maintenance management, etc.

Acknowledgements This research work was performed within the context of Sustain Owner ('Sustainable Design and Management of Industrial Assets through Total Value and Cost of Ownership'), a project sponsored by the EU Framework Program Horizon 2020, MSCA-RISE-2014: Marie Skłodowska-Curie Research and Innovation Staff Exchange (RISE) (grant agreement number 645733—Sustain-Owner—H2020-MSCA-RISE-2014); and the project "DESARROLLO DE PROCESOS AVANZADOS DE OPERACION Y MANTENIMIENTO SOBRE SISTEMAS CIBERO FISICOS (CPS) EN EL AMBITO DE LA INDUSTRIA 4.0", Ministerio de Economía y Competitividad del Gobierno de España, Programa Estatal de I+D+i Orientado a los Retos de la Sociedad. (DPI2015-70842-R). Financed by ERDF (European Regional Development Fund).

References

1. Zille, V., Berenguer. C., Grall, A., & Depujols, A. (2010). Simulation of maintained multicomponent systems for dependability analysis. In A. Faulin, S. Martorell, & J. Ramirez-Márquez (Eds.), *Simulation methods for reliability and availability of complex systems* (pp. 253–272). Berlin: Springer (Chapter 12).
2. Rizzolo, L., Abichou, B., Voisin, A., & Kosayyer, N. (2011). Aggregation of health assessment indicators of industrial systems. In *The 7th Conference of the European Society for Fuzzy Logic and Technology, EUSFLAT*.
3. Abichou, B., Voisin, A., & Iung, B. (2015). Choquet integral capacity calculus for health index estimation of multi-level industrial systems. *IMA Journal of Management Mathematics, 26*(2), 205–224.
4. Abichou, B., Voisin, A., Iung, B., Do Van, P., & Kosayyer, N. (2012). Choquet integral capacities-based data fusion for system health monitoring. *IFAC Proceedings*, *45*(20), 31–36.

5. Hjartarson, T., & Otal, S. (2006). Predicting future asset condition based on current health index and maintenance level. In *ESMO 2006 - 2006 IEEE 11th International Conference on Transmission & Distribution Construction, Operation and Live-Line Maintenance*. https://doi.org/10.1109/TDCLLM.2006.340747.
6. Naderian, A., Cress, S., Piercy, R., Wang, F., & Service, J. (2008, July). An approach to determine the health index of power transformers. In *Conference record of the IEEE International Symposium on Electrical Insulation* (pp. 192–196). https://doi.org/10.1109/ELINSL.2008.4570308.
7. Naderian, A., Piercy, R., Cress, S., Service, J., & Wang, F. (2009). An approach to power transformer asset management using health index. *IEEE Electrical Insulation Magazine, 25*(2), 2. https://doi.org/10.1109/MEI.2009.4802595.
8. Azmi, A., Jasni, J., Azis, N., & Kadir, M. Z. A. (2017). Evolution of transformer health index in the form of mathematical equation. *Renewable and Sustainable Energy Reviews, 76,* 687–700. https://doi.org/10.1016/j.rser.2017.03.094.
9. Subject Specific Guidance. (2017). *Capital investment, operations and maintenance decision making*. IAM.
10. Scatiggio, F., & Pompili, M. (2013). Health index: The TERNA's practical approach for transformers fleet management. In *IEEE Electrical Insulation Conference (EIC)*, 178–182.
11. Deloitte/Canadian Electricity Association. (2014). *Asset health indices: A utility industry necessity*.
12. Vermeer, M., Wetzer, J., van der Wielen, P., de Haan, E., & de Meulemeester, E. (2015). Asset management decision support modeling, using a health and risk model. In *PowerTech, IEEE. Eindhoven*, 1–6. https://doi.org/10.1109/PTC.2015.7232556.
13. UK DNO Common Network Asset Indices Methodology (2017, January 30). Health and Criticality. Version 1.1.
14. Scatiggio, F., Rebolini, M., & Pompili, M. (2016). *Health index: The last frontier of TSO's asset management* (pp. 1–9). Terna Rete.
15. Pompili, M., & Scatiggio, F. (2015). Classification in iso-attention classes of HV transformer fleets. *IEEE Transactions on Dielectrics and Electrical Insulation, 22*(5), 2676–2683.

Part II
Focusing on Value-Based Asset Management

Chapter 5
A Framework to Embed Asset Management in Production Companies

Irene Roda and Marco Macchi

Abstract The aim of this chapter is to investigate how to embed Asset Management in production companies. A framework is defined based on literature analysis and focus groups findings, in which the fundamentals to guide the integration of Asset Management are systematized. Two dimensions are identified—the asset life cycle and the hierarchical level of the asset-control activities—and four founding principles—life cycle, system, risk and asset-centric orientation—as levers to integrate Asset Management within an industrial organization. An empirical investigation is then developed through multiple case-study involving eight production companies in Italy, with the purpose to map the elements of the framework against the real mechanisms in the industrial practices. This allows testing the relevance of the framework itself and demonstrating its potential as a support for companies to implement gap analysis on AM practices. Empirical evidence on current practices of AM in production companies is contextually unveiled.

Keywords Asset management · Life cycle engineering · Modelling/simulation · Production system asset management · Production system maintenance

1 Introduction

Nowadays physical Asset Management (AM) is established and recognized as a holistic process involving different functions within an organization in the management of the assets along their entire lifecycle. In particular, AM is intended as a paradigm shift in focus with respect to maintenance management. The asset is brought at the center of the management process towards value creation and maintenance is one of several activities to be done along the lifecycle of the asset [1, 2].

I. Roda (✉) · M. Macchi
Department of Management, Economics and Industrial Engineering, Politecnico di Milano, Piazza Leonardo da Vinci, 32, 20133 Milan, Italy
e-mail: irene.roda@polimi.it

M. Macchi
e-mail: marco.macchi@polimi.it

A. Crespo Márquez et al. (eds.), *Value Based and Intelligent Asset Management*,
https://doi.org/10.1007/978-3-030-20704-5_5

In the industry, AM finds its roots in terms of application in most capital-intensive sectors such as the oil & gas, energy and network utilities sectors due to the inherent criticality of the impact of the assets on the business. Nevertheless, current challenges and opportunities in the industrial world such as globalization, aging of assets, emerging of new technologies, etc., are more and more bringing the attention of production companies towards AM, both in the manufacturing and process production sectors [3, 4]. More and more, companies are aware of the need to break down their knowledge departmental silos and to take a more holistic approach for generating value from the management of their own assets (i.e. production systems). Nevertheless, AM is still scarcely adopted by production companies as a business process. Managers are finding difficulties in understanding how to implement AM in their own reality [5] and some dimensions of AM decisions are based more on intuition and visions rather than structured and well-tooled analyses [6]. Nowadays, the consensus has been reached about AM as an essential process, contributing to an organization's objectives [7] and the publication of the body of standards ISO5500X on AM in 2014 supported reaching such consensus and understanding AM concepts at general level. Nevertheless, the standards provide general guidelines adaptable to any kind of asset and organization, but still lack of clear indications on how to tailor AM in the specific context of production companies.

Looking at the scientific body of knowledge about AM and, in particular, about AM in the production sectors, this is still fragmented and the collaboration between organizations and academic researchers is still underway to extend it [7, 8]. Tailoring the AM discipline to establish an AM business process that fits the needs of specific sectors is still an open challenge [9].

For these reasons, the main objective of this chapter is to propose a framework that can guide managers in production companies to embed AM in their company. This framework should remain in alignment with the business objectives and be integrated with the rest of the organization.

The research question is: "How to support production companies to embed AM in their organization?" The answer cannot depend solely on the indications given by the standards, independently of the target application sectors. Therefore, based on literature review and focus groups, the chapter firstly aims to systematize the fundamentals to embed AM in production companies within a framework. Then, an empirical investigation is developed through a multiple case-study in order to map the elements of the framework against the actions undertaken in the actual practices, to explain and validate the framework itself and to demonstrate its potential as support for companies for implementing gap analysis. This also unveiled empirical evidence on current practices of AM in production companies.

Section 2 of the chapter describes the methodology adopted for the research. Section 3 illustrates the framework built based on literature analysis and focus groups in order to synthesize the fundamentals for embedding AM in production companies. Section 4 is dedicated to present and discuss the findings of the cross-case analysis, as emerged from the multiple case study by mapping the real mechanisms through the framework's elements. Finally, Sect. 5 is dedicated to the conclusions and discussion of scientific and industrial contributions.

2 Methodology

The methodology used in this research was designed based on the works on research methods in AM by El-Akruti and Dwight, Kusumawardhani et al. [10, 11]. In particular, the retroductive research strategy was selected in accordance to the research's objective (Fig. 1).

Retroduction represents an attempt to overcome the pitfalls of purely inductive or deductive research processes [12]. Theory is considered as conceptualization and not as ordering framework; this allows consulting and redeveloping theory in a close relationship with empirical information. Basically, retroductive research strategy involves the building of hypothetical models as a way of uncovering the real structures and mechanisms which are assumed to produce empirical phenomena. The model, if it were to exist and act in the postulated way, would, therefore, account for the phenomena in question [13]. Literature analysis and focus groups allowed defining the model. The model provides the basis for the observation and collection of qualitative and quantitative data [10]. Testing the relevance of the postulated model can be done through case study methods. In fact, as stated in [10, 11], due to the multi-disciplinary nature of AM research, the non-experimental and quasi-experimental methods (like case study) are the most suitable. Case study allows exploring evidences and testing the relevance of the postulated framework and model according to the retroductive approach's objective [10].

With respect to the research steps, the first conceptualization step was based on literature analysis combined with focus groups developed with industrial exponents. In accordance to the applied research approach, not only a state of art review in the international scientific literature was conducted, but also an investigation on the

	Inductive	Deductive	Abductive	Retroductive
Objective	To establish universal generalizations from observations	To test theories, to eliminate false ones and corroborate the survivor	To describe and understand a phenomenon in terms of the actor's motives and understanding	To discover underlying mechanisms to explain the observed regularities
Start	Accumulate observations or data	Identify a regularity to be explained	Discover everyday common concepts, meaning and motives	Document and describe a regularity
	Produce generalizations based on empirical data	Construct a theory and deduce consequences		Construct a hypothetical model of a structure or mechanism
Finish	Use these "laws" as patterns to explain further observations	Test the consequences by matching them with data	Develop a theory and test it successively	Find the real mechanism by observations and/or experiments
Applications	To discover generalizations	To test theories	To understand people's actions	To verify the causal structure or mechanism

Fig. 1 The four types of reasoning in AM research Kusumawardhani et al. [11]

industrial current practices was done. The search for related publications was mainly conducted as a keyword search, using as main keywords: «Engineering Asset OR Physical Asset OR Asset» AND «Management OR Management System» AND «Production System OR Production Line OR Equipment OR Industrial System OR Industry OR Manufacturing». Both library services such as Scopus or Google Scholar and a wider surfing in the web were addressed. Moreover, different focus groups were organized in order to collect insights from industry. In particular, one focus group was organized in the context of TeSeM Observatory (www.tesem.net) and around thirty companies' representatives, mainly operations managers and maintenance managers, from different sectors (among which the mechanical, the food & beverage, the chemical and the oil & gas sectors) participated in a roundtable session. Focus groups at different companies' premises were also organized involving a smaller audience to openly discuss about AM and its challenges and meanings. In particular, three smaller focus groups were organized involving AM responsibles of three companies from the chemical, food & beverage and mechanical sectors. Each focus group lasted 3 h in average and opinions were collected about the findings from the literature analysis. The result of this research phase is a framework for AM integration in production companies.

Secondly, an exploratory phase was implemented. In fact, based on the framework, the case study method was used in order to find the real mechanisms and verify the framework validity by observation. The study targeted eight production companies in Italy, selected among companies with a medium or high maturity level in Maintenance Management practices. In fact, it was assumed that only companies with certain level of maturity in Maintenance Management—one of the precursors of AM [14, 15]—are ready enough to talk about and implement the wider concept of AM. The selection was possible thanks to the survey of the TeSeM observatory made with the purpose to benchmark the maturity in Maintenance Management practices, and counting on a sample of more than 300 companies up to 2015 (the maturity assessment of maintenance management practices adopts the method presented in Macchi and Fumagalli [16]). The selected companies belong to different industrial sectors in order to avoid biases and to cover a broader scope of the production industry. Table 1 shows the panel of companies selected for the case study.

The main source of the primary data for this research phase was a face-to-face semi-structured interview. Semi-structured interview brings the opportunity for the interviewee to share information relatively freely with the interviewer since such interviews possess some degree of flexibility in content [17]. Additionally, semi-structured interviews have the ability to light the road for the researcher by permitting practitioners to cooperate in the interview guidance to some extent, while still researcher as the interviewer keeps the control over the process. The nature of the interviews was chosen in order to allow starting the dialogue from the hypothesized framework and model in the previous research step. The face-to-face interview also allowed to observe not only the firm's environment, but also to notice any "weak signal" on the respondents' side—e.g. hesitation, irritation, non-verbal messages or non-given answers—that could provide a further dimension for the overall interpretation of the investigated phenomenon. The interviews were all recorded digitally

Table 1 Case study: involved companies

Case	Type	Sector	Core business[a]		People interviewed
A	Large	Chemical	2010	Manufacture of basic chemicals, fertilizers and nitrogen compounds, plastics and synthetic rubber in primary forms	• Maintenance and technical materials Executive
B	Large	Appliances	2751	Manufacture of electric domestic appliances	• Site Industrial Engineering Manager
C	Large	Steel	2420	Manufacture of tubes, pipes, hollow profiles and related fittings, of steel	• Maintenance Manager
D	Large	Steel	2400	Manufacture of basic metals	• Technical Director • Maintenance Manager
E	Large	Petrol-chemical	1920	Manufacture of refined petroleum products	• Maintenance Manager
F	Large	Machine tools	2849	Manufacture of other machine tools	• Technical Functions Manager
G	Large	Food & beverage	1100	Manufacture of beverages	• Global Maintenance Director • Real estate and Energy Management
H	Large	Tyre	2211	Manufacture of rubber tyres and tubes; retreading and rebuilding of rubber tyres	• Corporate Maintenance Coordinator

[a]Statistical Classification of Economic Activities in the European Community, Rev. 2 (2008)

and each of it lasted around 1 h and a half in average. The interviews protocol was composed of eleven open questions. It was designed addressing the elements of the framework through different questions about the main issues that were identified through the literature analysis and focus groups in order to allow triangulation when analyzing the findings. The chosen unit of analysis was the company from the perspective of the maintenance, technical services or industrial engineering function. The data collected from the case studies were then analyzed using a uniform approach, interpreting the transcripts according to the coding technique, in order to denote the relevant concepts emergent during the interviews [18]. Overall, 75 first-order codes were identified and the highest order themes reflected the elements of the framework. The analysis of the interviews and the main findings allowed confirming the elements within the defined framework and validating its application as a guide to integrate AM as well. In fact, the framework worked as a support for the companies involved in the case study to identify the AM level of integration in their organization and the existing gaps. Moreover, the case study implementation contextually allowed to unveil empirical evidence on current practices of AM in production companies.

3 Framework Definition

3.1 Literature Analysis on Existing Frameworks

As it is stated in Sect. 1 of this chapter, the scientific body of knowledge about AM and, in particular about AM in the production sector, is still fragmented, and the collaboration between organizations and academic researchers is still underway to extend it [7, 8]. A complete overview over the development process of the AM concepts and existing frameworks is provided in [19]. The authors show that, at general level, several AM frameworks have been defined so far in the literature. Nevertheless, the available frameworks that have been considered for AM are generally not comprehensive and do not consider a holistic approach, primarily focusing on just one stage of the lifecycle of an asset (commonly, the middle of life stage, addressing asset maintenance management) [19–21].

Some authors addressed the need to shift the focus from the technical aspects of physical assets to a more business-oriented AM approach. The recently published ISO 55000 standard goes in the same direction and defines guidelines that can be followed to develop an AM system. Nevertheless, the scientific and technical literature about AM is either generic, not addressing any specific application sector (see for example the ISO5500X body itself), or dedicated to specific sectors such as the energy, building and constructed facilities, transportation infrastructures or intangible assets (human factor). In fact, 80% of papers found in the search engine Scopus, by looking at physical Asset Management as keyword, are related to those sectors. Very few papers can be found dealing with AM in production companies. Nevertheless, several recent technical reports can be found about the topic [5, 22], showing the interest about AM in such industry.

Based on these findings, the aim of this research is to develop a literature analysis specifically focused on works illustrating AM as a consolidated discipline and, in particular, journal papers and books chapters proposing an AM framework addressing the production sector. The aim is to identify the main elements to define a framework to guide companies to embody AM in their realities.

The search for papers fitting the requirements of such analysis led to the identification of seven papers (Table 2). The findings of the cross-analysis of the papers is showed in the table. The main topics that are treated by each paper in describing the AM framework or scope have been identified under six main categories (Asset life

Table 2 Main aspects in the existing AM frameworks

		Life-cycle	Control-levels	Related organizational functions	Asset knowledge	Risk	System
El-Akruti et al. [7, 23] E-Akruti and Dwight [19]	2013, 2016	x	x	x	x		
Komonen et al. [6, 24]	2006, 2014	x	x			x	x
Amadi-Echendu et al. [1]	2010	x	x	x			
Schuman and Brent [20]	2005	x		x	x		
Frolov et al. [8]	2010			x	x		
Tam and Price [25]	2008		x		x	x	
Tranfield et al. [4]	2004	x	x		x		
		5	5	4	5	2	1

cycle, Control levels, AM related organizational functions, Asset knowledge, Risk, System).

The first two elements—asset life cycle and control levels—are the ones that are cited by most of the selected publications together with asset knowledge. Indeed, the first two aspects are considered as relevant to establish asset-related decision-making and they are not easily separated when analyzing AM, being intimately connected when a decision is taken [19]. The asset life cycle includes three relevant stages, Beginning of Life (BoL), Middle of Life (MoL), and End of Life (EoL) stages, wherein asset-related decision making is required [26]. All selected papers highlight the importance of asset lifecycle management in their proposed frameworks except for [25] that focus on the middle of life stage only and [8] who don't stress the lifecycle aspect in their research.

Hierarchical levels of the asset-control activities (control levels) comprise the strategic, tactical, and operational levels wherein asset-related decisions are allocated, in terms of accountabilities, within the company organization. In their framework, El-Akruti et al. [7] define a set of planning and control activities that exist at the three organizational levels, i.e. strategic, tactical or aggregate and operational. Komonen et al. [6] highlight the three asset management levels in their framework for holistic management of production assets, and the three control levels can also be found in the works by Amadi-Echendu et al., Tranfield et al., Tam and Price [1, 4, 25].

The other aspect that is cited by more than a half of the selected papers is the asset knowledge. The key issues pointed out by Tranfield et al. [4] are the need to clarify the relationships between the asset a company owns and the achievement of the company's strategic purpose, and which are the overriding purpose and function of the asset base. Moreover, according to the authors, companies need to have clear information about what asset they own, what type, how many and their location. Knowledge of the assets can be gained through asset inspection and examination routines, condition assessment and monitoring [4]. Existing asset knowledge and asset performance evaluation are also two important elements in the work by Frolov et al. [8]. The relevance of asset conditions is also highlighted by Schuman and Brent [20]. According to them, the requirements with regard to system effectiveness in terms of reliability, availability and maintainability are of equal importance to the functional requirements of throughput, quality, capital cost, schedule, etc. In this same direction, El-Akruti et al. [7] state that the value creation process expected by AM, can be detected in terms of the asset performance. In their paper, Tam and Price [25] discuss about the relevance of an asset database that provides basic reference to information regarding assets' physical properties for strategic decisions at senior managerial level. According to the authors, the availability of useful data is paramount to making the best decision in asset management. Indeed, the role of information technology and systems is crucial to facilitate it [4, 25].

Other aspects that have been identified through the analysis of the paper are AM related organizational functions, risk and system. Regarding the first one, some papers advocate the importance of identifying and involving several organizational functions in the management of the assets. El-Akruti et al., identify functions involved in the asset life cycle activities and in supporting activities as well. The importance of a multi-disciplinary approach is identified by Amadi-Echendu et al., Hastings, Frolov et al., Schuman and Brent [1, 2, 8, 20] in order to enable an AM holistic approach, through the coordination among different kinds of actors. With this regard, clarifying accountability on assets is essential to ensure its sustainability in the company [19, 21].

The risk dimension is considered crucial in the framework by Tam and Price [25] in which risk relates to potential hazardous events caused by failures. Risk analyses (both considering the risk related to the asset reliability and performance but also taking into account the business objectives, changes in the business environment, viewpoints of various joint-parties, the balanced governance of potential opportunities and versatile risks), is considered as an important input for the AM decision-making process, together with the technical and economic analyses, in the framework by Komonen et al. [24].

Only the work by Komonen et al. [24] among the selected publications, explicitly highlights the importance of the system perspective. In fact, the authors divide the AM framework they propose into three levels: corporate, plant and equipment level. Noticing that the plant and corporate levels have often been narrowly treated so far. According to the authors, different objectives and analysis should be performed at the different asset levels: "for example, the corporate executives should determine the role of each production unit. The plant-level management should model the production system, carry out criticality assessment […]. At the production line, subprocess or equipment level, the important management processes would be, e.g., modeling of processes or technical functioning of equipment and taking care of criticality assessment". This aspect is not addressed in the other selected publications for this analysis, El-Akruti et al. [23] introduce some reference to the importance of referring to systems when implementing AM.

3.2 Focus Groups Findings

Based on the focus groups organized with industrial exponents, the different elements identified through the literature analysis were discussed and their importance based on direct industrial opinions was collected.

First of all, it was evident that a full integration of asset life cycle and control levels is the heart of AM. On one hand, it was agreed that any time a decision is taken about assets, the whole life cycle should be considered analyzing what is inherited from the past in term of influencing variables, and how the future will be affected by the decision in case it is taken. On the other hand, it was apparent that a relevant factor for decision-making is the company organization; in particular, all three hierarchical levels of asset-control activities within the organization need to be involved, ensuring alignment through the implementation of feedback loops between the levels. Overall, the ability of a company to implement AM stands in the capability to integrate the two aspects into a robust and clearly defined AM system in its organization. This confirms what discussed in literature.

Moreover, the importance of adopting a long-term perspective in the management practice emerged from the focus groups. This means, on one side, to work on the integration among different functions that have to take decisions on the asset along the various stages of its lifecycle. On the other side, the need to have and use decision-making tools to support orientation towards a lifecycle perspective such as Life cycle cost or Total Cost of Ownership methodologies emerged in the discussions.

Among the other aspects identified, an important issue that was discussed is the need of deep and consolidated knowledge of the asset portfolio that must be managed by the company. Developing better knowledge about the asset base is considered as an urgent matter to build an AM system.

Interestingly, the stress of the industrial people involved in the focus groups was on the need to consider that industrial asset in the production sectors (both for process production and manufacturing) are typically very complex systems, composed by several machines and components connected among each other. For this reason, the system aspect was stressed more than what is discussed in the literature. Indeed, as a result of the focus group, the need to adopt a hierarchical approach, considering different aggregation levels when managing assets (e.g. corporate, plant and equipment level), was evident, as fundamental element to take informed decisions.

Finally, the focus groups allowed to identify the risk dimension as a relevant aspect in the AM process. AM should allow managing assets but also the risk associated with the asset and with the decision-making processes on the assets. In particular, the focus is put on the operational risks related with uncertainty, i.e. the risk associated with the future asset behavior and its expected performance.

3.3 Framework Proposal

Provided the findings of the literature analysis and of the focus groups, it was possible to define a framework (Fig. 2) including the main elements required for AM embodiment in production companies.

In particular, the framework is composed by two dimensions and four founding principles. On one hand, the two dimensions are the asset control levels and the asset lifecycle. As it emerged from the literature analysis, these two aspects are relevant to establish asset-related decision-making and they are not easily separated, being intimately connected when a decision is taken [19]. This is the reason why they are inserted in the framework as two dimensions: they represent the elements needed to position an asset-related decision within the AM process. On the other hand, the four principles are: life cycle orientation; system orientation; risk orientation; asset-centric orientation. These are called principles since they define and guide the approach, methods, and systems to be adopted in order to integrate AM (and related decisions) within a company. As an example, in order to ensure that AM is integrated in a production company, the company has to be able to position any asset-related decision in terms of control level (strategic, tactical or operational), by simultaneously checking the alignment with the other levels, and in terms of asset

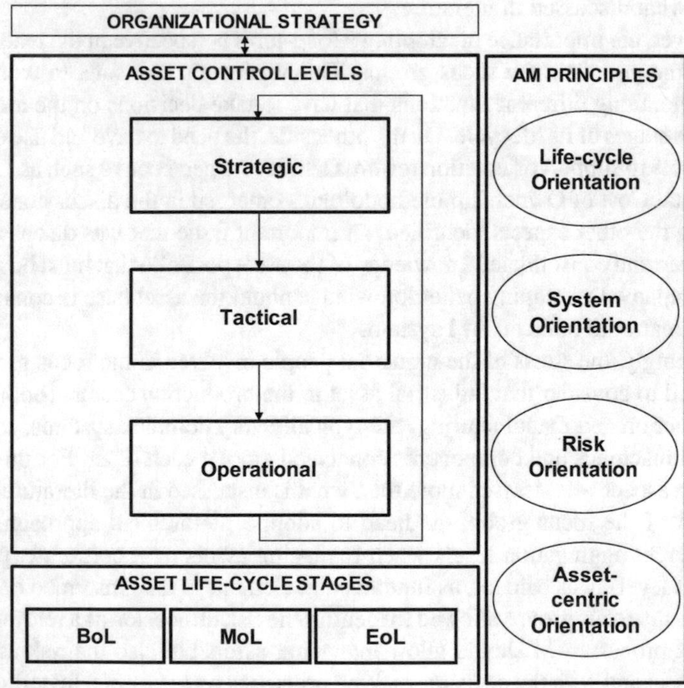

Fig. 2 Framework for embedding AM in production companies

lifecycle stages. Moreover, it has to approach the decision ensuring it is following the four principle. For example, let's consider, as decision, the investment in a new equipment within an existing plant. First of all, it is a strategic decision that must be aligned with the organizational strategy and guide activities and tactical and operational level. Then, it is a decision positioned in the MOL of the plant, as the plant is an existing one. Moreover, proper indicators and decision-making tools should be adopted to ensure lifecycle orientation (for example, considering the impact of the decision on total cost of ownership), system orientation (for example, considering the impact of the local decision on the performance of the whole pant), risk orientation (considering the operational risks connected with the investment decision) and asset-centric orientation (considering, as essential, the knowledge of the plant in which the new equipment will be installed and the kind of data that are required to be analyzed to guide the decision). More in detail, each founding principle is described hereafter.

The adoption of life cycle orientation in decision-making means that the AM process should incorporate long-term objectives and performances to drive decision-making. Supporting tools can be adopted by the company to aid the achievement of this objective, such as the LCC (life cycle cost)/TCO (total cost of ownership) [27, 28]. Moreover, given that the three stages of the life cycle of the assets—BOL, MOL, EOL—differ in the scope of decisions, different organizational functions need to collaborate in the AM process, covering all organization's hierarchical levels [19].

System orientation is another essential principle to ensure focusing efforts and resources on the right decisions. In particular, criticality of the assets at system level is fundamental for decision-making: as it is expressed in the ISO 5500X, "an asset is defined critical if it has potential to significantly impact on the achievement of the organization's objectives". As industrial assets are complex systems composed by various components interrelating among themselves, and such interactions, together with the state of each component, affect the state and performance of the system itself, the criticality of the assets at system level and the systemic effect of any local decision have to be considered to have a robust AM business process. Furthermore, the need to realize value from the asset systems motivate the relevance of a holistic consideration of asset systems in their entirety, and not merely of the individual components [29].

Risk orientation should be also applied in decision-making. This should generally rely on the definition of risk given in ISO 31000:2009. Amongst the different types of risk, considering the operational risk in the industrial assets is then essential: the failure of critical assets proved to be the risk that is recognized by companies to have the biggest impact on business (according to the results of the industrial survey on operational risk management 30). Being aware of such importance, leading companies use analytical tools to gain better visibility into the risks within their operations. Moreover, they establish a risk culture and empower the workforce with the information to be "predictive decision-makers" [30]. Eventually, a risk orientation inevitably leads to the realization of value from assets: risk enables to take into account the likelihood and consequence of not fulfilling stakeholders' expectations, which is relevant to preserve the target values generated by the assets [20, 31].

Asset-centric orientation is another essential principle as the management of assets is dependent on knowledge about the organization's assets, in terms of both current equipment, business role of the assets and future prospects. In other words, asset managers need to have a practical working knowledge of the major assets so to be able to make sound business decisions [2]. Thus, it is advocated that it is necessary for AM implementation to have an asset common database where all the data about each asset and its components are stored together [32, 33]. The asset database would provide basic reference to information regarding assets' properties, usable for strategic, tactical and operational decisions. Besides, tracking of changes during the life cycle of the asset is facilitated by a common database: it is an essential information for integrating the asset-life cycle dimension; it allows knowledge sharing about the assets along the different life cycle stages.

3.4 Concluding Remarks

Overall the defined framework brings the four principles, as well as the two dimensions (asset control levels and life-cycle stages) together, considering them the elements that have to be considered by a company aiming to integrate AM in its organization. In fact, the two dimensions allow giving a frame in which to position asset-related decisions by keeping a holistic perspective and providing a shared basis to the different functions and actors participating to the AM decision-making process. Those dimensions are also highlighted in the ISO5500X body of standards as fundamental for an AM system. Moreover, the four principles, guide companies to adopt the proper systems, tools, methods to ensure implementing an integrated AM process when taking asset-related decisions.

4 Cross Case Study Findings Analysis

The fundamentals defined in the framework were used as the basis for studying the actual level of integration of AM business process in the management systems of production companies and to subsequently understand the main gaps. The findings coming from multiple case study development are herein synthesized, focusing on each founding principle and the orientation towards it of each analyzed company.

4.1 Life Cycle Orientation

Based on the framework, life cycle orientation means: (i) promoting an integrated organizational structure in which all the necessary competencies and organizational functions are involved at each stage of the life cycle of the asset; (ii) adoption of long-term performance objectives and indicators in managing assets.

As far as the first issue regards, the findings from the case study analysis allowed assessing if companies present a proper level of integration among functions to manage the assets. In particular, by keeping the point of view of the maintenance function, a general trend towards integration emerged from the analysis. Nevertheless, there are still gaps at the organizational level to achieve a complete integrated system for the management of the assets.

At the early stage of the life cycle of the assets (BOL), dealing with design, construction, and commissioning, the desired condition is to get to closer cooperation among the various organizational functions such as design, purchasing, and maintenance. A certain trend towards this direction was detected in all the analyzed companies; however, the integration cannot be considered complete. What clearly emerged from the majority of the cases is the desire of the maintenance function to have a more active role in the BOL stage. As for the intermediate stage of the life cycle of the assets (MOL), dealing with use and maintenance, awareness of the role of maintenance as an "evolved" fundamental function that must work in an integrated manner with the various functions, and in particular with the production function, emerged. Nevertheless, in some cases, a certain "suffering" by the maintenance function is still perceptible that would like to participate more to decision-making, and that instead is often confined to managing assets in terms of reliability and availability in a still partially isolated way. Looking at the end of life stage of the assets life cycle (EOL)—dealing with renewal, disposal, recycling, reuse, as well as with potential life cycle extensions by, e.g., retrofitting or revamping (which means deciding to move the end of life ahead in time)— it is the stage in which there is the lowest level of integration among the functions. In particular, the maintenance function in most of the analyzed companies mainly takes executive role, without participating in the decision-making process. It leaves wide rooms to achieve better integration.

As for the second issue of implementation of AM guided by long-term objectives and performances, interesting findings have emerged evaluating the tools and indicators used by companies to support decision-making. Concentrating on the investments planning and, in particular, the investments assessment, traditional methods are mainly adopted like ROI (Return on Investment), NPV (net present value) and IRR (Internal Rate of Return). Although these indicators theoretically imply the adoption of a long-term perspective, in practice, in the majority of cases, their calculation method consists of an accurate assessment of only CAPEX (the costs recognized in the capital of the company, i.e. Capital Expenditures) and of only including a rough estimate of OPEX (the operating costs for the year, i.e. Operational Expenditures). In particular, it is evident that the approximate estimate of OPEX is likely to underestimate the impact of the investment decision to the future performance of the asset that can generate inefficiencies and, therefore, hidden costs due to the inefficiencies.

Few are the cases where the investments assessment (made with the above methods) are flanked by other methods. Among them, it is worth citing: (i) the use of a TCO model oriented to model the OPEX based on an accurate engineering estimate of the performance of the asset system, (ii) the assessment of the satisfaction of the stakeholders, and/or (iii) a RAM analysis (Reliability, Availability, Maintainability) for a provisional estimate of industrial plant performance losses (as failures are considered as relevant events leading to inefficiencies).

Overall, investments still seem a decision majorly featuring a financial problem, while all the companies recognize the need to increase the contribution of an engineering viewpoint to be integrated with the financial analysis. This would enable to evaluate the convenience of choices enriching the financial indicators with models capable of synthesizing the technical and operating characteristics of the industrial assets, with the ultimate goal to obtain adequate performance estimations, which are at the basis of informed financial analyses and robust decisions in the long-term.

4.2 System Orientation

Regarding the system orientation, the case studies allowed investigating whether the complexity characterizing the industrial assets is taken into account in the decision-making processes, or not. In fact, industrial assets are typically systems composed of multiple components with their own RAM characteristics; and those components interact with each other to perform the requested function of the industrial asset.

All analyzed cases showed an awareness of the importance of adopting a system performances analysis, with the ultimate goal to take into account the effect that every local decision—hence, also the RAM characteristics of individual assets—inevitably has got at global level. This means, for example, making decisions for improvement of productivity in an industrial plant on the basis of a careful analysis of the criticality of the individual equipment with respect to the function they have for the production flow at system level.

Although awareness was shown by companies, today analyses keeping the systemic perspective are not implemented in a systematic way. This may be justified by various contingent reasons. In some cases, that is the case of a production line with no (or of limited capacity of) inter-operational buffers among the equipment, the asset (as a system) requires the operation of all the equipment (components) for its operation. In this specific case, each equipment assumes the same criticality at local and systemic levels, namely the OEE (Overall Equipment Effectiveness), measure centered at equipment level, would be sufficient to have an accurate approximation of the effect of local problems on systemic performances. In other cases, there is not yet a full integration of engineering techniques or tools, within the reliability and maintenance engineering systems, so to enable that the global effect of local decisions is fully analyzed; in fact, some companies are still relying on traditional KPIs like OEE keeping the local perspective: it is clearly not enough, when the complex interdependencies of individual assets affect the achievement of the performance of the system of assets (e.g. production capacity of the system). Finally, in one case, the systemic analysis is restricted to a limited number of decisions in the life cycle of the asset, and the use of some specialized engineering techniques but with no systematic approach throughout the life.

Overall, system orientation generally requires an enriched application of techniques through the extensive use of advanced engineering tools capable of system-level performances analysis, with particular concern to system effectiveness in terms of reliability, availability and maintainability as well as other functional requirements of throughput, quality, etc.

4.3 Risk Orientation

As far as risk orientation regards, when making decisions related to assets, every company applies all the needed measures required in the legislative framework of the specific sector to reach the level of compliance for the management of critical risks related to safety and environmental impact. A differentiation then emerges in regard to how those operational risks—such as the risks linked to the effect of future behavior of assets on the expected technical performance—are managed. In this regard, the sectorial contingency has an impact on the practices adopted by the companies. In some sectors, the most capital-intensive ones, ensuring asset integrity is a priority due to its high impact on business. It is the case of the most advanced companies in terms of the systematic integration of typical approaches of RAM analysis and methods for operational risk management in AM. In general, the attention to the risks over performance and operation of the assets has been growing as a critical element in all analyzed cases. Concerning the uncertainty related to a risk, the aspect that stimulates a greater reflection is if the expected technical performance of the asset, and any inefficiency expected from its operation, are quantified in terms of cost to support decision-making by companies. As emerging evidence, what in general is still missing is an alignment between the system's technical performance measures and the financial indicators, which are the ones that are taken into account in the company to make decisions. Nonetheless, companies believe that an engineering analysis should support the final choices, aiming at a reduction of operational risks related to the losses in production efficiency, which eventually ensures a more informed decision-making.

Overall, the risk of not fulfilling stakeholders' expectations in terms of production efficiency and, then, related costs of the losses is still not fully analyzed, even if its relevance for the target values generated by the assets is apparent in all the cases.

4.4 Asset-Centric Orientation

The adoption of an asset-centric management approach appeared to be influenced by the sectors of the companies. In the most capital intensive sectors, the role of the assets and their performance is definitely recognized as central to ensure the production and the achievement of business objectives. It follows the high emphasis on ensuring clear ownership of the assets and collecting data referring to each asset.

In all cases, the relevance of the definition of a clear ownership of assets to ensure control and commitment to AM is recognized by the companies. The various analyzed companies, while proving to have all shared this need, have made different organizational decisions. In some cases, a centralized ownership at the level of maintenance/technical direction/industrial engineering functions has been chosen; in a few cases, the ownership was instead given to an executive belonging to the top management board.

In terms of information systems, in order to support an asset-centric management approach, information systems are seen with a central role, and in particular the maintenance information systems appear to play a relevant role for the MOL. Even so, the maintenance information systems are still partially integrated with other enterprise information systems and the need to move towards the definition of a better system, in which different kinds of data (technical and economical) related to the assets are collected and analyzed in an integrated manner, is recognized. This is considered an enabling element for the effective implementation of AM.

Overall, the issues addressed by asset-centric orientation are two-fold. On one side, from an organizational point of view, companies addressed the need to set clear responsibilities and commitment for AM implementation. On the other side, the need for an integrated information system to support AM along the lifecycle of the asset emerged.

4.5 Concluding Remarks and Discussion on Existing Gaps

The analysis of the interviews and the main findings allowed confirming the elements within the defined framework and validating its application as a guide to integrate AM. In fact, the framework worked as a support for the companies involved in the case study to identify the AM level of integration in their organization and the existing gaps.

Based on the findings described in the sections above, it was possible to make an overall diagnosis. Table 3 visualizes the level of integration of each principle within each company. Information are qualitative and are presented to have an overview on the AM integration level in each of the companies analyzed. Moreover, the last column of the table provides an overall value for the level of integration of each principle.

The main findings are illustrated in the reminder and are shown in the table in a qualitative way.

- Life-cycle orientation is the principle that is most integrated by companies even if there are still some cases where it needs to be improved.
- The least integrated aspect is the system orientation, in fact just one company uses a model in order to consider the global effect of local decisions when approaching decision-making, and yet not systematically.

Table 3 Cross case study findings

	Companies								Principles integration level
	A	B	C	D	E	F	G	H	
LIFE-CYCLE ORIENTATION									
SYSTEM ORIENTATION									
RISK ORIENTATION									
ASSET-CENTRIC ORIENTATION									
Overall AM integration level									

- The company that resulted with the highest level of AM integration given the orientation towards the four principles is company F. Company F presented some weaknesses regarding risk orientation since future uncertainty is considered through average values based on estimations or historical values and not through a what if approach (scenario analysis) nor through a probabilistic approach.

To conclude, it is worth remarking that the case studies were developed by considering companies belonging to different industrial sectors to have a general overview, and validating the framework as a guide for AM integration in production companies. The framework proved to be useful to assess the level of AM integration in a company, hence it can be used as a guide to identify activities to integrate AM in any production company. Even if the companies selected for the case study development are companies with high level of maturity in terms of Maintenance Management, when widening up the perspective over AM, gaps to be filled in have been identified. None of the companies resulted to fully incorporate the four founding principles for AM implementation. Lifecycle orientation is the principle that more companies are looking at in order to tend towards it (by means of re-organization, testing of new tools, etc.), while system orientation is still quite weak in all companies. Moreover, it is worth noticing that the gaps that have been identified regarding the level of integration of the AM in the companies not only are due to contingent reasons (industry, types of assets to manage etc.) but also to the low development level of the necessary technologies/methodologies. For example, the availability of a standard model for total cost of ownership, as a methodology to support the decision-making process, or the availability of performance indicators at system level, were pointed out in the majority of interviewed companies.

5 Conclusions

In this research, the AM integration process is addressed within the context of production companies.

From a scientific perspective, the contribution of this chapter is the proposal of a framework for AM integration specifically thought and developed for production companies. The framework is derived from an analysis of existing frameworks in the literature, by integrating the main elements that were identified during the time by different authors and confirmed by industrial experts through the development of focus groups. The case study developed involving eight production companies allowed exploring evidences in industry and testing the relevance of the proposed framework.

From a practical point of view, the research findings are interesting for managers and engineers within this context (i.e. technical managers, operations managers and/or top managers), proposing a framework illustrating the principles to be considered in order to embed AM in their companies (Fig. 2). Production sectors differ among each other and each single company is characterized by its own peculiarities. This is the reason why the framework is intended to provide the main principles to guide the AM integration in a company without indicating specific tasks to be implemented or tools to be used, making it useful for different companies.

Based on the lessons learnt from the case studies, we believe that companies have to get more and more aware of the importance of addressing AM by reflecting the four founding principles and by accordingly structuring an Asset Management system. Moreover, we would assert that the founding principles are actually a required background upon which the decision-making process can be developed, building both on the life cycle perspective and through alignment among the strategic, tactical and operational control levels within the organization. By adopting these measures, which are actually summarized in the AM framework presented in this chapter, sustainable value creation from assets can be ensured.

Acknowledgements The research work was performed within the scope of TeSeM Observatory (www.tesem.net). Moreover, the research work was performed within the context of SustainOwner ("Sustainable Design and Management of Industrial Assets through Total Value and Cost of Ownership"), a project sponsored by the EU Framework Programme Horizon 2020, MSCA-RISE-2014: Marie Skłodowska-Curie Research and Innovation Staff Exchange (RISE) (grant agreement number 645733).

References

1. Amadi-Echendu, J. E., Brown, K., Willet, R., & Mathew, J. (Eds.). (2010). *Definitions, concepts and scope of engineering asset management*. Springer-Verlag. Epub ahead of print 2010. https://doi.org/10.1007/978-1-84996-178-3.
2. Hastings, N. A. J. (2009). *Physical asset management*. Springer Science & Business Media.
3. Korpi, E., & Ala-Risku, T. (2008). Life cycle costing: A review of published case studies. *Managerial Auditing Journal, 23,* 240–261.
4. Tranfield, D., Denyer, D., & Burr, M. (2004). A framework for the strategic management of long-term assets (SMoLTA). *Management Decision, 42,* 277–291.
5. Aberdeen Group. (2006). *The asset management benchmark report: Moving toward zero downtime* http://www.aberdeen.com [online].

6. Komonen, K., Kortelainen, H., & Räikkönen, M. (2014). Corporate asset management for industrial companies: An integrated business-driven approach. In *Asset management: The state of the art in Europe from a life cycle perspective* (pp. 47–63).
7. El-Akruti, K., Dwight, R., & Zhang, T. (2013). The strategic role of engineering asset management. *International Journal of Production Economics, 146,* 227–239.
8. Frolov, V., Ma, L., Sun, Y., & Bandara, W. (2010). Identifying core functions of asset management. In J. E. Amadi-Echendu, K. Brown, R. Willet, J. Mathew (Eds.), *Definitions, concepts and scope of engineering asset management* (pp. 19–30). Springer-Verlag.
9. Komonen, K., Despujols, A. (2013). Maintenance within physical asset management : A standardization project within CEN TC319. In: *COMDEM*. Helsinki.
10. El-Akruti, K., & Dwight, R. (2010). Research methodologies for engineering asset management. In: *ACSPRI Social Science Methodology Conference 2010. Sydney, Australia*. http://ro.uow.edu.au/engpapers/5001/. Accessed September 28, 2014.
11. Kusumawardhani, M., Gundersen, S., & Tore, M. (2017). Mapping the research approach of asset management studies in the petroleum industry. *Journal of Quality in Maintenance Engineering, 23,* 57–70.
12. Sæther, B. (1998). Retroduction: An alternative research strategy? *Business Strategy and the Environment, 7,* 245–249.
13. Atkinson, D. (2014). Approaches and strategies of social research. *Essay for Reasearch Methods Class ST700.*
14. Al-Turki, U. (2011). A framework for strategic planning in maintenance. *Journal of Quality in Maintenance Engineering, 17,* 150–162.
15. Liyanage, J. P., & Kumar, U. (2003). Towards a value-based view on operations and maintenance performance management. *Journal of Quality in Maintenance Engineering, 9,* 333–350.
16. Macchi, M., & Fumagalli, L. (2013). A maintenance maturity assessment method for the manufacturing industry. *Journal of Quality in Maintenance Engineering, 19,* 295–315.
17. Bryman, A. (2006). Integrating quantitative and qualitative research: How is it done? *Qualitative Research, 6*(1), 97–113. Epub ahead of print 2009. https://doi.org/10.1177/1468794106058877.
18. Corbin, J., & Strauss, A. (2014). *Basics of qualitative research: Techniques and procedures for developing grounded theory*. Sage Publications.
19. El-Akruti, K., & Dwight, R. (2013). A framework for the engineering asset management system. *Journal of Quality in Maintenance Engineering, 19,* 398–412.
20. Schuman, C. A., & Brent, A. C. (2005). Asset life cycle management: Towards improving physical asset performance in the process industry. *International Journal of Operations and Production Management, 25,* 566–579.
21. Amadi-Echendu, J. E. (2004). Managing physical assets is a paradigm shift from maintenance. In: *IEEE International Engineering Management Conference* (pp. 1156–1160).
22. Eriksen, L., & Steenstrup, K. (2017). *Market guide for asset performance management*. Gartner.
23. El-Akruti, K., Dwight, R., & Zhang, T. (2016). Exploring structure and role of engineering asset management system in production organisations. *International Journal of Strategic Engineering and Asset Management, 3,* 1–22.
24. Komonen, K., Kortelainen, H., & Räikkonen, M. (2006). An asset management framework to improve longer term returns on investments in the capital intensive industries. In: S. London (Ed.), *Engineering asset management* (pp. 418–432).
25. Tam, A. S. B., & Price, J. W. H. (2008). A generic asset management framework for optimising maintenance investment decision. *Production Planning and Control, 19,* 287–300.
26. Ouertani, M. Z., Parlikad, A. K., & McFarlane, D. C. (2008). Towards an approach to select an asset information management strategy. *IJCSA, 5,* 25–44.
27. El-Akruti, K., Dwight, R., Zhang, T., et al. (2015). The role of life cycle cost in engineering asset management. In *Engineering asset management-systems, professional practices and certification* (pp. 173–188). Springer International Publishing.
28. Roda, I., Garetti, M. (2014). TCO evaluation in physical asset management : Benefits and limitations for industrial adoption. In B. Grabot et al. (Eds.), *APMS 2014, Part III, IFIP AICT 440* (pp. 216–223).

29. Xu, Y., Elgh, F., & Erkoyuncu, J. (2013). Cost engineering for manufacturing: Current and future research. *International Journal of Computer Integrated Manufacturing* 37–41.
30. Aberdeen Group. (2007). *Operational risk management*. Epub ahead of print 2007. https://doi.org/10.1057/9780230591486.
31. Roda, I., Parlikad, A. K., Macchi, M., et al. (2016). A framework for implementing value-based approach in asset management. In *Proceedings of the 10th World Congress on Engineering Asset Management* (pp. 487–495).
32. Al-Najjar, B. (1996). Total quality maintenance: An approach for continuous reduction in costs of quality products. *Journal of Quality in Maintenance Engineering, 2,* 4–20.
33. Kans, M., & Ingwald, A. (2008). Common database for cost-effective improvement of maintenance performance. *International Journal of Production Economics, 113,* 734–747.

Chapter 6
An Approach to Value-Based Infrastructure Asset Management

Rengarajan Srinivasan and Ajith Kumar Parlikad

Abstract Effective management of infrastructure assets requires intricate considerations with regards to safety, serviceability, reputation and cost. Additionally, infrastructure assets have different requirements from various stakeholders and have a longer service life. However, traditional asset management decisions focussed predominantly on cost and there is an inherent need to understand the value of an infrastructure asset to various stakeholders and to utilise this value to drive asset management decisions. In this chapter, a systematic approach to make value-based asset management decisions is proposed. The proposed process provides an efficient method to map the stakeholder's requirements to the value provided by the asset. This map can then be used to assess the value and make effective decisions. The developed approach is demonstrated through a case study involving transportation tunnels. The essential consideration of value is expected to allow organisations in evaluating the balance between cost, risk and performance and thereby allowing better informed decisions.

Keywords Infrastructure planning · Maintenance & inspection · Mathematical modelling

1 Introduction

Managing infrastructure are of vital importance for countries and have considerable socio-economic impact. The longer service life and ever-changing user demands poses distinct challenges in maintaining adequate service levels and reducing risks. Additionally, the infrastructure assets are characterized by complex deterioration and interdependent systems, which makes decision making a challenging task. Furthermore, infrastructure assets involve multiple stakeholders ranging from governments to end-users. Reducing levels of available budget and the need to maintain high level of service performance places considerable onus on asset owners to effectively

---

R. Srinivasan · A. K. Parlikad (✉)
Department of Engineering, University of Cambridge, Cambridge, UK
e-mail: aknp2@cam.ac.uk

A. Crespo Márquez et al. (eds.), *Value Based and Intelligent Asset Management*,
https://doi.org/10.1007/978-3-030-20704-5_6

manage the infrastructure. Therefore, it is crucial for organizations to understand the value deliver by such infrastructure assets and to use this value to make asset management decisions.

Current asset management decisions are predominantly based on the principles of life cycle costing and are mainly focused on the asset owners. These cost-based approaches lack consideration of various stakeholders associated with the asset. The value of an infrastructure asset depends on:

- The benefits arising from the asset to the stakeholders by providing efficient service and effective performance
- The risk posed by the asset based on the operation and condition
- Expenditure occurred by the asset over its life.

Systematic consideration of benefits realized through the asset whilst minimizing risks and associated costs leads to value driven asset management decisions. In the current economic climate and shortage of funds are forcing asset owners to justify the maintenance needs and also to look at innovative ways of managing the assets and extend the life. Especially for infrastructure assets, the asset management solutions can include asset specific or non-asset specific approaches. For example, asset specific solutions consist of repairing and refurbishing the asset, while non-asset specific solutions for a bridge might include speed and weight restrictions. In order determine the optimal strategy, it becomes imperative to understand the value.

Infrastructure assets seldom provide value on their own but contribute to the value generated at the system or network level. However, individual assets have the ability to affect the value generated by the system depending on their criticality to the service. For example, a bridge on its own does not deliver value, but the associated road network generates value for the users and the owners. The infrastructure assets are also highly dependent on other assets or systems. The impact of one asset will have an influence on the operation of other assets. This needs to be understood to develop effective asset management solutions.

Existing work in asset management focused on sector specific issues such as water, pavement and municipal assets. Halfway describes the challenges associated with municipal infrastructure asset management and presents an integrated framework to manage the process and data fragmentation associated with multiple asset management silos [1]. Management of pavements from a network level perspective and the optimization of pavement maintenance strategies have also been given considerable attention in the literature [2, 3].

Research has also been focused in understanding asset specific solutions such as modelling deterioration of assets and determining optimal replacement or repair type/time. Micevski developed a Markov model to represent structural deterioration of storm water pipes and uses Bayesian techniques to convert data into model parameters [4]. Similarly, Ana and Bauwens propose statistics-based techniques to model the deterioration of urban drainage pipes [5]. Han et al. apply Bayesian estimation technique and Markov hazard model to predict the deterioration of infrastructure assets [6].

Most of the existing works have focused on particular aspects of asset such as deterioration while others have considered optimizing the cost to determine optimal maintenance policy or schedule. The importance of value-based asset management and what constitutes value are well defined [7]. There has been lack of attention with regards to developing approaches focused on multiple stakeholders and also to understand the value generated by the assets. The recent developments in the ISO 55000 standard describe the need for value-based approaches [8]. However, there are no systematic approaches to identify value and to make value-based decisions.

There have been considerable interests in multi criteria decision making for infrastructure management [9]. These approaches allow consideration of various stakeholders' interest into a decision-making criteria. On the other hand, infrastructure asset not only have multiple stakeholders but also have various interdependencies between other assets that need to be taken into account. This has not been given considerable attention in the existing literature. In this chapter, we propose a systematic approach to capture the value generation process and to utilize this value to drive asset management decisions. We demonstrate the proposed approach using a real case example involving transportation tunnels. The material presented in this chapter was first published by the authors in the Infrastructure Asset Management journal.

The chapter is structured as follows: Sect. 2 provides an illustration of the value generated by infrastructure assets. Section 3 describes the systematic process of value driven asset management decision making. Section 4 presents the application of the proposed process in an industrial context and Sect. 5 concludes the chapter.

2 Value of an Infrastructure Asset

In order to develop a value-based asset management, it is essential to understand what constitutes a "value" of an infrastructure asset. This can be understood from the top down and bottom-up perspectives. Top-down value is associated with organization level aspects such as business model, while bottom-up value is attributed to the asset or the asset systems and its functionality which allows in value generation.

Infrastructure organizations' business model and objectives drives the different type of value required by various stakeholders. From the top-down perspective, the value generated by an infrastructure asset is attributed to the ability to deliver the intended functionality at a required level of performance, while satisfying various stakeholders' objectives. It is important to note that the achievement of functionality needs to take place at acceptable level of expenditure with clear understanding of the impending risks. Therefore, value-based asset management is about finding the optimal balance between cost, risks and the associated performance over the life-cycle of the infrastructure.

On the other hand, the infrastructure asset or the network delivers value by meeting stakeholders' requirements enabled by safe and reliable services. One distinguishing factor related to infrastructure asset is that a single asset seldom provides value on

Table 1 Comparison of cost-based and value-based asset management approaches

	Cost based (traditional)	Value-based (recommended)
Core focus	Cost	Cost, risk and performance
Management philosophy	Minimize expenditure while maintaining satisfying performance requirements	Maximize performance while satisfying budgetary constraints
Stakeholder focus	Decision maker or asset owner	All stakeholders of the asset
Impact on service	Maintain minimum service levels	Explore innovative approaches to improve service levels
Difficulty	Well established body of knowledge	Concepts not well understood
Decision focus	Generally focusses on asset specific issues	Focusses on system level dependencies and business value

its own. The value is provided by the infrastructure asset playing an effective role in a wider system or network of assets. For example, a bridge on its own does not deliver value, but the associated road network delivers value to different stakeholders. Additionally, infrastructure assets affect the value through its interactions with other assets in the system. Thus, it is essential to understand from the bottom up of how asset affect value directly or indirectly through its interactions with system level assets. Table 1 compares the differences between traditional cost-based approach with the value-based approach.

Whole Life Value can be defined as "*the benefits, costs and risks associated with an asset over its whole-life by taking into account of the interests of all stakeholders affected by its construction, existence and usage and its wider economic, social and environmental impacts*" [10].

The key to obtaining value is to find the critical balance between the costs and benefits of different renewal, maintenance and disposal interventions [11]. The term whole-life value is attributed to finding the best trade-off between short term considerations and the conflicts of longer-term interests and objectives of various stakeholders. ISO 55000 standard proposes that realization of the value will involve balancing costs, risks, opportunities and performance [8]. Furthermore, it is critical to interpret the term "whole-life" of an infrastructure. Typical life-cycle consists of a number of stages such as requirements, design, build/procure, operate and maintain and disposal or end-of-life stages. From an infrastructure context "life" in general refers to the design life of the asset as defined by the designer or the manufacturer. However, many assets are operational for longer duration than its designed life through effective interventions during their life. The decision-making horizon is influenced by various aspects and general consideration is to take into account of the period of responsibility of the asset [11]. The decision-making horizon depends on the nature of the asset and the business model of the organization. For instance, a contractor managing a set of assets will be interested in managing it for the contract duration.

On the other hand, a county council managing a bridge might want to manage the asset over a longer period of 30 years or so. Therefore, it is essential to have a realistic understanding of the expected life time of the assets or the duration of the decision planning horizon. This could be:

- Design life of asset
- Expected life of asset
- Planning horizon
- Decision making horizon
- Regulatory cycle
- Contract duration.

Consequently, it is essential to better understand the term "whole-life", depending on the nature of the decision, business driver of the organisation, regulatory or service contract requirements and the type of the asset. The next section introduces a systematic process for making value-based asset management.

3 Process for Value Driven Asset Management Decision Making

The value driven decision-making process consists of three stages and is shown in Fig. 1. The three stages are:

Stage A: Establish the context
Stage B: Value Mapping

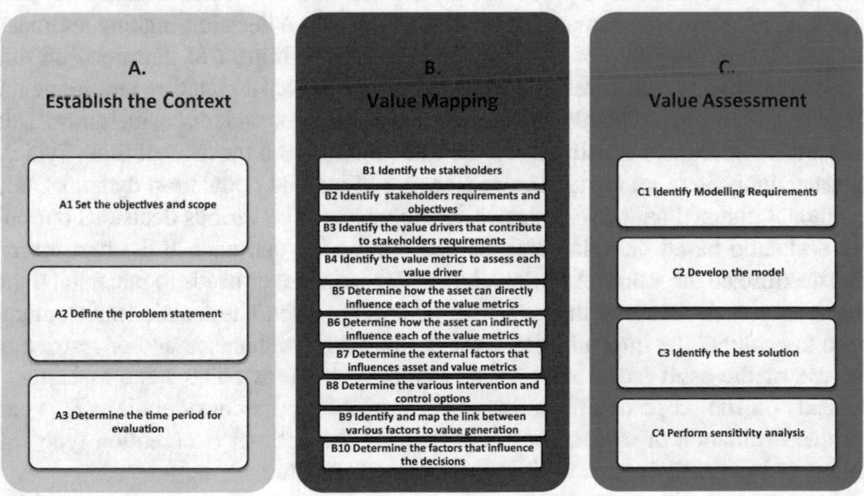

Fig. 1 Systematic approach to value-based infrastructure asset management

Stage C: Value assessment.

First stage is related to establishing the context under which value-based decisions need to be made. This is important as there are different types of assets and problems associated with them. Typical asset management decisions vary widely depending upon the context, asset type and their functionalities. This includes determining optimal intervention decisions such as time to repair, inspection and type of repair to carry out. Other problem types include justification of funding requirements for regulatory compliance, scheduling maintenance activities, portfolio management and prioritising maintenance tasks for a group of assets. Therefore, it is essential to establish the context. In this stage, the objective of the problem under consideration and the time period of evaluation are determined.

Stage B forms the core stage of the process as it establishes the value creation mechanisms. Each infrastructure asset generates value by providing functionality on its own or through interaction with other systems in the network. Consequently, any failure or disruption will have an impact on the value generated. The main objective of this stage is to understand the value and the associated risks contributed towards the various stakeholders' requirements. The inclusion and analysis of stakeholders' interest in asset management is one of the key guiding principles of ISO 55001 standards (see clause 4.2) [12]. Additionally, the impacts on the value either directly or indirectly associated with the asset are identified along with factors that can be used to control this value such as intervention options. The output of this stage is in the form of a map depicting how value is generated and the various interconnections that impact this value either positively or negatively. This value map will aid asset owners in understanding the various dependencies that need to be considered effectively when making decisions. Additionally, this value map will also indicate the typical information requirements for asset management.

Stage C is dedicated to value assessment and the associated decision-making aspect. It provides a systematic framework to develop a decision-making approach for calculating the value of the asset and the related impact of decisions on this value. Specifically, for infrastructure assets with long-term planning requirements, there is a need to conduct demand analysis for the asset under consideration. This will inform additional constraints on the value metrics as a function of time. Typical demand analysis of an infrastructure asset could include population dynamics and additional planned new infrastructure developments. The various decisions options are evaluated based on value and the decision maker can choose the best option that maximized the value. Additionally, the decision maker needs to understand the implication of decisions on the long-term value of the asset. The decision maker might need to evaluate the impending risk for postponing maintenance and/or restricting the use of the asset (such as speed or weight restrictions). The value assessment depends on the scope and objective of the problem under consideration. This can include evaluation of options or in determining the optimal intervention type and timing or in prioritizing the work schemes based on value.

In the next section, a case example is presented to illustrate the application of the proposed process in an industrial context.

4 Case Study

4.1 Introduction

The case example concerns with the development of appropriate repair strategy to address seepages that occur in tunnels used for underground transportation. The proposed process is applied to this example to determine the repair option that provides the best value.

The seepages need not have an immediate impact on the structural integrity, however could have an impact on other assets such as signaling systems or rails. The impact of seepage depends on the location and extent of seepage, and can have varying degrees of consequences. In order to mitigate the risks posed by seepage, the asset manager has various repair options, each of which have varying degree of success in minimizing the seepage and will be effective for fixed period of time. Current practice in the organization to address the seepage problem was predominantly based on expert opinion and was subjective in nature. The aim was to develop an objective method that can be consistently used across the portfolio of tunnels rather than being case specific. The next sub-section will illustrate the application of the value-based decision-making process to this problem.

4.2 Stage A: Establish the Context

The main objective was to determine the best repair option for tunnel seepage problem that would achieve the best value for money and has the least business impact. The repair option that yields the lowest service disruption, low safety risk and low financial impact is the optimal strategy. The time period of evaluation was 30 years, consistent with the long-term planning horizon of the asset owner.

4.3 Stage B: Value Mapping

In the second stage, the value generation process is captured in the form of a map depicting the factors that influence the value generated by the tunnels to the various stakeholders. The value mapping process was carried out through a series of workshops with experts from different parts of the organizations such as operations, maintenance and civil assets group.

The first step in this stage is to identify the various internal and external stakeholders associated with tunnels. The vital internal stakeholders were the different business units of the organization such as operations, maintenance and internal regulators. The external stakeholders were the customers using the tunnel, government regulator and the mayor of the city. The next step was to identify the requirements and objectives for each of the identified stakeholder. The most requirements from the stakeholders were in relation to the effectiveness of the service provided, safety and the costs incurred. From the asset owners' perspective, further to service delivery, reputation of the organization was one of the important requirements. Additionally, the mayor of the city was interested in the reputation and also the ambience was paramount.

The various value drivers were identified based on categorizing the various stakeholders' requirements. The essential categories for the value drivers are related to service delivery, safety, costs, sustainability, reputation. In order to assess the value drivers, various metrics were established. To assess safety, accident frequency and passenger fatalities and causalities were identified as metrics. Similarly, for assessing the service risk, lost customer hours and service reliability were used as metrics. The amount of media coverage, measured in column inches was used to assess the reputation value driver. It was further possible to convert all the value metrics into monetary terms for the purposes of asset management decisions and this was part of risk quantification to drive intervention decisions.

The next step was to identify the factors that influence the value generated by the tunnels. In other words, the main aim of this step is to identify how the value generated by the system is affected by seepages in the tunnel. This is related to the various ways in which the functionality provided by the tunnels is affected. This was identified by existing maintenance practices and guidance developed by the asset owner. The tunnel can directly affect the value through normal structural deterioration to complete collapse. The seepage related problems affect the value-based on the location in the tunnel and includes tunnel lining, lining joints and through head walls. In this particular example, the focus is on seepages occurring in tunnel lining joints. Seepages in general need not affect the value drivers directly, but can influence them through interaction with other systems. For instance, seepage occurring in tunnel lining joints will impact signaling systems, which in turn can cause service delays and will impact on the reputation of the organization. On the other hand, seepages can also cause corrosion of rails and in this case will lead to service disruptions and have safety implication for the tunnel operator.

Additionally, external factors such as weather and geology will have an impact on how seepages occur in the tunnel. For example, depending on the soil type and excessive rain fall will cause seepages in the tunnels. Furthermore, external construction activities and other events in the vicinity of the tunnel can cause excessive ground movements leading to structural integrity issues and can lead to tunnel weakening.

The next step is to identify the various intervention and control options that can be used to reduce the risk and maximize the value. This includes different inspection techniques, repair options and temporary mitigation solutions. Principal and special inspections were identified as mechanisms to assess the condition of the tunnel and can be used to determine the optimal time to repair. The main repair strategies for seepage in tunnels were face sealing, acrylic grouting, polyurethane grouting and lead caulking. Temporary solutions such as water management systems can be used to mitigate the risk of seepage affecting other assets. Finally, the factors affect the decisions were identified. In the specific case of seepage in tunnels, location of the tunnel in the network and the criticality of the network (based on number of passengers carried) will have an impact on the repair strategy. Further, heritage status and political factors need to be considered when deciding the optimal repair strategy for seepage problem. Figure 2 shows the typical value map for the tunnel focussing on the seepage problem occurring in lining joints.

4.4 *Stage C: Value Assessment*

The next stage in the process is to assess the value to determine the best repair option for the seepage problem. The first step in the assessment stage is to identify the essential modelling requirements from the value map. The key value influencing factors that need to be modelled are the signalling disruptions and the corrosion of rails. For the particular case of seepage related problem, it is reported either during inspection or during the maintenance of other assets. Therefore, this is a reactive maintenance strategy and the objective is to choose the best repair option that would either eliminate the seepage problem or minimise it. For each of the factors that influence value, suitable value metrics are identified based on the value map. Signalling disruptions has an impact on service reliability, reputation and costs, whereas corrosion of rails impacts service reliability, safety and costs. This needs to be taken into account when deciding the repair option depending on what other assets are being affected. Additionally, we assume the demand on the tunnel (number of passenger) to be constant over the time period under consideration.

In order to assess the overall value for each repair strategy, weighted sum of individual value metrics is used to determine the best choice. The combined weighted sum indicates the total business impact to the organization and the best repair option is the one which has the least business impact. Based on the modelling requirements, the factors that need to be modelled are:

- Impact of signaling disruptions and rail corrosion due to seepage on service, safety, reputation and costs
- Impact of various repair options on the different value metrics.

4.5 *Quantifying Service Disruption*

Service disruption is attributed to full or partial closure of the line and is determined by the duration of disruption (D) and the impact of disruption I_D. Additionally, the impact of disruption is loss in revenue per customer L times the average number of customers (N_c)

$$S_D = DI_D = DLN_c$$

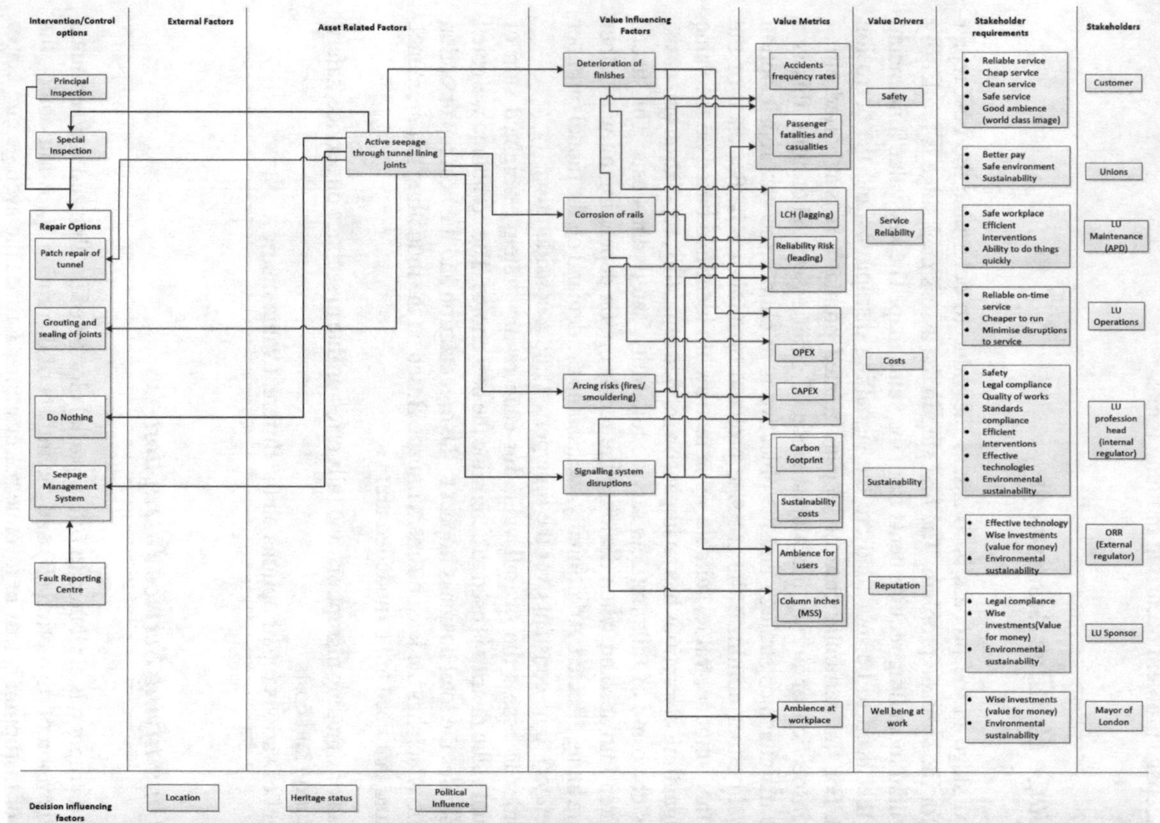

Fig. 2 Value map for the seepage problem occurring in tunnels. Opex, operating expense; Capex, capital expenditure

The impact of disruption will depend on the location of the tunnel in the network and its strategic importance. In this particular case example, L is £8.82 per customer and the average number of customers per hour was 3260. Therefore, the impact of one hour of service disruption is £28,750.

4.6 Quantifying Safety Risk

Safety issues arise due to derailments attributed to corrosion on tracks and depends on the average number of fatalities (F) for the network associated with the tunnel and the financial impact per fatality (I_R).

$$S_R = FI_R$$

In this particular example $F = 0.0054$ was calculated using historical data and I_R was £1.728 M. Therefore, the safety risk was $S_R = £9439$.

4.7 Quantifying Reputation

The impact of reputation due to signaling disruptions is based on the amount of media coverage for the incident. In this particular example, it was assumed that the reputation was linearly proportional to the service impact at about 1/10th.

$$Rep = \frac{S_D}{10}$$

4.8 Modelling the Intervention Options

In order to address the seepage problem, four different repair options were considered and is presented in Table 2. Active seepage in tunnel lining joints requires the repair to be carried out on average between 3 and 5 joints and the total cost is indicated in the second row. Each of the repair option is effective only for a certain fixed time period as shown in the fourth row of the table. Based on the time period of evaluation, this will have an impact on the number of repairs that needs to be done and is shown in the fifth row. The last row represents the likelihood of seepage not occurring in the next year.

4.9 Assessing Total Business Impact

The whole life value is assessed as a measure of the total business impact due to disruptions causes by seepages in the tunnels. The total business impact is the weighted sum of individual value drivers. The total business impact due to signalling disruption for a given year t is calculated as:

$$BI_{Sig}(t) = w_1 S_D(t) + w_2\, Rep(t) + w_3\, Cost(t)$$

Similarly, the total business impact due to corrosion of rails caused by seepage for a given year t is given as:

$$BI_{Cor}(t) = w_1 S_D(t) + w_3\, Cost(t) + w_4 S_R(t)$$

where w_1, w_2, w_3, w_4 are the relative weights for service, reputation, cost and safety respectively. Each of the repair options will have a different impact on the various value metrics and therefore the business impact will be different for all the repair options. This reflects the amount of risk reduced by a particular repair option and the resulting reduction in the associated total business impact.

The whole life value for an evaluation period of $N = 30$ years, the total business impact can be calculated as:

$$BI_{sig}(OPT) = \sum_{i=1}^{N} \frac{BI_{sig}(i)}{(1+r)^i}$$

$$BI_{Cor}(OPT) = \sum_{i=1}^{N} \frac{BI_{Cor}(i)}{(1+r)^i}$$

where OPT refers to each of the repair options shown in Table 2 and r is the discount rate. The above formula represents the discounted sum of business impact for N years and $r = 3\%$ is used in this particular example. The above formulation can be used to assess the business impact per year and the total business impact for the decision

Table 2 Intervention options

	Option 1	Option 2	Option 3	Option 4
Repair type	Face sealing	Acrylic grouting	PU grouting	Lead caulking
Cost per joint (£)	500	1050	1050	3000
Total cost (£)	2500	5250	5250	15,000
Effectiveness (Years)	1	10	100	1000
No. of repairs	30	3	1	1
Risk reduction (%)	50	70	75	90

Table 3 Results for signalling disruptions

	Option 1	Option 2	Option 3	Option 4
Service (£)	290,209	174,125	145,105	116,084
Reputation (£)	29,020	17,412	14,510	11,608
Cost (£)	50,471	12,603	5250	15,000
Total impact (£)	369,701	203,601	164,865	**142,692**

Table 4 Results for corrosion of rails

	Option 1	Option 2	Option 3	Option 4
Service (£)	156,461	93,876	78,230	62,584
Safety (£)	95,281	57,168	47,640	38,112
Cost (£)	50,471	12,603	5250	15,000
Total impact (£)	312,800	169,461	136,414	**119,432**

horizon. Tables 3 and 4 shows the relative impact of service, safety, reputation and cost for signaling disruptions and rail corrosion due to seepage. Equal weights for each of the value drivers have been used in this case example. It can be seen that in both cases the lead caulking is the best repair option. This because of the fact that lead caulking reduces the risk of seepage occurring consequently (about 90%).

4.10 Sensitivity Analysis

The impact of service reliability on the repair options vary depending upon the location of the tunnel in the network. Furthermore, service impact has significant business consequences for the organization and therefore the service impact is varied and the total business impact is calculated for each of the option. The change in the service impact is directly related to criticality of the tunnel in the network and indicates the different lines that operate in the various tunnel networks. This is based on the location of the tunnel and the number of passengers carried in a particular line. Figure 3a illustrates the impact of various operating lines on the different repair options. For signaling disruptions due to seepages occurring in line 6 and 5, option 2 (acrylic grouting) is the preferred choice. For all other lines, option 4 (lead caulking) is the best option because the lines are critical to the network. However, from Fig. 3b, it can be seen that lead caulking is the best option for seepages causing corrosion on rails, as it has significant impact on service and safety risks. Using sensitivity analysis on service impact, the organization can understand the relation between the criticality of the network and the best repair option for tackling seepages.

One of the other challenges in infrastructure management is the need for understanding short term decision impact against long-term planning. To understand this effect, the time period of analysis is varied and the resulting business impact for the

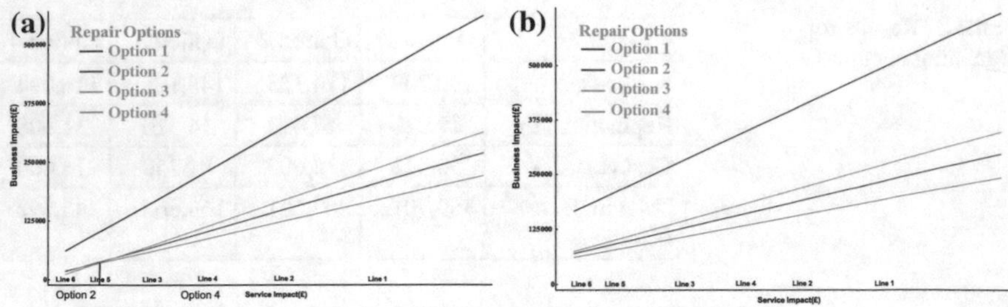

Fig. 3 Sensitivity analysis on service reliability: impacts of service on **a** signalling disruption and **b** corrosion of rails

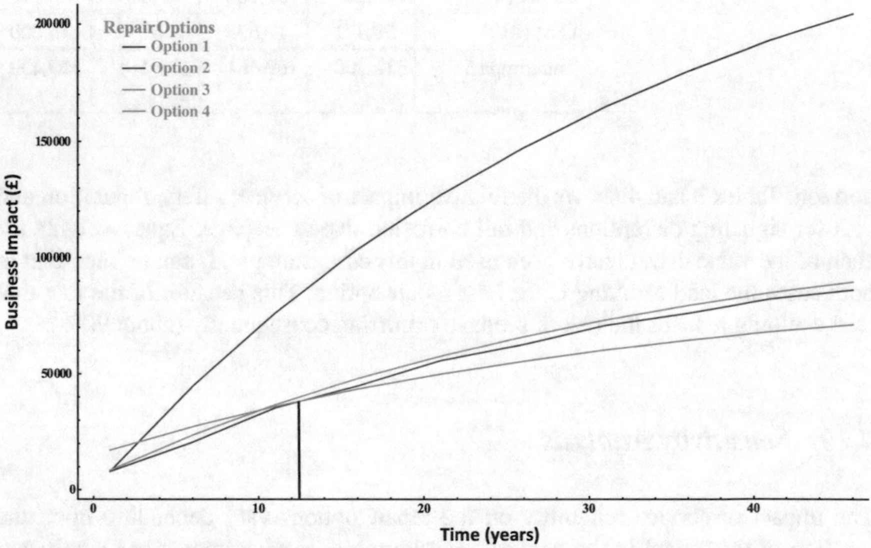

Fig. 4 Impact of time period of analysis on signalling disruptions due to seepage

various repair options for the signaling disruption is calculated and is shown in Fig. 4. From the figure, it can be seen acrylic grouting is the best repair options when the timer period of analysis is less than 12.2 years. When the decision horizon is longer, then lead caulking is the preferred repair option that will have the least business impact for signaling disruption due to seepages. Using this analysis, the organization can plan for the best repair option depending upon the funding available and offers flexibility in managing the seepage problems. On the other hand, the organization can use such analysis to justify the need for long-term planning and the associated budget requirements for regulators.

5 Conclusions

Infrastructure asset management is becoming more critical for organizations, especially to provide high quality service with an increasingly stringent budget. Additionally, there is a need to manage infrastructure assets to meet the varying demands of various stakeholders. A value-based infrastructure asset management decision making process is proposed, which can be used to systematically identify the value generation process and to incorporate this value into asset management decisions. The proposed approach takes into account the various stakeholders' requirements and a map representing the flow of value is developed. This value map can then be used to develop innovative ways to manage assets with clear understanding of the interdependent systems and the implication of intervention decisions. The proposed process was demonstrated through a case example concerning the identification of best repair option to mitigate the risks posed by seepages that occur in tunnels.

Acknowledgements The research work was performed within the context of SustainOwner ("Sustainable Design and Management of Industrial Assets through Total Value and Cost of Ownership"), a project sponsored by the EU Framework Programme Horizon 2020, MSCA-RISE-2014: Marie Skłodowska-Curie Research and Innovation Staff Exchange (RISE) (grant agreement number 645733—Sustain-Owner—H2020-MSCA-RISE-2014). The research was funded by the EPSRC Innovation and Knowledge Centre for Smart Infrastructure and Construction at the University of Cambridge.

References

1. Halfawy, M. R. (2008). Integration of municipal infrastructure asset management processes: Challenges and solutions. *Journal of Computing in Civil Engineering, 22*(3), 216–229.
2. Zhang, H., Keoleian, G. A., & Lepech, M. D. (2012). Network-level pavement asset management system integrated with life-cycle analysis and life-cycle optimization. *Journal of Infrastructure Systems, 19*(1), 99–107.
3. Buttlar, W. G., & Paulino, G. H. (2015). *Integration of pavement cracking prediction model with asset management and vehicle-infrastructure interaction models, Technical report*.
4. Micevski, T., Kuczera, G., & Coombes, P. (2002). Markov model for storm water pipe deterioration. *Journal of Infrastructure Systems, 8*(2), 49–56.
5. Ana, E. V., & Bauwens, W. (2010). Modeling the structural deterioration of urban drainage pipes: The state-of-the-art in statistical methods. *Urban Water Journal, 7*(1), 47–59.
6. Han, D., Kaito, K., & Kobayashi, K. (2014). Application of Bayesian estimation method with Markov hazard model to improve deterioration forecasts for infrastructure asset management. *KSCE Journal of Civil Engineering, 18*(7), 2107–2119.
7. The Institute of Asset Management. (2016). *Subject specific guidance: Capital investment, operations and maintenance decision making*. The Institute of Asset Management.
8. ISO. (2014a). *ISO 55000: Asset management*. Geneva, Switzerland: ISO (International Standards Organization).
9. Kabir, G., Sadiq, R., & Tesfamariam, S. (2014). A review of multi-criteria decision-making methods for infrastructure management. *Structure and Infrastructure Engineering, 10*(9), 1176–1210.
10. Hooper, R., Armitage, R., Gallagher, A., & Osorio, T. (2009). *Whole-life infrastructure asset management: Good practice guide for civil infrastructure*. CIRIA.

11. The Institute of Asset Management. (2015). *Asset management—An anatomy.* The Institute of Asset Management.
12. ISO. (2014b). *ISO 55001: Asset management—Management systems—Requirements.* Geneva, Switzerland: ISO (International Standards Organization).

Chapter 7
Exploiting EAMS, GIS and Dispatching Systems Data for Criticality Analysis

Improving Maintenance and Felling and Pruning Management in Power Lines

Adolfo Crespo Márquez, Antonio Sola Rosique, Pedro Moreu de León, Juan F. Gómez Fernández, Antonio González Diego and Eduardo Candón Fernández

Abstract This Chapter deals with the process of criticality analysis in overhead power lines, as a tool to improve maintenance, felling and pruning programs management. Felling and pruning activities are tasks that utility companies must accomplish to respect the servitudes of the overhead lines, concerned with distances to vegetation, buildings, infrastructures and other networks crossings. Conceptually, these power lines servitudes can be considered as failure modes of the maintainable items under our analysis (power line spans), and the criticality analysis methodology developed, will therefore help to optimize actions to avoid these as other failure modes of the line maintainable items. The approach is interesting, but another relevant contribution of the Chapter is the process followed for the automation of the analysis. Automation is possible by utilizing existing companies IT systems and databases. The Chapter explains how to use data located in Enterprise Asset management Systems, GIS and

A. Crespo Márquez (✉) · P. Moreu de León · J. F. Gómez Fernández · E. C. Fernández
Intelligent Maintenance System Research Group (SIM), Department of Industrial Management, School of Engineering, University of Seville, Camino de Los Descubrimientos s/n, 41092 Seville, Spain
e-mail: adolfo@us.es

P. Moreu de León
e-mail: moreu@us.es

J. F. Gómez Fernández
e-mail: juan.gomez@iies.es

E. C. Fernández
e-mail: eduardocandon@gmail.com

A. S. Rosique
INGEMAN. Association for the Development of Maintenance Engineering, School of Engineering, Camino de los Descubrimientos, Seville 41092, Spain
e-mail: asrasrasr@telefonica.net

A. G. Diego
Viesgo Electrical Distribution. PCTCAN, Santander, Spain
e-mail: antonio.gonzalez@viesgo.com

A. Crespo Márquez et al. (eds.), *Value Based and Intelligent Asset Management*,
https://doi.org/10.1007/978-3-030-20704-5_7

Dispatching systems for a fast, reliable, objective and dynamic criticality analysis. Promising results are included and also discussions about how this technique may result in important implications for this type of businesses.

Keywords Criticality analysis · Asset and maintenance management · Felling and pruning management in power lines · Life cycle management and sustainability · Decision support systems · Risk management

1 Introduction

Within the OPEX budget, felling and pruning work is commonly the most important activity for electricity distribution companies [1], which are capital-intensive companies with a long-term return on investment (ROI) [2]. As a general rule, the power line corridors are treated at fixed intervals along the line as a whole, which leads to low levels of efficiency, given the varied nature of both the vegetation, with its very different growth rates, and the distances from its conductors along the line. In addition, the new Spanish regulatory framework forces distributors to seek maintenance optimisation tools that focus on "asset management". Therefore, defining a proper methodology to increase the efficiency and effectiveness of felling and pruning maintenance plans, involving a transition from a cyclical maintenance model to a predictive maintenance model, has become a relevant issue for electrical distribution companies in Spain.

Most innovative companies are in the process of implementing advanced strategies, which rely on the possibility to have access to a high quantity and quality of information. Basically, using these strategies maintenance efficiency is improved because of the following four key factors:

1. *Improved knowledge of the network and the vegetation underneath it*. Transportation and distribution companies spend millions of Euros every year on vegetation management, but do not have sufficient information about it to maximise the efficiency of this treatment. With the right information, it is possible to find out where the vegetation is within the network, the area it occupies and its growth rate, and based on these details it is possible to calculate the optimum frequency for the treatments. Furthermore, an enhanced knowledge of the network and the vegetation for the providers of the felling and pruning services will lead to a reduction in the cost of their operations.
   The areas involved in improving the knowledge of the network aimed at improving the competitiveness of the felling and pruning services, to reduce the financial risk of their operations, are as follows (see Fig. 1):
   - Vegetation cartography (for instance using high-resolution LiDAR data capture (density >15 pts/m^2 and high resolution images −3 cm—along the high and medium voltage lines and 2,000 km of medium voltage lines; LiDAR-PNOA data provided by the ©National Geographic Institute of Spain (0.5 pts/m^2) and

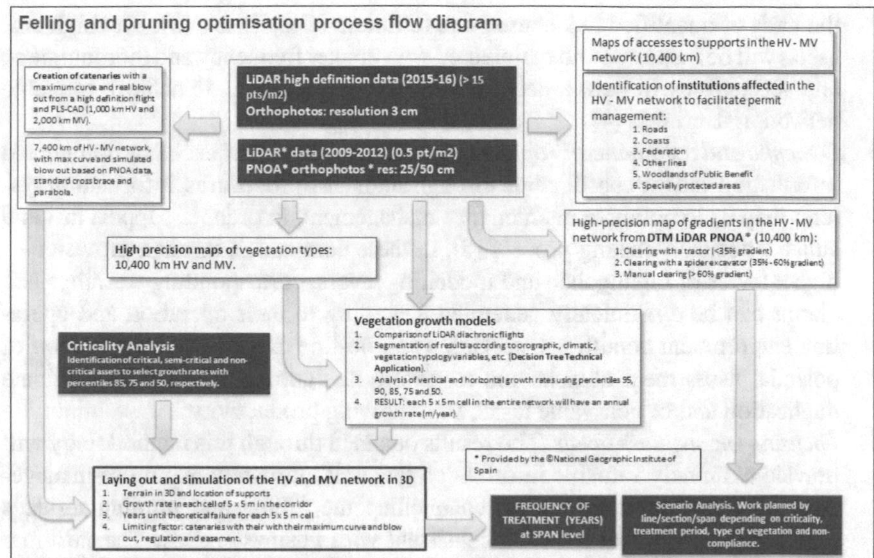

Fig. 1 Maintenance optimization process for overhead power lines [1]

PNOA orthophotos provided by the ©National Geographic Institute of Spain (25–50 cm) used to provide complementary information to the LiDAR high definition data, where these are not available) to quantify the intensity of the work to be done.

- Maps of accesses to the supports in the high and medium voltage network.
- Identification of the institutions affected by the high and medium voltage network, to facilitate the management of permits.
- High-precision map of gradients in the entire high and medium voltage network, to quantify the type of work (manual or mechanised) to be carried out depending on the terrain. The types of gradient that have been defined are as follows: <35%, 35–60%, >60%.
- Calculation of the frequency of felling and pruning at span level (objective to achieve a predictive maintenance), based on the vegetation growth rates obtained, orography, the network in 3D and the criticality analysis of each single line span.

2. *Prioritisation of work based on the asset's criticality.* The criticality analysis is considered a prerequisite or a necessary stage to review the existing maintenance programs, as well as the felling and pruning programmes associated with the assets (overhead power lines spans in this case). The level of indenture selected is the maintainable item, for which the maintenance plans are developed, resulting in a massive number of assets (around 200,000 assets for the entire network in the case study, to this level). Later, inspection and maintenance activities on these assets, plus suitable frequency of vegetation treatment, will be prioritized on

the basis of quantified risk caused due to failure of the assets [3]. The high-risk assets will be inspected and maintained with greater frequency and thoroughness, and vegetation will have a deepen treatment and analysis, to achieve tolerable network risk criteria [4].

3. *Dynamic and automatically optimization*. Due to the proper exchange of updated information and the coordination through automatic procedures, information systems help us to optimize maintenance management, in order to support fast and suitable decision making process [5]. In these companies, through provision of higher levels of intelligence and modelling layers, corresponding warnings and alarms can be dynamically generated according to their operation and operation environment conditions, serving as an on-line diagnostic or prediction of potential assessment of risks and as priority assignment in order to eliminate duplication and/or non-value tasks, so improving productivity.
4. *Focusing on business needs*. The results obtained through this methodology will provide extremely valuable information that will ultimately maximise management efficiency in the network, channelling the felling and pruning services provider in a way that must be consistent with business needs. The information will be managed on a centralised basis by means of a so called *"Felling and Pruning Management System*" based on GIS technology, which will use a multi-variable analysis to produce optimised maintenance plans for the short and medium term, minimising expenses, monitoring risks, making the work done by the contractor carrying out the work sustainable, and complying with the applicable Spanish and autonomous regional legislation. The implementation of the strategy presented in this Chapter was expected to provide the business with an annual saving of 33% in felling and pruning budget resulting in a dramatic efficiency improvement.

All these elements are relevant, but in this Chapter we concentrate on the process followed to provide a very dynamic analysis for the determination of the criticality of the assets. This analysis was used, as mentioned, to update the preventive maintenance plans, in general, and to reassign the frequency of vegetation treatment at a power line span level, in particular. In the sequel, the Chapter is organized as follows: Sect. 2 presents the existing requirements for the criticality analysis, conditioning the selection of the technique to be used for that purpose. Section 3 describes very precisely each step of the methodology implementation process using specific examples. Section 4 presents most relevant results obtained, their discussion and implications for the improvement of the management of the felling and pruning works. Finally Sect. 5 summarizes conclusions of the work and outlines aspects of further interest and research.

2 Requirements for the Analysis and Method Selected

The criticality assessment process to deal with the problem of this Chapter requires a specific methodology, which must cope with the following requirements:

- The process must be applicable to a large scale of in-service systems within the network (around 200,000), for which PM plans are designed and surrounding vegetation treatment is derived;
- The analysis should support regular changes in the scale adopted for the severity effects of the functional losses of the assets (this is a must to align maintenance strategy in dynamic business needs in current environments).
- The process must allow easy identification of new maintenance needs for assets facing new operating conditions, for instances new network developments, new demand of services, etc.;
- Connection with the company *Enterprise Asset Management System (EAMS), the GIS and the Dispatching System* should be possible, in order to automatically reproduce the analysis, with a certain cadence, over time.
- Connection with the *Felling and Pruning Management System* of the company, for on-line updates of vegetation status, treatment and budget control;
- The process should be tested in the network showing good practical results.

After considering all these needs, in this Chapter we have selected the methodology developed by Crespo Márquez et al. [6] because it fits properly for this problem resolution. The methodology was developed to prioritize the assets within an industrial/infrastructure context, where the maintenance organization has important amounts of data for complex in-service assets, for which a certain maintenance strategy has been previously developed and implemented. The criticality analysis is accomplished with the purpose of adjusting assets maintenance strategies to dynamic business needs over time.

Notice that most of current quantitative techniques for assets criticality analysis use a weighted scoring method defined as variation of the RPN method used in design [7]. This time, however, a very precise procedure must be considered when determining factors, scores and combining processes or algorithms [8] for criticality assessment. Unlike FMECA now we do assess assets criticality, not failure modes criticality. The referred methodology can be applied to the problem of this Chapter if we properly develop the following steps:

1. Determine frequency levels and the frequency factors;
2. Determine criteria and criteria effect levels to assess functional loss severity;
3. Determine non-admissible functional loss effects;
4. Determine criteria weights in the functional loss severity;
5. Determine severity scales per criteria effect;
6. Determine criticality limits.

The methodology has been developed with the premise that the results derived from the criticality analysis must be aligned with the priorities of the company.

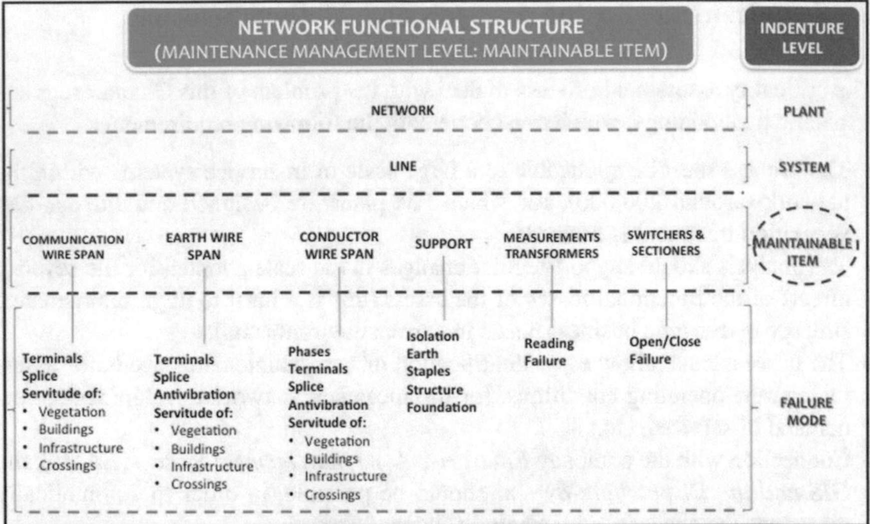

Fig. 2 Selected functional structure of the network for the analysis

It implies that the methodology must serve to the company target, and not in the opposite way. As a result, we will remark some aspects of the methodology that have been slightly adapted, with the aim of the results show, as faithfully as we can, the reality of the business.

At the same time, as it was mentioned above, this time the analysis requires a very precise level of indenture in the functional structure of the network (Fig. 2), resulting in a massive number of assets. Notice that, besides the needs of ranking the different spans for the felling and pruning work improvement, the organization will use the same information of the rest of the maintainable items, for general maintenance optimization purposes.

3 The Criticality Analysis Process

3.1 Determine Frequency Levels and Frequency Factors

The criticality concept (CR) is defined as the product of the failure frequency of and item (FF) times the possible consequence of its functional loss (C) as in Eq. 1.

$$\mathrm{CR} = \mathrm{FF} \times \mathrm{C} \quad (1)$$

Therefore, the first step in the process is to determine the frequency levels and the frequency factors. Frequency levels allow us to differentiate the assets by its failure

Table 1 Frequency levels and frequency factors

Annual frequency failure	Classification	Management definition	Frequency factor
$2 \leq f$	Very high	Inadmissible failures	2
$1 \leq f < 2$	High	Recurrent failures	1.5
$0.5 \leq f < 1$	Medium	Acceptable failures	1.2
<0.5	Low	Possible failures	1

recurrence. The frequency factor is the weight that we assign to each frequency level in order to use it within the criticality algorithm. Concerning the frequency levels, the most extended models define four levels: low, medium, high and very high failure frequency. In order to define the threshold among these frequency levels, a form of Pareto analysis is used, in which the elements are grouped into 4 frequency categories according to their estimated functional loss recurrence.

The use of Pareto approach guarantees that all items are properly distributed in the matrix spectrum, in order to maximize the sensitivity of the methodology. Thresholds values assigned must show the real management strategy of the company. Assuming that during the last years the company priority was more availability than efficiency. It is assumed that this fact led the assets to be a little over maintained, maintenance as well as felling and pruning work has been intense and equally carried out for all different lines, without prioritization, and as a consequence very low failure events are registered. With this in mind, we can clearly explain that the majority of assets will be located within the lowest failure frequency band. The frequency levels can be classified as follows (Table 1).

Once we have defined each level and the frequency failure thresholds, we must assign a failure frequency factor. This value will be given to each frequency in order to compute a criticality value. There are many different ways to assign the frequency factor to a level. If we follow the methodology, this value would be directly related to the numerical limits of the failure frequency defined for each level. In our case, we expect most of the items to be in the lowest level of frequency failure. We have given a value of "1" to this lowest frequency band. At the same time, we assume that more than two failures per year would be non-acceptable for an item, and that fact would cause a regular item (with average consequence of a failure) to become critical for the business. Value of "2" is then given to that case and the burdens in between are established proportionally (see Table 1).

3.2 *Determine Criteria and Criteria Effects Levels to Establish Functional Loss Severity*

To define a certain objective criteria to assess an asset functional loss, most theoretical models propose the consideration of two main arguments: integrity and sustainability. Integrity goes first, and issues like personal and industrial safety as well as environmental care, are considered under this argument. Sustainability is related to management efficiency and continuous improvement and it is based on assets integrity; aspects like availability, quality of service and maintenance costs, are included within this topic. It is important to remark that sustainability do not directly imply a certain monetary expense, even an estimated "profit loss" or "production loss", but can also be related to reputational or image lost, repercussion on the stakeholders or even hypothetical penalties for the loss of a certain service level.

3.3 *Determine Non-admissible Effects*

At this point, the process requires the definition of those functional loss effects that will have the consideration of "non- admissible", for the business. This first requires deciding in what criteria is this concept applicable. This consideration represents the allocation of the maximum punctuation, in total, in functional failure consequence to the asset (100 in our case), regardless its results in the rest of the criteria assessment. Looking back to the business asset management policy, it was decided to apply this "non-admissible" condition just for criteria related to Industrial safety, Environment and Quality of service (see Table 2, first three columns). We therefore have defined as non-admissible consequences, the maximum level of severity in industrial safety, environmental criteria.

3.4 *Determine Criteria Weights in the Functional Loss Severity*

Every single criteria criterion must have a specific weight in order to change subjective opinions of the criticality steering team members into a numeric value, ranking the asset according to how important is its function to meet business goals. AHP techniques helped to solve this problem and the reader is referred to Crespo Márquez [9] (Sect. 4.1 in Chap. 9, steps of the process 6 and 7, pages 121 and 122, concerning the Quantification of judgments on pair alternative criteria and the Determination of the criteria weighting and its consistency) for a detailed description of the utilization of the AHP in this specific process. For instance, we just limited the method utilization to the severity criteria level, not to the asset criticality classification level.

In the example of this Chapter, {wi}, weight given to the severity criteria i by experts, resulting from the AHP analysis are assume to be equal to $\{w_i\} = \{30, 12, 35, 14, 9\}$. This means, for instance, that the review team considers the impact on industrial safety to be almost two times more important than the impact of a failure on network availability.

Table 2 Selected criteria and criteria effects levels

Industrial safety: 30% Catastrophic: CA High: A Medium: M Low: B		*Environment*: 12% CATASTROPHIC: CA High: A Medium: M Low: B		*Quality service*: 35% Catastrophic: CA High: A Medium: M Low: B		*Availability*: 14% Very high: MA High: A Medium: M Low: B		*Maintenance costs*: 9% Very high: MA High: A Medium: M Low: B	
Very serious effect with damage to the network, or with fatalities or permanent disability, or severe damage to third parties (>1M€)	CA 100	Non-recoverable effect on vulnerable area (protected areas)	CA 100	Loss of continuity of electric service superior to 20,000 customers, critical essential facilities, tip lines or customers ≤30 kV	CA 100	Total loss of the capacity of the line with condition to the capacity of other lines, or line in branch without redundancy	MA 20	MC > 30,000 €	MA 5
High impact with damage to the net, or serious injury that causes a temporary disability, prolonged, or severe damage to third parties (>300 k€)	A 30	Recoverable impact on vulnerable area	A 12	Same for • 2,000–20,000 customers • Line loss ≥30 kV in N – 1 configuration	A 35	Total loss of line capacity in grid or branch network with redundancy	A 10	6,000 € < MC < 30,000 €	A 4

(continued)

Table 2 (continued)

Industrial safety: 30% Catastrophic: CA High: A Medium: M Low: B		*Environment*: 12% CATASTROPHIC: CA High: A Medium: M Low: B		*Quality service*: 35% Catastrophic: CA High: A Medium: M Low: B		*Availability*: 14% Very high: MA High: A Medium: M Low: B		*Maintenance costs*: 9% Very high: MA High: A Medium: M Low: B	
Average Impact on the Net, or damage to third parties (>50 k€)	M 18	Outside impact in non-vulnerable area	M 6	Same for: • 200–2,000 customers • Essential customers in medium voltage (<30 kV) • ≥55 kV, N − 1 lines	M 12	Partial loss of line capacity (>50%) with remote control manoeuvre points and with coupling	M 5	1,000 € < MC < 6,000 €	M 3
Slight impact or no impact	B 1	Slight impact or no impact	B 1	Any case not included in the above	B 1	Slight impact or no impact	B 0	MC < 1,000 €	B 1

Table 3 Criticality limits

Criticality	Criticality value
Not critical	90–200
Semi-critical	50–89
Critical	1–49

3.5 Determine Severity Scales Per Criteria Effect

The next step is to define the severity levels for each criteria effect. These levels will measure the severity of the consequences of a failure. In the same way that we have defined the failure frequency levels, the first step is to assess how many different levels must be defined for each criterion. In this project, the steering team decided that four levels was an optimum number to develop a precise and massive analysis. For each criterion, the consequences that a functional loss implies, in every level, must be determined. Each definition must be as simple and explicit as possible. If we are able to define it very simply, we will limit the possible debates later, in the working groups. The criteria effects scale defined is also included in Table 2.

3.6 Determine Criticality Limits

The determination of the criticality limits is a relevant business issue since it will later impact the number of assets for which a certain strategy will be addressed. In this Chapter example the limits considered were as in Table 3.

4 Retrieving Data to Easy Process Automation

At this point, the process would be ready to start, assessing asset by asset, for a massive number of assets (over 200,000 for high and mid voltage lines in our example). All the assets are registered in the company assets register of the EAMS that is connected to the GIS and therefore to the geo-referenced database of assets. An example of the codification of assets in the EAMS is presented in Table 4. An example of data concerning the geographical location of the asset is presented in Table 5, where data available in the different layers of the GIS is presented.

At the same time, fault location functionality of the dispatching systems can be used by a dispatch center to provide information about the potential number of customers to be affected when a failure takes place in a given location of the network. An electricity distribution grid contains a large number of power lines and equipment distributed over a wide area. A great number of these equipment are power protection equipment capable of detecting power faults as they occur, protecting consumers and

Table 4 Sample list of different types of assets

Network	MV line	Asset name	Element type	Asset code
Mid voltage	LMT1	01-10291|27-14906	Initial support	COD-35610
Mid voltage	LMT1	01-10291|27-14906	Final support	14906
Mid voltage	LMT1	7036_1|A41794	S&S	S38543
Mid voltage	LMT2	A15620_2|A26834	S&S	S38520
Mid voltage	LMT2	03|A15791	Span	03|A15791

Table 5 Data captured in the different GIS layers in the case study

Data layers in the GIS	Acronym	Content
Fire risk zone	ZRF	Yes/No
Place of public interest	LIC	Yes/No
Special protection zone (animals)	ZEPA	Yes/No
Natural park	EEDN	Yes/No
Vegetation fraction covered (%)	FCC	%
Railway crossing	FFCC	Yes/No
Main road crossing	CP	Yes/No
Populated zone	ZP	Yes/No
High frequency of persons area	AFP	Yes/No
Other network crossing AT, MT, BT, Phone	CoR	Yes/No

the grid itself from the consequences of these faults. When a fault is detected in a remote location, it is necessary to dispatch repair teams to the field to locate the place where the fault occurred. At the same time, in a smart grid the electricity distribution is managed through a communications network enabling remote monitoring and control of power equipment. If the number and location of sectionalizers and switches in a power line is known, and the number of customers served through that line is also known, the number of customers impacted by a fault of an asset of that line can be estimated [10].

According to previous information, we have found an important room for improvement when developing the criticality analysis process. If previous assets data is available, the criticality process can be automated following a set of simple rules that we explain in the next paragraphs.

4.1 Redefinition of Criteria Effects Levels According to Available Data

A first step in the automation process is to convert rules determining criteria effects levels (in Table 2) using now assets data that is available in the systems (GIS Geo Data

Table 6 Sample environment criteria effects level conversion (using GIS Data)

ORIGINAL		AUTOMATION	
Quality service: 35 % CATASTROPHIC: CA HIGH: A MEDIUM: M LOW : B		Quality service: 35 % CATASTROPHIC: CA HIGH: A MEDIUM: M LOW : B	
Loss of continuity of electric service superior to 20,000 customers, critical essential facilities, tip lines or customers < = 30Kv	CA 100	The failure of the equipment produces: ▪ Loss of non redundant line ▪ Loss of critical essential customers ▪ Loss of more than 20.000 customers ▪ Loss of customers in lines more than 30Kv	CA 100
Same for ▪ 2.000 - 20.000 customers ▪ Line loss> = 30KV in N-1 configuration	A 35	The failure of the equipment produces: ▪ Loss between 2.000 and 20.000 customers ▪ >=30Kv, N-1 lines	A 35
Same for: ▪ 200 - 2.000 customers ▪ Essential customers in medium voltage (<30Kv) ▪ >=55Kv, N-1 lines	M 12	▪ Loss of between 200 and 2.000 customers ▪ Loss of customers in medium voltage (<30Kv) ▪ >=55Kv, N-1 lines	M 12
Any case not included in the above	B 1	Any case not included in the above	B 1

Base or in the network dispatching systems). Computers, integrating the different types of information layers (services, cartography, network architecture, etc.), can interpret these converted rules, mapping locations and assigning severity to the assets automatically, for each specific criterion, saving an enormous time of analysis and producing a very robust and objective judgment. For instance, let's do that exercise to propose an equivalence of original criteria rules to new automated rules that are now based on assets GIS data, for the Environment criteria. We present that equivalence of rules in Table 6.

In Table 7, the same exercise is done for the equivalence of original criteria rules versus automated rules when using Dispatching data for the Quality of Service criteria.

Doing the same with all criteria effects levels, we can easily obtain Table 8, which is basically Table 2 adapted to compute data available in the business.

4.2 *Automatic Assessment of Assets Functional Failure Consequences*

At this time, failure consequences for all selected criteria, and for each single asset (maintainable item) can be assessed. To illustrate this point and to easy the generation of the algorithms that can be later implemented, it is convenient to generate a set of

Table 7 Sample quality of service criteria effects level conversion (Dispatching system Data)

ORIGINAL	
Environment: 12% CATASTROPHIC: CA HIGH: A MEDIUM: M LOW : B	
Non-recoverable effect on vulnerable area (protected areas)	CA 100
Recoverable impact on vulnerable area	A 12
Outside impact in non-vulnerable area	M 6
Slight impact or no impact	B 1

AUTOMATION	
Environment: 12% CATASTROPHIC: CA HIGH: A MEDIUM: M LOW : B	
Risk of fire in areas environmentally qualified as SCI, SPAB or natural spaces that have coverage fraction>= 70%.	CA 100
Risk of fire in areas environmentally qualified as SCI, SPAB or natural spaces	A 12
Coverage fraction ≠ 0	M 6
No impact or minimum impact	B 1

flow charts (one per criteria), describing graphically the automated process that will be then implemented. This process will generally be changed in the future, so it is convenient an up to date clear process documentation.

Once the flowcharts are generated, the corresponding pseudo-codes can be written. These codes describe, in IT language, the rules to be followed for each particular criterion during the automatic criteria consequences assessment. For instance, for the previous two cases, in Fig. 3 the flowchart for the assessment of the Environment criteria is shown, and in Fig. 4 the pseudo-code that was used in the case study for the automatic assessment of the criteria is presented. Continuing in the same way, in Fig. 4 the flowchart for the assessment of the Quality of Service criteria is presented, while in Fig. 5 the written pseudo-code that was used in the case study for the automatic assessment of that criterion is presented (Fig. 6).

4.3 *Automatic Evaluation of Assets Criticality*

Once the assessment for each criterion is completed, the criticality of the assets, as a result of multiplying the frequency factor times the consequence of the functional loss can be computed. See for instance the calculation in Table 9, assuming a certain level of criteria consequences for each asset.

Table 8 Adaptation of the criteria effects levels table to existing data

Industrial safety: 30% Catastrophic: CA High: A Medium: M Low: B		*Environment*: 12% Catastrophic: CA High: A Medium: M Low: B		*Quality service*: 35 % Catastrophic: CA High: A Medium: M Low: B		*Availability*: 14% Very high: MA High: A Medium: M Low: B		*Maintenance costs*: 9% Very high: MA High: A Medium: M Low: B	
Area of high frequency of people	CA 100	Risk of fire in areas environmentally qualified as SCI, SPAB or natural spaces that have coverage fraction ≥70%	CA 100	The failure produces loss: – of non-redundant line • of essential customers • of >20,000 customers – Loss of customers in lines more than 30 kV	CA 100	The failure of the equipment produces: failures in lines: With multiple circuits or lines in tip (of any voltage)	MA 14	The failure of the equipment produces: failures in lines: 55 and 132 kV underground lines	MA 9
· Crossing with road (Motorway, dual carriageway, link, national route) · Crossing with railway (Ave, conventional train or station) · Populated areas · On high, medium or low voltage line and telecom line	A 30	Risk of fire in areas environmentally qualified as SCI, SPAB or natural spaces	A 12	The failure of the equipment produces: • Loss between 2,000 and 20,000 customers • ≥30 kV, N-1 lines	A 35	The failure of the equipment produces: failures in lines: Of 132 and 55 kV in meshed net, or of meshed medium voltage without remote control	A 7	• 55 and 132 kV overhead lines • 30 kV underground lines	A 5

(continued)

Table 8 (continued)

Industrial safety: 30% Catastrophic: CA High: A Medium: M Low: B		*Environment*: 12% Catastrophic: CA High: A Medium: M Low: B		*Quality service*: 35 % Catastrophic: CA High: A Medium: M Low: B		*Availability*: 14% Very high: MA High: A Medium: M Low: B		*Maintenance costs*: 9% Very high: MA High: A Medium: M Low: B	
Crossing with an autonomic road	M 18	Coverage fraction $\neq 0$	M 6	• Loss between 200 and 2,000 customers • Loss customers in medium voltage (<30 kV) • $\geq$55 kV, N − 1 lines	M 12	The failure of the equipment produces: failures in lines: Medium voltage with remote control elements	M 5	Overhead lines of medium voltage or underground lines of less than 30 v kV	M 2
Any case not included in the above	B 1	No impact or minimum impact	B 1	Any case not included in the above	B 1	Any case not included in the above	B 1	Any case not included in the above	B 1

Table 9 Sample criticality results

Asset Code	Frequency score	Industrial safety	Environment	Quality of service	Availability	Maint. cost	Criticality CR = FF × C
COD-35610	L	A	B	A	A	A	1*(30 + 1+ 35 + 14 + 5) = 85
14906	VH	B	CA	B	B	A	2*100 = 200
S38543	H	B	B	B	A	B	1.5*(1 + 1 + 1 + 10 + 1) = 21
S38520	H	B	B	B	B	B	1.5*(1 + 1+ 1 + 0+1) = 6
03lA15791	L	CA	A	B	B	B	1*100 = 100

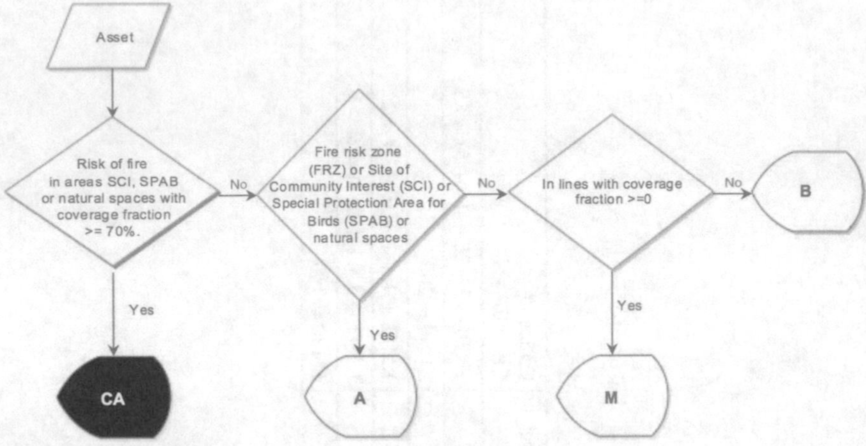

Fig. 3 Flow chart for the assessment of the criteria: Environment

```
Function: Value EN

Define EN, Fire_risk_zone, SCI_zone, SPAB_zone, natural_spaces_zone like character;
Define Coverage fraction like whole number;
Read Fire_risk_zone, SCI_zone, SPAB_zone, natural_spaces_zone, Coverage fraction
        Algorithm:
            If (Fire_risk_zone="YES" |SCI_zone="YES"| SPAB_zone="YES" |natural_spaces_zone="YES") & covera
            ge fraction>60%
            Then
                    Write "CA"
            Else
                    If (Fire_risk_zone ="YES" | SCI_zone ="YES" | SPAB_zone ="YES" | natural_spaces_zone ="Y
                    ES")
                    Then
                    Write "A"
                    Else
                            If (Fire_risk_zone ="NO" | SCI_zone ="NO" | SPAB_zone ="NO" | natural_spaces_
                            zone ="NO") &       coverage fraction >0%
                            Then
                                    Write "M"
                            Else
                                    Write "B"
                            ENDIF
                    ENDIF
            ENDIF
    Endfunction
```

Fig. 4 Pseudo-code for the assessment of the creiteria: environment

4.4 Automatic Obtaining of the Network Criticality Matrix

A suitable representation of criticality, as obtained in Table 9, can be the criticality matrix in Fig. 7, and a real production criticality matrix, as obtained in Fig. 7, is shown in Fig. 8. It is very important to develop certain features that may results very practical elements of a real production criticality matrix. For instance we call readers attention about the following ones:

- List the items within a certain matrix location

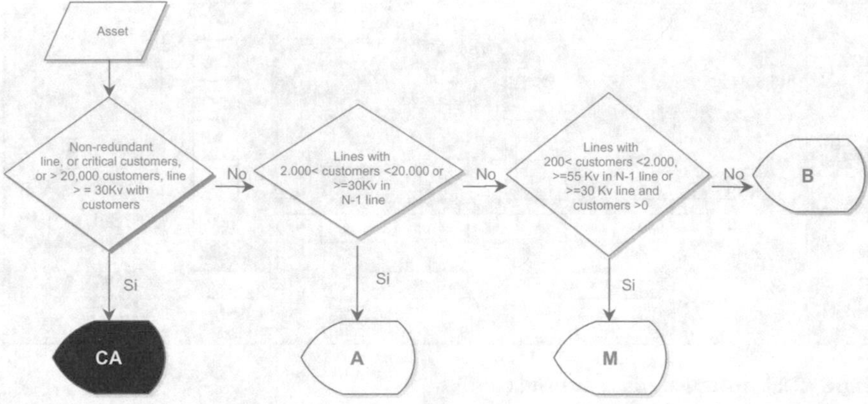

Fig. 5 Flow chart for the assessment of the quality of service criteria

```
Function: Value QS

Define QS, line_non-redundant, critical_customers, N-1 like character;
Define customers, line like whole number;
Read line_non-redundant, critical_customers, N-1, customers, line;
  Algorithm:
        If (line_non-redundant = "YES" | customers >20.000 | critical_customers >0 | line >=30 & customers >0)
        Then
                    Write "CA"
        Else
              If (2.000< customers <20.000 | line >=30 & N-1 = "YES")
              Then
                    write "A"
              Else
                    If (200< customers <2.000 | line >=55 & N-1 = "YES" | line >=30 & customers >0) Then
                              Write "M"
                    Else
                              Write "B"
                    ENDIF
              ENDIF
        ENDIF
Endfunction
```

Fig. 6 Pseudo-code for the assessment of the quality of service criteria

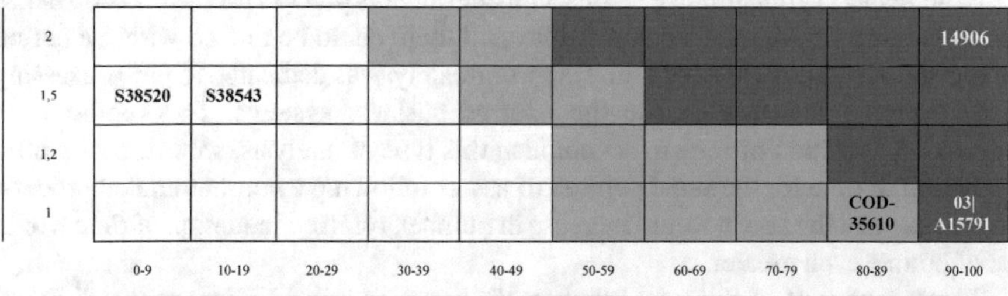

Fig. 7 Sample criticality matrix

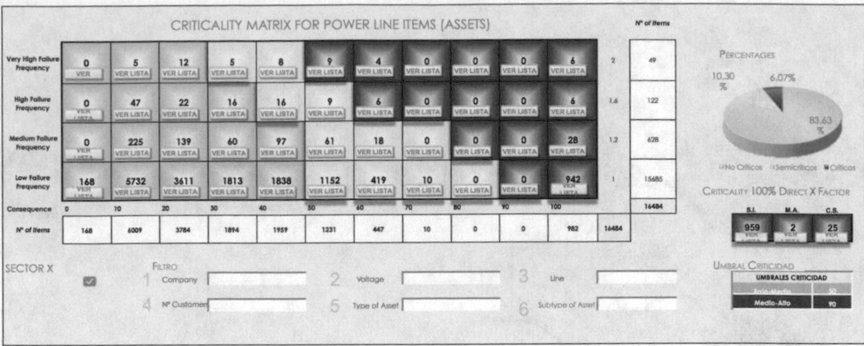

Fig. 8 Real production criticality matrix

- List the item of a certain criticality level
- Possibility to filter items per type of maintainable item
- Possibility to filter items per line
- Possibility to filter items per voltage of the lines
- List of item with non-admissible functional loss effects and per time of effect
- % of items under each criticality category

These features are used on a daily basis and can seriously increase efficiency in day-to-day operations analysis.

5 Final Results of the Methodology Application

In this Section we review main results obtained through the use of the methodology described above, some of these results are quantitative results, but some other are related to organizational aspects of the process and implications to the business.

With respect to quantitative results, considering this process for the 200,000 maintainable items of the selected power lines, all them could be ranked within a period of one month. And 50% of this time, approximately, was dedicated to pre-processing and arranging data available in the referred business systems. This represent an enormous reduction of time to accomplish this type of analysis, we estimate a 80% reduction of time for the same number of assets following a non-automated process. In the case that the assets would increase in number, referred reduction of time would be of course even greater.

Percentage (out of the total number of assets) of critical items in the different categories are listed in Table 10, showing a very important amount of assets resulting non-critical (close to 70%), these items could immediately be subjected to a risk-cost-benefit analysis to discard preventive maintenance tasks. At this point it was important to focus attention on:

Table 10 Percentage of items per each criticality category

Type of asset	Critical	Semi-critical (%)	Non-critical (%)
Supports	6.4	12.8	41.6
Aerial spans	1.9	4.9	14.2
Underground branches	0.4	0.3	7.0
Maneouvert elements (S&S)	1.0	2.7	6.9
Total	9.7	20.7	69.7

Table 11 Percentage of items per each criticality category

Type of asset	Critical	Semi-critical	Not critical
Aerial spans	9%	23%	68%

- Task accomplished with a higher frequency than stated in the legal directives;
- Task that were designed beside legal tasks, with the initial intention to have a better control of systems dependability;
- Task that when discarded really represented cost savings for the business (many tasks do not really represent cost savings when discarded because of similar parallel tasks that mast be accomplished).
- Task that when discarded do not represent early deterioration of the items.
- Concerning the impact of these results on the felling and pruning work, the percentage of spans per category are listed in Table 11, showing also results of a 68% of spans resulting non-critical spans, and a 23% of semi critical, while only 9% spans resulted to be critical.

This information could be crossed or combined with the vegetation growth models that were developed for each cell of 5 × 5 m of the entire network, and which provide an annual growth rate [meters/year] of that cell. The vegetation growth models are not part of this Chapter but very interesting tools because they also allow a 3D simulation of the network providing values for:

- The maximum arrow catenaries and real and simulated blow out.
- 3D terrain and location of supports.
- Application of growth rate in each cell of 5 × 5 m in the corridor.
- Years until theoretical default for each cell of 5 × 5 m.
- Limiting factor: catenaries of maximum clearance and blow out, regulation and servitude.

The combination of models: Span criticality versus Vegetation growth per span corridor, allows again a risk-cost-benefit analysis to discard felling and pruning tasks per line span (so now with much more detailed level of indenture than before) and improves dramatically the effectiveness and efficiency of felling and pruning treatments, to ensure regulatory distances not to be invaded by vegetation (see Fig. 9).

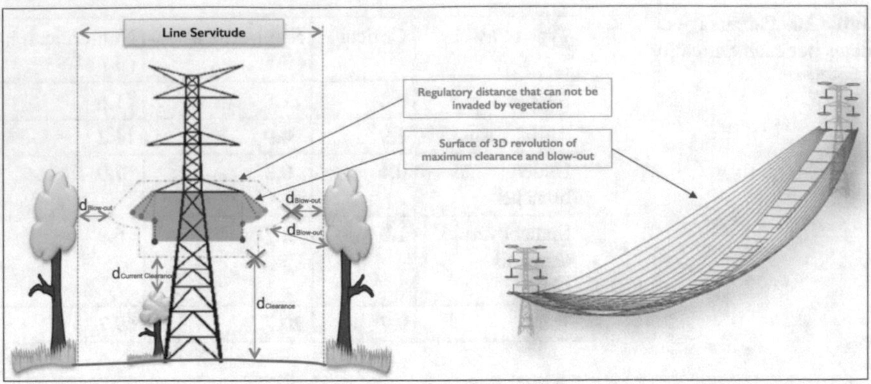

Fig. 9 Regulatory distances not to be invaded by vegetation

In this case the suggested period of treatment per span (SPTS) was obtained as in Eq. 2

$$\text{SPSP} = \text{Min}\ (\text{T}_\text{i}),\ \text{with i} = 1 \ldots \text{n number of cells per span} \tag{2}$$

where T_i is the time for the vegetation of cell i to reach the above mentioned limiting factors (Fig. 9) of the cell, considering the analysis of vertical and horizontal growth rates, and taking percentiles of, for instance, 95, 75 and 50%. Of course the more critical the span the more conservative we are in our estimations (the higher admissible vegetation growth rate).

The implementation of the methodology did generate a reduction of 33% in expenses in felling and pruning treatments to a minimum increase of risk for the business, and right for the next economic business cycle.

6 Conclusions

In this Chapter we show a practical way to implement criticality analysis in a power distribution network, we exemplify the concepts and procedure using several maintainable items of the lines.

We demonstrate the importance of the selection of a suitable methodology, allowing the study of assets criticality to the required indenture level. We explain how, in this digital era, this process can be automated thanks to assets existing data in business systems like EAMS, GIS and dispatching systems. Automation requires simple rules translation and algorithm development.

Results in the application of the method to extensive number of assets in power lines were considered relevant by different businesses, because of the extent of the savings, but also because of the "easy-to-implement" technique. In most of cases a relevant decrease of the budget assigned, specially to felling and pruning task, but also to preventive maintenance, was reaching significant values (most of the times exceeding the 30% of the original budget).

Acknowledgements This research work was performed within the context of Sustain Owner ('Sustainable Design and Management of Industrial Assets through Total Value and Cost of Ownership'), a project sponsored by the EU Framework Program Horizon 2020, MSCA-RISE-2014: Marie Skłodowska-Curie Research and Innovation Staff Exchange (RISE) (grant agreement number 645733—Sustain-Owner—H2020-MSCA-RISE-2014); and the project "Desarrollo de Procesos Avanzados de Operacion y Mantenimiento sobre Sistemas Cibero Fisicos (CPS) en el Ambito de la Industria 4.0", Ministerio de Economía y Competitividad del Gobierno de España, Programa Estatal de I+D+i Orientado a los Retos de la Sociedad. (DPI2015-70842-R). Financed by ERDF (European Regional Development Fund).

References

1. Cuasante, D., González, A., Garañeda, R., Soto, G., Crespo Márquez, A., & Sola A. (2018). Innovative techniques for the predictive maintenance of overhead power lines. *Practical application in the improvement of efficiency in felling and pruning in Viesgo (Spain)*. Cigre. Paris Session on Overhead Lines and Information Technology.
2. Newbery, D. M. G. (2002). *Privatization, restructuring, and regulation of network utilities*. Cambridge: MIT Press.
3. Arunraj, N. S., & Maiti, J. (2007). Risk-based maintenance -Techniques and applications. *Journal of Hazardous Materials, 142,* 653–661.
4. Brown, S. J., & May, I. L. (2000). Risk-based hazardous protection and prevention by inspection and maintenance. *Transfer ASME Journal of Pressure Vessel Technology, 122,* 362–367.
5. Fernández, J. F. G, & Márquez, A. C. (2012). *Maintenance management in network utilities: Framework and practical implementation. springer series in reliability engineering*. ISBN 978-1-4471-2757-4.
6. Crespo Márquez, A., Moreu de León, P., Sola Rosique, A., & Gómez Fernández, J. F. (2016). Criticality analysis for maintenance purposes: A study for complex in-service engineering assets. *Quality and reliability engineering international, 32,* 519–533.
7. Duffuaa, S. O., Raouf, A., & Campbell, J. D. (2000). *Planning and control of maintenance systems*. Indianapolis: Wiley.
8. Moss, T. R., & Woodhouse, J. (1999). Criticality analysis revisited. *Quality and Reliability Engineering International, 15,* 117–121.
9. Crespo Márquez, A. (2007). *The maintenance management framework: Models and methods for complex systems maintenance*. London: Springer.
10. U. S. Department of Energy. (2014). *Fault location, isolation, and service restoration technologies reduce outage impact and duration*. Smart Grid Investment Grant Program.

Part III
Intelligence in Operational Decision Making—CBM/PHM and Predictive Analytics

Chapter 8
A CPS for Condition Based Maintenance Based on a Multi-agent System for Failure Modes Prediction in Grid Connected PV Systems

Jesús Ferrero Bermejo, Juan F. Gómez Fernández, Antonio J. Guillén López, Fernando Olivencia Polo, Adolfo Crespo Márquez and Vicente González-Prida Díaz

Abstract Failure control based on condition monitoring needs to be sustainable and well-structured, and to rely on procedures fulfilling international standards, in order to keep and improve solutions over time. Failure detection and prediction in networks of assets demands an even more sophisticated approach and a clear conceptual framework, to be able to consider individual assets degradation behaviours and corresponding integrated effects on the network. In these cases, logic of failure control has to manage not only reliability data but also operation and real time internal and locational variables. Cyber-Physical Systems (CPS) approach easies the integrations of physical processes, network of assets and intelligent computation; CPS may enable co-operation among autonomous and distributed intelligence. Because of all these reasons this chapter sustains that failure detection and prediction in networks of assets can seriously benefit of CPS. However, CPS implementation needs a conceptual framework allowing the permanent development of current and new algorithms for advanced asset degradation and production forecasting. Multi-Agent System (MAS) architecture complies with these framework requirements from the scalability point of view, but in order to cope with these solutions adaptation to loca-

J. Ferrero Bermejo · F. Olivencia Polo
Magtel Operaciones, Seville, Spain
e-mail: jesus.ferrero@magtel.es

F. Olivencia Polo
e-mail: fernando.olivencia@magtel.es

J. F. Gómez Fernández · A. J. Guillén López · A. Crespo Márquez (✉) · V. González-Prida Díaz
Intelligent Maintenance System Research Group (SIM), Department of Industrial Management, School of Engineering, University of Seville, Camino de los Descubrimientos s/n, 41092 Seville, Spain
e-mail: adolfo@us.es

J. F. Gómez Fernández
e-mail: juan.gomez@iies.es

A. J. Guillén López
e-mail: ajguillen@us.es

A. Crespo Márquez et al. (eds.), *Value Based and Intelligent Asset Management*,
https://doi.org/10.1007/978-3-030-20704-5_8

tional and operational changes, artificial neural networks (ANN) are developed in this chapter on top of the legacy supervisory control and data acquisition system, to implement an innovative failure detection and power generation forecasting method. The model and method is demonstrated in grid connected solar photovoltaic power plants.

Keywords Renewable energy · Solar photovoltaics · Multi-agent system · Condition based maintenance · Cyber-physical systems · Artificial neural network · Failure detection

1 Introduction

Photovoltaic (PV) systems are rapidly gaining acceptance as one of the best alternative sources of energy. This source of energy, also known as clean energy, is included in the tariff plans of many countries. However, the power output of the PV system can highly fluctuate depending on weather conditions. Consequently, it is essential to develop accurate short-term PV energy supervisory and control systems, models for energy prediction and operation planning, reserve planning, peak load matching of power systems, and failure detection. A grid connected (PV) photovoltaic system is composed of all essential elements to generate electric power, the following ones being fundamental, see Fig. 1:

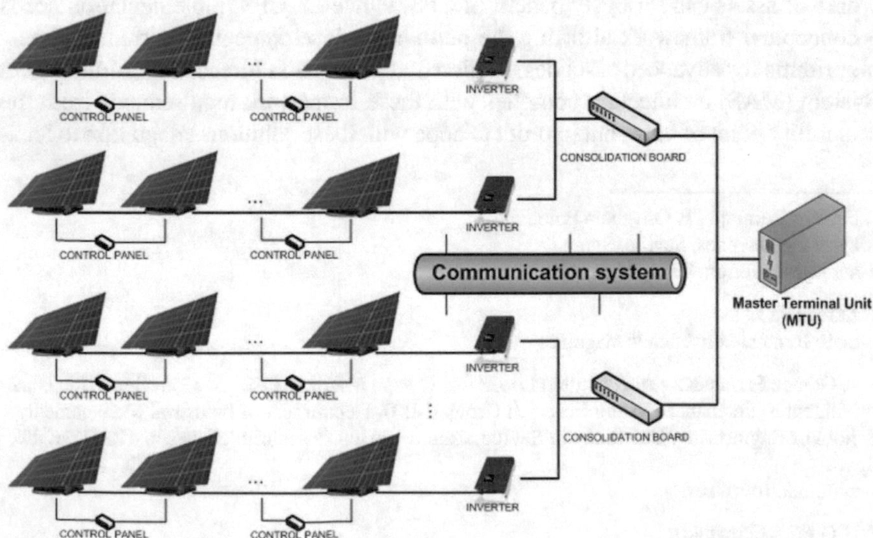

Fig. 1 Physical layer architecture

- Solar panels (trackers): Receive the solar power and transform it into continuous electric current.
- Inverter: Transforms the generated continuous electric current by group of trackers into alternating current, which after being conditioned appropriately in the centre of transformation, is injected into the electrical distribution network.
- Counting/Register: This element takes into account the electric power produced by the solar plant, and using the stored registers, produces the invoicing for the distributing company of electricity.

Prediction models for PV energy can be classified into two categories, Physical Models and Time Series Models:

- Physical Models which are based on the physical properties of the individual elements involved in the system (for example: a PV cell, the inverter, etc.), and aim to develop mathematical models of PV performance as a function of independent variables, including the characteristic parameters of photovoltaic modules, solar radiation, ambient temperature, etc. As example, representing the efficiency of PV cells with dependence upon solar radiation and temperature of the cell [1, 2]. These models are commonly based on equivalent circuit models of a diode or models with four or more parameters [3–6].
- The time series models are generally used for the statistical forecast of solar energy. Autoregressive and autoregressive moving average are among the models commonly used [7, 8]. Nonlinear methods, such as fuzzy model [9] and wavelet based methods [10], has proven they are better to linear models. Artificial neural networks (ANN) can reach a good performance to predict the solar energy [1, 11].

The usual approach to monitor PV installations is to use a centralized Supervisory Control and Data Acquisition (SCADA) system, also known as Power Management System. This approach has deficiencies in flexibility and adaptability and is no longer sufficient for certain control operations [12]. Conventionally, power management is ensured by a central facilitator whose program is based on long series of conditional branches. While this solution achieves a constant supply of the load, cannot meet other objectives such as fault tolerance of an element [13]. The proper exchange of updated information and the coordination for maintenance is a core activity [14] in order to create company competitive advantage through excellent maintenance management. Therefore, finding a control strategy becomes more and more difficult with increasing system size. Moreover, monitoring, analysis, and control technologies are sometimes too slow, limited to just the protection of specific components. The system restoration process is based mainly on the experience and results of operators' offline studies [15].

The changing demands of power distribution networks require the application of intelligent methods to maintain the system state within the suitable permitted limits. These kind of advanced strategies converge with the developments of Industry 4.0, particularly with CPS (Ciber Physical System). CPS is defined as transformative technologies for managing interconnected systems between its physical assets and computational capabilities [16, 17] combine two interesting approaches describing

CPS: (i) CPSs refer to ICT systems (sensing, actuating, computing, communicating, and so on) embedded in physical objects, interconnected through several networks, including the Internet, and providing a wide range of innovative applications; (ii) CPSs are also ubiquitous embedded cyberphysical applications that are surfacing (emerging) and are now bridging the physical and virtual worlds and share all kinds of collaborative networks.

CPSs provide an optimum framework for high level integration approaches [18], including systems integration and processes integration, as it is proposed in this chapter. Based on the results of four European innovation projects (i.e. SOCRADES, IMC-AESOP, GRACE and ARUM) focused in this topic, MAS (Multi-Agent Systems) and SOA (Service-Oriented Architectures) are identified as key CPS technologies.

Multi-Agent Systems (MAS) are considered as an approach able to perform this functions [19–21]. MAS provide a convenient way to deal with the dynamics in large complex systems, so the system control is performed in a decentralized manner, thereby reducing the complexity and increasing the flexibility and fault tolerance. Agents have the ability to coordinate different devices, such as power sources, loads or switches in a decentralized manner and are able to strike a technically and economically optimal operation under consideration of all types of constraints [22]. In this context, a distributed architecture allows processing data locally and minimizes the need for massive data exchanges. A distributed system allows the high performance necessary to prevent or contain the rapidly changing adverse events [23]. In photovoltaic networks, multi-agent systems were developed for the management of distributed power sources, storage batteries and programmable power sources [13, 24, 25]. In addition, multi-agent systems were implemented to determine the optimal network operation point [26] and restore the system to a safe point after a failure [11, 21].

Potential PV plant re-investments must be evaluated; incorporating equipment reliability analysis in future operating and environmental conditions in order to avoid future production disruptions. This type of photovoltaic plan was usually built modularly, each 100 kW may represent over 600,000 € of investment, therefore the possibility to replicate this CMB based on MAS is great helpful if it has the ability to dynamic update before changes.

Risk of failures could even reach ten times the purchase equipment cost [27]. In our case study, failure mode analysis adds enormous value for protecting a production (at an initial price of 0.4886 €/kWh) of 513,030 €/year in an opened in 2008 plant with 37,180 photovoltaic panels in groups of 100 kW for each inverter, aggregating its own energy (210,000 KWh/year) jointly into a transformer. Therefore, the inverter is relevant equipment in the plant to be analysed by MAS in the CBM framework, searching early detection of degradation and a quantitative measure of risk.

This Chapter is structured as follows: Sect. 2 presents the way to comprehend the failure prediction searching sustainability and adaptability of solutions, using a structured framework. Section 3 provides the developed architecture and method with the proper data integration. In Sect. 4, the implementation is shown, describing the specific case of power inverters. Finally Sect. 5 concludes the chapter.

2 Optimizing CBM and Predictive Maintenance with MAS

The problem of predicting solar power is becoming extremely relevant because inaccurate prediction often lead to substantial economic losses and limit the expansion of renewable energies. Predicting photovoltaic power becomes to a challenging problem if we consider that this is dependent on weather conditions.

Moreover, it is necessary to develop solutions for failure prediction in an appropriate manner, considering real-time and non-real-time data from different sources. The main difficulty for processes efficiency, is data combination in a structured and sustainable way. The three main sources of information are CMMS (computerized maintenance management systems), RCM (reliability centred maintenance) and CBM. The purpose of CMMS is to manage, plan and document the maintenance activities, and it encompasses the historical knowledge of solutions and learning. CBM and RCM approaches deepen into the knowledge based on the physical condition (real-time status of equipment, using alarms, thresholds and degradation patterns) and on reliability (real-time risks, predictions of residual life, survivability and resiliency studies) respectively.

The recent development of the PHM discipline (Prognosis and Health Management) is promoting a new CBM concept or "extended CBM", providing powerful capabilities for physical understanding of the useful life of a system through dynamic pattern recognition [28]. According to EN 13306, Predictive Maintenance (PdM) is the part of CBM that is focused on the prediction of failure and prevents the failure or degradation. So CBM can be considered as the most general term that includes the rest.

Condition-Based Maintenance (CBM) is defined by EN 13306:2010 as "Preventive maintenance that includes a combination of condition monitoring and/or inspection and/or testing, analysis and subsequent maintenance actions". ISO 13372:2012 standard defines CBM as "Maintenance performed as governed by condition monitoring programmes". CBM monitors the condition of components and systems in order to determine a dynamic preventive schedule. In recent decades, the emergence of cheaper and more reliable ICT-Information and Communication Technologies (intelligent sensors, personal digital devices, wireless tools, etc.) has allowed an increase in the efficiency of CBM programs [29].

For a wide comprehension of the CBM potential is needed to analyse its three components [30]: monitoring, diagnosis and prognosis. Guillén et al. [28] describe these components in terms of failure mode control:

- Detection/monitoring is associated with the system states (for example the transition from function state to fault state) and, in general, with normal behaviour-anomalies distinction (in reference to defined baseline data).
- Diagnosis is associated with the location of the failure mode and its causes.
- Prognosis is associated with the evolution of the failure mode or its future behaviour (risk of failure and remaining useful life in a moment).

This disaggregation of the CBM concepts also appear in the approach of the ISO 13374-1:2003 "Condition monitoring and diagnostics of machines—Data processing, communication, and presentation". This standard, jointly with OSA-CBM [31], are ones of main international references, detailing information models and processing architecture requirement.

PV systems contain a large number of sub-systems and components. This case generates a wide variety of maintenance situations and the challenge is to employ escalating to its surrounding assets of the plant. The challenge is how structure the information sustainably and interrelated properly and how present it in a form which they can assimilate, decreasing mismatches between perceived and real risk, improving rapid decision on critical incidents.

The use of MAS architecture for CBM allows a very specific treatment for each component following these approaches. This aids to understand the different uses of each of them and their requirements, optimizing the technical design of the solutions (hardware, software, algorithm, etc.) and its uses within the maintenance function.

In order to standardize and make scalable the MAS application, we have based this development in the CBM framework of Guillén et al. [28] as in standard conceptual computing model, serving as templates for its implementation from a computational context. In this framework the abstract representation of entities and their relationships are implemented by Unified Modelling Language (UML). This CBM framework manages the three types of monitoring outputs (detection, diagnosis, and prognosis) simultaneously and in uniform way, and describes five consistent blocks for the characterization of any CBM solution, see Fig. 2. Thus any future change, modification, improvement, management of the solution will be very much facilitated and understood.

- Block 1. Physical description according to the system belongings to a plant, an installation, an industry, etc.
- Block 2. Functional description about the action and activity assigned to, required from or expected from a system. In fact, a functional failure is the way in which a

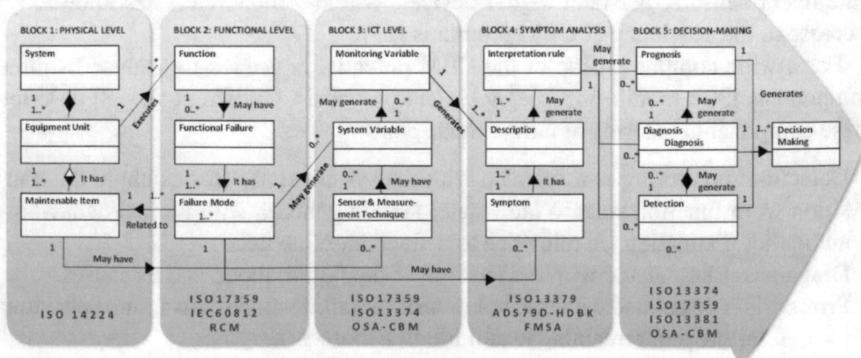

Fig. 2 CBM/PdM framework (Guillén et al. [28])

system is unable to fulfil a function at the performance standard that is acceptable for the user and a failure mode is the effect by which a functional failure is observed [ISO 13372:2012, Condition monitoring and diagnostics of machines—Vocabulary].

- Block 3. Information sources, organizing the different information for symptoms treatment and producing monitoring variables from sensors, software systems (control, operation and maintenance), monitoring devices and techniques, data bases, data warehouses, etc.
- Block 4. Symptoms analysis management in relation to the failure mode which is linked to symptoms as qualitative description of the latter.
- Block 5. Maintenance decision-making, linking detection, diagnosis and prognosis to different maintenance decisions. The maintenance tasks and general actions that are triggered as consequences of the monitoring outputs can be registered, listed and catalogued to be used.

Maintenance management is focus on failure mode in order to preserve system functions [32]. In conclusion, the key of this CBM framework is the failure mode determination, linking the physical structure elements and functional logic elements and identifying parameters required to predict it.

3 CBM/PdM Based on Multi-agent System for Prediction

As we have seen before, a CBM approach is focused on identifying physical degradation patterns of assets, according to their operation and operation environment conditions, searching failure reduction due to triggering warning or alarm messages.

Therefore, access to many disparate elements and advanced knowledge generation are crucial for remote intelligent maintenance management in order to support fast and suitable decision making. This intelligent management has to include end-to-end and flexible performance monitoring, and considering decentralization, as low as possible, by software agents programed for each case depending on the variety of the decisions [33] mapping relationship between product performance variation and asset degradation and considering operational environment throughout its entire life cycle. In this sense, concurrently with the operation, based on monitoring information and predictive simulations, we could obtain automatically warnings about levels of risk to optimize maintenance, or to generate corrective and preventive tasks. All the maintenance knowledge has to be created, validated, represented and available by the CMB/PdM system, with the intention of being applied to the activities.

Now, the developed MAS architecture (see Fig. 3) for the CBM framework is presented for failure prediction from an energy production point of view, searching early warnings and a quantitative measure of the risk.

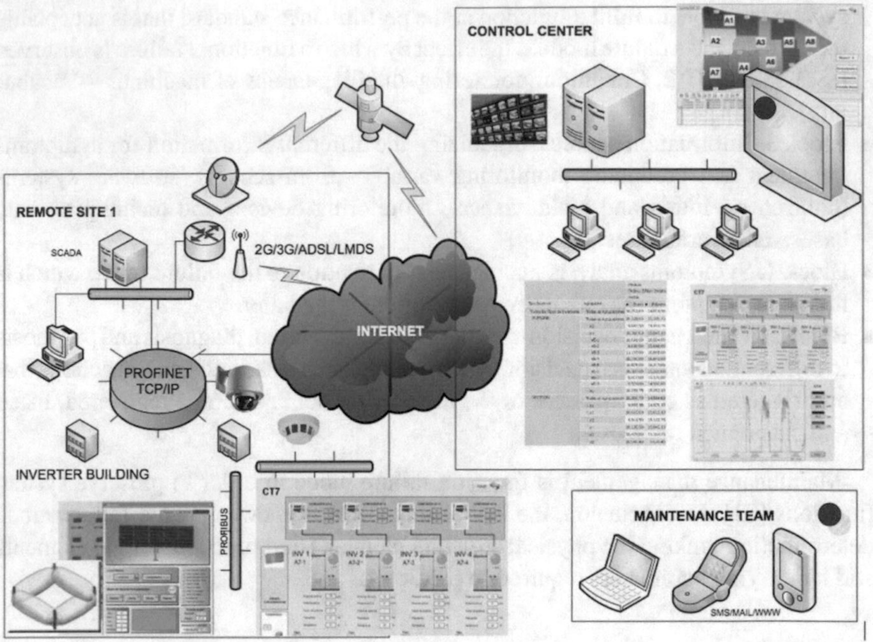

Fig. 3 System architecture

3.1 Developed MAS Physical Architecture for CBM/PdM

In order to develop the CBM/PdM of the PV system by a multi-agent control system, its physical architecture is composed of the following modules:

- Distributed sensors: A set of anemometers, temperature sensors, meteorological stations and irradiance sensors distributed in the solar park.
- Master-slave control panels: Every couple of trackers shares a small control panel to make in situ the data acquisition (remote site) and to send control commands to the trackers.
- Consolidation board: They are located in the huts of the inverter and concentrate the data from the solar trackers, from the inverters and from the sensors associated with this area.
- Master Terminal Unit (MTU): It has the function of collecting the data from all the consolidation boards. It opens as a high accessibility cluster.
- Communication system: It is composed of five networks depending on the employed protocols of each PV element of the system and in order to make scalable the solution.
- MODBUS as distribution network 2, interconnecting the consolidation boards with the inverters and sensors.

- MODBUS as Master-Slave communications network for supervision and control every couple of solar trackers.
- PROFIBUS DP as distribution network 1, interconnecting the consolidation board with the solar panels.
- Ethernet TCP/IP in the main network, which is constituted by a ring of optical fibers that interconnects the MUT with the consolidation boards.
- 3G-Link, allowing remote access to the MTU using a console session or a web access; with this network it is possible to send SMS and e-mails to the operators indicating the system alarms.

3.2 Developed MAS Logical Architecture for CBM/PdM

From the point of view of information technologies, the physical elements described in the previous section, for the CBM/PdM of the PV system by a multi-agent control system, are logically organized according to the hierarchical levels of organization defined in ISO 18435:

- Sensors/effectors (environment).
- Agent organization 1: legacy SCADA. This subsystem is composed of the following agents:
  - Real time logger. Collects data from the sensors network.
  - Historian. Archives raw and processed data from the real time logger.
  - Human Machine Interface (HMI). Allows human interaction with the SCADA application.
  - Alarm. This agent throws events to other system components.
- Agent organization 2: CBM/PdM subsystem. The agents that make up this organization are the following ones:
  - Modeller. Collects data from the sensors network and generates the power generation forecasting model.
  - Energy Predictor. Evaluates the forecasting model and outputs the predicted generated power.
  - Failure Detection. Compares the predicted generated power with the real time production.
  - Failure Diagnostics. Gathers and analyses the information available in the SCADA along with the failure detection output to determine the point of failure.
  - Failure Prognosis. Provides estimation to time to fail or RUL that will be use to programming the maintenance action.
- Interaction protocols. Communication protocols between agents are based on FIPA Agent Communication Language (ACL).

At the bottom the assets and the physical processes are included. The processes and the state of the assets are measured and the control is executed (level 1, in the work unit) by sensors, actuators and devices (either by wire or wireless). In level 2 (work center), the data is collected and processed in any case, both mainly on real time. There are different types of data transmissions in sensors and devices depending on the supplier; and different types of connection with them (PLC, fieldbus, and distributed systems or servers), then a server is needed to operate all of them and to communicate data, commands and events to the upper level based on FIPA Agent Communication Language (ACL).

3.3 Failure Detection and Energy Prediction Method

Starting from a RCM approach, the MAS platform intends to integrate the use of the most advanced techniques of acquisition and treatment of signal (thermography, vibration, ultrasound, and in general any that can be developed in the future) with the development of new multivariate algorithms for the prediction of the future state and the evolution of the condition of the systems, from a reliability and economic risk point of view, providing simulations of the systems lifecycle according to [34] recommendations.

We have employed a formal method to support our MAS CBM framework based on the main standards that are a reference in the field of maintenance, trying to guarantee the basic objectives of all of them in a coherent way. ISO 17359 "Condition monitoring and diagnostics of machines— General guidelines" is the industrial reference for the implementation of Condition Monitoring programs.

Therefore, it must provide the control of the condition of the systems on real time by monitoring variables directly related to their failure modes for reliability and risk management. Then it searches a descriptor matrix relating failure modes with an array of monitoring variables that allow us to define the status and evolution of the specific failure mode.

The proposed prediction method is based on training the ANN with the records stored in the historian component, including the main variables involved in the asset degradation and produced photovoltaic energy. An ANN is a cognitive information processing structure based training or learning algorithms, connecting three types of neuron layers: an input layer, an output layer and generally, one or more hidden layers. In a more formal engineering context, before continuous initial inputs, with the topology of a directed graph of neurons, it carries out information processing by means of neurons state. The aptitudes of ANN are to replicate reality self-adaptively in complex and noised operating conditions with enough non-formal failure data and complex covariates interaction.

The selected ANN type is a backpropagation multi-layer perceptron with a single hidden layer, using initial randomly weights, a hyperbolic tangent function as activation function of each hidden neuron, and linear hyperplane as network function. The learning algorithm is an error correction supervised minimizing the penalized mean square error through the Quasi-Newton method. The error optimization algorithm selected is the Levenberg-Marquardt algorithm with gradient times to avoid falling into a minimum relative error in the free software R. The algorithm stops when the maximum number of cycles is reached or the maximum number of failures occurred in the verification process (early stopping).

The main effort consists on identifying the monitoring variables required to predict failure modes (when that was feasible). Our failure detection and prediction framework has innovative features compared to previous works in the literature, structuring MAS based on ANN models, in a formalized way and combining not only failure mode degradation solutions but also energy generation predictions depending on the different operating and environmental conditions.

This Chapter focuses on failures resulting as a consequence of equipment deterioration and useful life reduction due to operational and geographical (environmental) features that could have a great influence on the equipment and consequently in the generated energy. Now we will focus on the definition of the basic monitoring variables in order to support this detective/predictive intelligence on information processing. The monitoring variables should be categorized and codified globally for all the architecture, and particularizing each one for an asset as failure mode descriptors [35, 36] through linking them with failure mode symptoms of the asset. Then, thanks to our formal descriptor matrix, the study can be applied:

- For all the assets of the grid connected PV system, because one attribute could be employed in different assets or its value could be transferred to another and;
- For adaptation to different states or conditions, one monitoring variable could be selected with some thresholds levels in a given descriptor and differently in another one.

The identification of the required descriptors to detect/predict a failure mode is an inductive process, because the effect of equipment conditions could be dissimilar among them. Through descriptors adjustment we can be exhaustive in failure mode detection/predictions per asset. Accordingly, this has to be accomplished per each critical failure mode, recommending a Failure Mode Effect and Criticality Analysis—FMECA in each critical equipment of the photovoltaic plant.

4 Developed CP CBM Based on MAS. The Power Inverter Case

Therefore, in order to introduce the developed CBM/PdM methodology, we have selected as critical failure mode the insulation problem in a power inverter due to the consequences on energy disruptions and the frequency of its occurrence.

Monitoring Variables (SCADA)											Failure Mode – Insulation Problem (Inverter)		
V1- Ambient T (ºC)	V2- Internal T (ºC)	V3- Global horizontal radiation (W/m2)	V4 - Operation Time (hours)	V5- Signal of insulation fault	V6- Accumu-lated active energy	Descriptors		D1	D2	D3	Detec-tion	Diagno-sis	Progno-stic
			Direct Derivation	●		D1	Insulation Alarm				✓		
●	●	●	●		●	D2	Estimated energy power					✓	
Estimated by ANN			●		●	D3	Derived RUL estimation		●				✓
			ANN + COX model										

Fig. 4 Descriptors-failure mode assignation for detection/diagnosis/prognosis agents in insulation failure mode-power inverter case

4.1 Structure Definition for Insulation Failure Mode

Our CBM/PdM system has been developed in order to make easy the knowledge generation per each individual failure mode without disruptions with other implemented failure modes and in a sustainable way, recycling defined and present knowledge in the system.

In Fig. 4, reader can see a matrix where maintenance technicians can make assignments easily, that is based on a combination of descriptors (in middle of the matrix). Three main descriptors have been designated by maintenance experts: Insulation Alarm (D1), Estimated energy power (D2), RUL estimation (D3). The foundations of descriptors are shown in left side of the matrix. First descriptor, D1, is derived directly from SCADA signal (V5) for insulation fault. D2 is derived indirectly using an ANN based on the SCADA signals from (V1) to (V4), difference of T (internal vs external), Global horizontal radiation and Operation hour of the inverter, and (V6) based on Accumulated active energy. And the third descriptor is a RUL estimation of inverter using (V6) and comparing along operation time (V4) with (D2), that is, comparing real with estimated energy power using ANN + COX model. Then, descriptors can use monitoring variable and other descriptors in a scalable perspective. Finally, the reason of each descriptor is presented in right side of the matrix, with marks corresponding with Detection, Diagnosis and Prognosis agents in each case (D1, D2, D3 respectively).

In order to do this, on-real time monitoring of the PV plant has to detect failure modes of critical elements, and even to predict produced energy according to adaptability objective. Thus, for any decision about failure mode detection/prediction, the agents interact in the following sequence (see Fig. 5):

- In agent organization 1, real-time logger agents insert records with new data into the historian database agents.
- In agent organization 2, modellers periodically recover data from the historian database, updating the model and notifying the new model to the predictor agent.

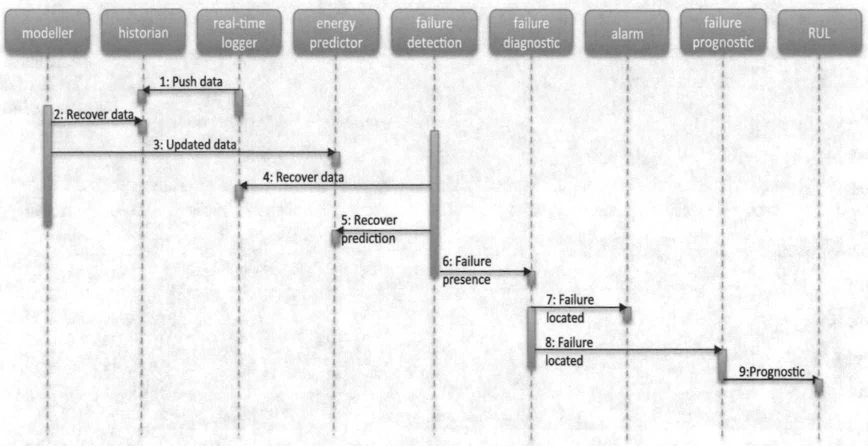

Fig. 5 Multi-agent collaboration diagram

- Failure detector agents ask predictor and real-time logger agents for the actual and forecasted states, comparing both values and in case of failure presence, and generating an event to Failure diagnostic.
- Failure diagnostics using information available in the SCADA and in failure detection output tries to determine the elements with failure.
- If prognosis is developed, Failure prognosis provides estimation of RUL to facilitate accuracy in the decision making and programing maintenance actions.
- Moreover, alarms are triggered to alert the maintenance team in three levels management: operational point of view, due to failure presence and failures located alarms, and in a tactical point of view due to RUL prognostic.

Consequently, MAS offers on-real time warnings about state and a RUL estimation of each failure mode of the power inverter (see Fig. 6). MAS shows the sequence of warnings on time with the respective level of criticality, red for severe (from 0 to 0.2 of RUL) and yellow for mild (from 0.2 to 0.35), and overlapping with the maintenance activities such as preventive and corrective (solid circle and x respectively). In this way, the decision of maintenance staff is facilitated toward effectiveness in these maintenance activities.

In Fig. 6, reader can see the RUL estimation according to operation time and based on our MAS for four of the modelled failure modes of the inverter:

- A. Insulation Problem;
- B. Cooling Problem;
- C. Electronic Protection;
- D. Wiring Problem.

In Fig. 6, a failure mode has recovered its RUL by a preventive activity, but when RUL was in a severe state. RUL of B failure mode was similar and even more critical than A, and taking advantage of preventive action it was re-established to 100% of

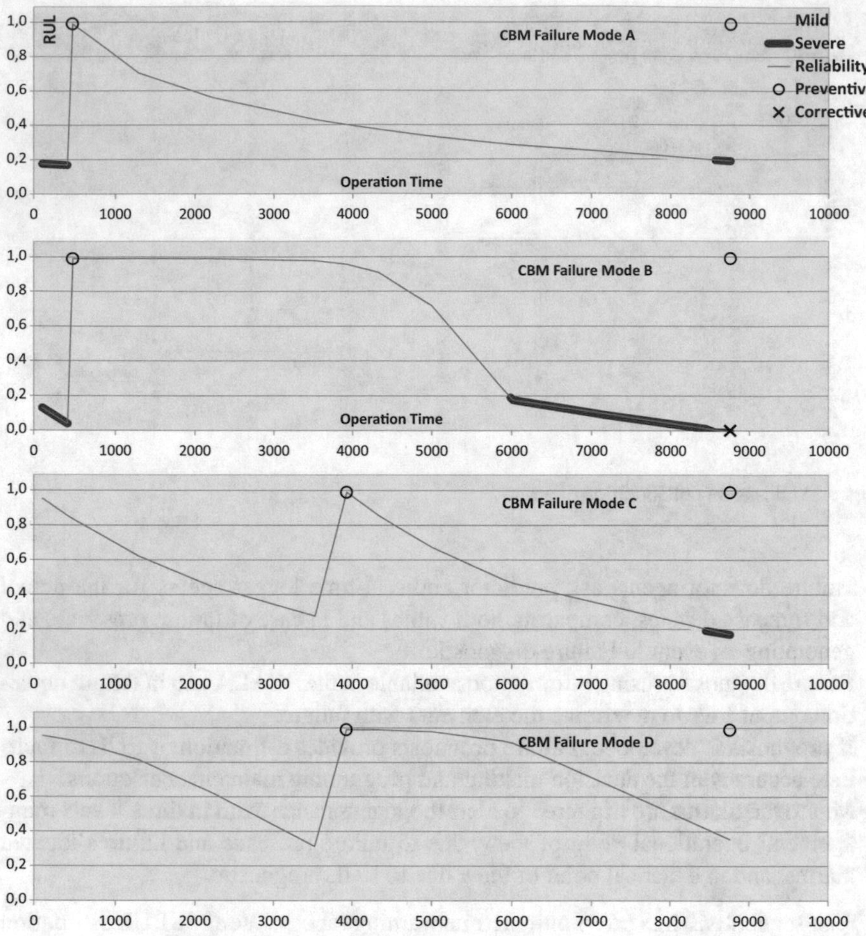

Fig. 6 Sequence on time of warnings in MAS according to maintenance activities

RUL. Although C and D failures weren't attended by the same preventive activity, their RULs were decreasing till 3000 h of operating time, when they began to reach mild level. Consequently, a second preventive activity was executed to re-establish C and D RULs in 3929 h of operation time, but without attending A and B failure modes. The following preventive action occurred in advance due to the generated corrective activity to solve the B failure mode, and in the same moment A, C and D failure modes were recovered to 100% of RUL.

These results are provided by experimental studies per agent. From now on, the practical case of diagnosis agent of estimated energy power is shown.

4.2 Experimental Diagnosis Agent Implementation for Insulation Failure Mode

The previous introduced diagnosis agent for insulation problems in power inverter is presented here with the experimental activities taken into account. This diagnosis agent detects insulation problems through differences on accumulated energy among obtained and estimated in real time. The experimental implementation is based on training a multi-layer perceptron ANN with the records stored in the historian of power inverters, including the above mentioned monitoring variables. This ANN has an input layer with four input neurons, corresponding to the difference among ambient and internal temperature (°C) of the power inverter, the global horizontal radiation (W/m^2), and the operation time (hours) and the accumulated active energy (kWh). All inputs were normalized in the range $[-1, 1]$. The output layer contains a single output neuron corresponding to the estimated active energy accumulated (kWh). ANN training has been done using historical records between 01/05/2012 and 31/05/2013, with a temporal resolution of both, hourly and quarter of an hour for production data and minutely for sensors data. There have been created two similar neural network models with temporal resolutions according with the available data.

Previously to model generation, it has been done a detailed quality filtering, in a way that data labelled as wrong has not been included in the model. It has also been excluded from modelling process data correspondent to hours without production, in order to not alter the statistical nature of results.

Calibration data was divided in three sets: training, validation and testing (70%, 15 and 15%, respectively). The learning algorithm parameters are as follows: (a) maximum number of cycles = 200, (b) maximum validation failures = 40, (c) min_grad = 1.0e−10, (d) goal = 0, (e) mu = 0.005, (f) mu_dec = 0.1, (g) mu_inc = 10.

For every zone in the photovoltaic plant equipped with an inverter it was created a neural model in 4 stages:

- Generating a thousand ANN models with the same architecture varying the initial weights in backpropagation algorithm.
- Doing a statistical analysis on each model output, considering the mean error, the mean absolute error and the root mean square error, being Ed the real active energy, Êd the simulated active energy, n the number of observations and d the temporal resolution.
- Selecting the 10 best models (best MAE and RMSE in the independent data set).
- Combining the outputs of the 10 best models by means of a simple average.

$$MAE = \frac{1}{n}\sum_{d=1}^{n}\left|E_d - \hat{E}_d\right| \tag{1}$$

$$RMSE = \sqrt{\sum_{d=1}^{n}\left(E_d - \hat{E}_d\right)^2/n} \tag{2}$$

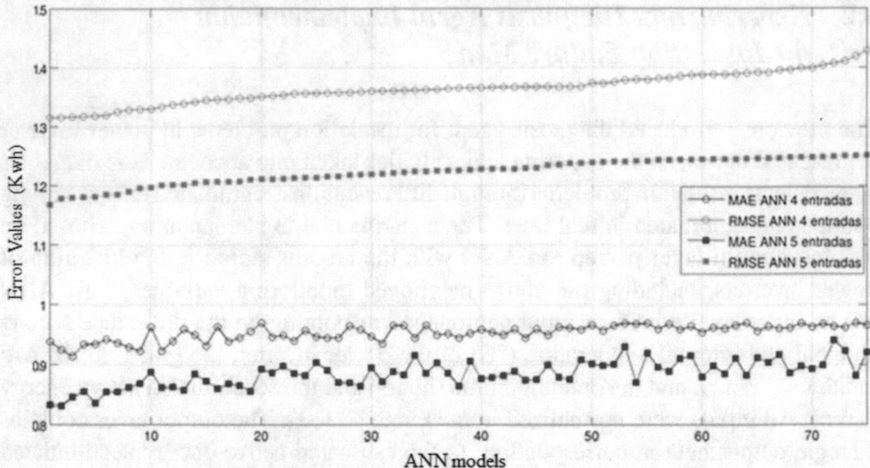

Fig. 7 Observed errors in test set for best 75 ANN quarter-hourly models

The ANN has been investigated considering Quarter-Hourly Temporal Resolution. 1000 neural networks were created. The best 10 networks were selected in both cases with respect to MAE and RMSE values in the test set, and combined in an average model. Figure 7 represents the best 75 neural models for one zone, in RMSE ascendant order, showing the obtained improvement including the input variable operation time as fifth input. It is worthy to highlight that precision in quarter-hourly data for measured energy production variable is ± 1 kWh, being the error of the simulated values of the same order of magnitude.

Moreover, the averaged neural network model has found out the relationships between input and output variables, despite the existing noise and system error present, see Fig. 8. Next graphs in Fig. 9 provide an accurate image of the model output precision.

A fragment of the simulated and real energy production time series is introduced in Fig. 10, where the good obtained results in the test data set confirm the capability of this ANN to predict the active energy in future dates. It achieves errors in the same order of magnitude than system errors produced by measure devices (± 1 kWh).

5 Conclusions

This chapter discusses about the necessity of support our CBM management in a structured framework in order to address a sustainable knowledge, clarifying the steps and concepts based on international standards, and encompassing a variety of knowledge matters with high specialization and, combining and correlating them through multi-agent systems thanks to huge available information. The independence

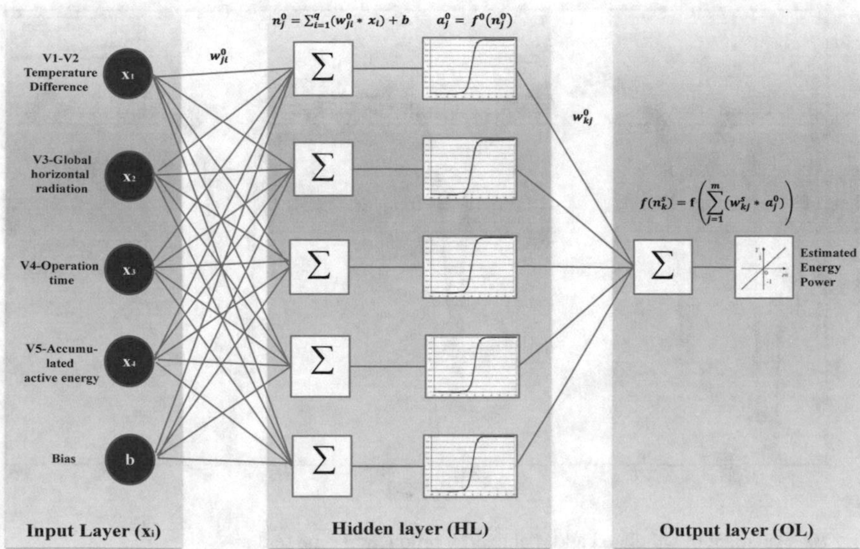

Fig. 8 Developed artificial neural network

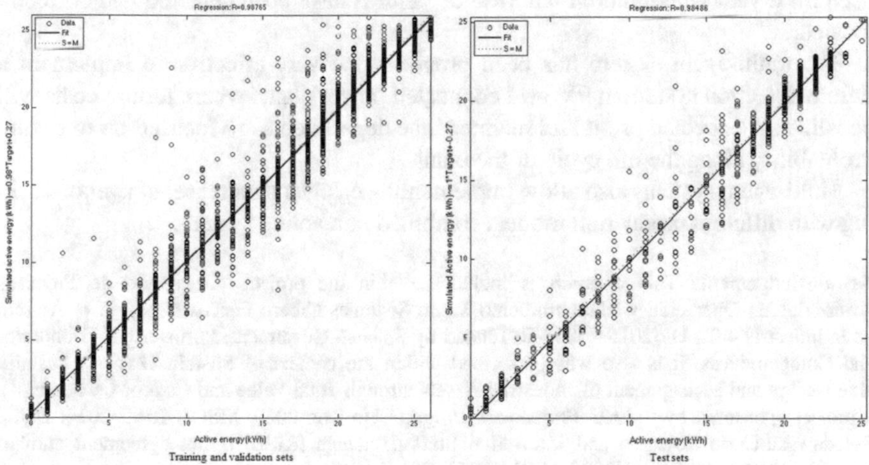

Fig. 9 Observed errors in training & validation sets in left side and, in test set

of the agents allows for a continuous update of the model, without affecting the processing of the rest of the agents.

With the presented structure, knowledge will be very much facilitated and understood using the provided template; capturing and training new information as monitoring variables and descriptors for improve decision making with a risk modelling perspective. This has been exemplified for a power inverter in a PV plant, which has been validated in practice in prototype software. The system has been successfully

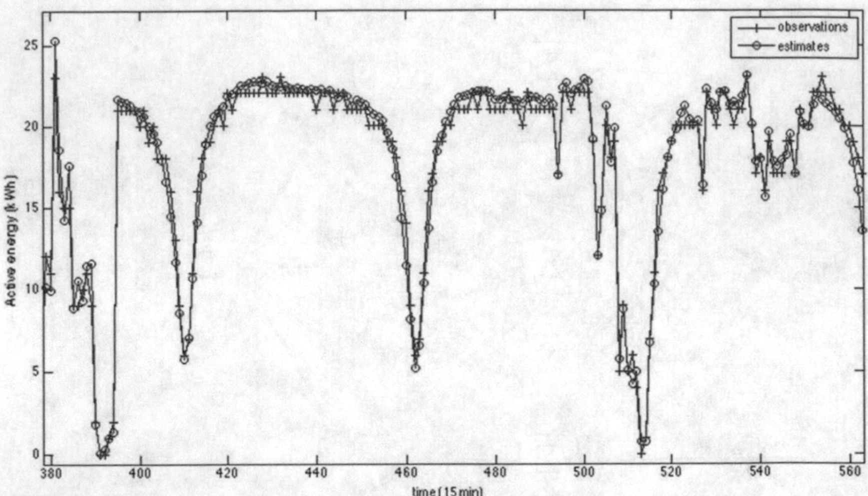

Fig. 10 Fragment of simulated and real energy production time series

used for a year where more than 98% of failures have been detected and correctly located.

The multi-agent system has been proved to be very effective to implement a failure detection system in PV grid connected power plants where failure costs will penalize the expected profit and maintenance departments are focused on to ensure profitability along the life cycle of the plant.

Multi-agent systems also allow implementing easily a parallel set of agents working with different production models combined in a voting system.

Acknowledgements This research is included within the project "Desarrollo de Procesos Avanzados de Operación y Mantenimiento Sobre Sistemas Cibero Físicos (Cps) en el Ámbito de la Industria 4.0", DPI2015-70842-R, funded by Spanish Goverment, Ministery of Economics and Competiviness. It is also was performed within the context of Sustain Owner ('Sustainable Design and Management of Industrial Assets through Total Value and Cost of Ownership'), a project sponsored by the EU Framework Program Horizon 2020, MSCA-RISE-2014: Marie Skłodowska-Curie Research and Innovation Staff Exchange (RISE) (grant agreement number 645733—Sustain-Owner—H2020-MSCA-RISE-2014).

References

1. Yacef, R., Benghanem, M., & Mellit, A. (2012). Prediction of daily global solar irradiation data using Bayesian neural network: A comparative study. *Renewable Energy, 48,* 146–154.
2. Benghanem, M., & Mellit, A. (2010). Radial Basis Function Network-based prediction of global solar radiation data: Application for sizing of a stand-alone photovoltaic system at Al-Madinah, Saudi Arabia. *Energy, 35,* 3751–3762.

3. Mellit, A., Benghanem, M., Arab, A. H., Guessoum, A., & IEEE. (2003). Modelling of sizing the photovoltaic system parameters using artificial neural network.
4. Mellit, A., Benghanem, M., Arab, A. H., Guessoum, A., & Moulai, K. (2004). Neural network adaptive wavelets for sizing of stand-alone photovoltaic systems.
5. Mellit, A., Benghanem, M., Arab, A. H., & Guessoum, A. (2005). An adaptive artificial neural network model for sizing stand-alone photovoltaic systems: application for isolated sites in Algeria. *Renewable Energy, 30,* 1501–1524.
6. Hiyama, T. (1997). Neural network based estimation of maximum power generation from PV module using environmental information—Discussion. *IEEE Transactions on Energy Conversion, 12,* 247.
7. Ashraf, I., & Chandra, A. (2004). Artificial neural network based models for forecasting electricity generation of grid connected solar PV power plant. *International Journal of Global Energy Issues, 21,* 119–130.
8. Mellit, A., & Shaari, S. (2009). Recurrent neural network-based forecasting of the daily electricity generation of a Photovoltaic power system. *Ecological Vehicle and Renewable Energy (EVER), Monaco, March* (pp. 26–29).
9. Moubray, J. (1997). *Reliability-centered maintenance*. Industrial Press Inc.
10. Rausand, M., & Høyland, A. (2004). *System reliability theory: Models, statistical methods, and applications* (vol. 396) Wiley.
11. Mellit, A., Benghanem, M., Bendekhis, M., & IEEE. (2005). Artificial neural network model for prediction solar radiation data: Application for sizing, stand-alone photovoltaic power system. In *2005 IEEE Power Engineering Society General Meeting* (Vols. 1–3, pp. 40–44).
12. Orioli, A., & Di Gangi, A. (2013). A procedure to calculate the five-parameter model of crystalline silicon photovoltaic modules on the basis of the tabular performance data. *Applied Energy, 102,* 1160–1177.
13. Kostylev, V., & Pavlovski, A. (2011). Solar power forecasting performance–towards industry standards. In *1st International Workshop on the Integration of Solar Power into Power Systems, Aarhus, Denmark*.
14. Patton, J. D. (1980). *Maintainability and maintenance management*. Research Triangle Park: Instrument Society of America.
15. Guasch, D., Silvestre, S., & Calatayud, R. (2003). Automatic failure detection in photovoltaic systems. In *Proceedings of 3rd World Conference on Photovoltaic Energy Conversion* (Vols. A–C, pp. 2269–2271).
16. Lee, J., Bagheri, B., & Kao, H. A. (2015). A Cyber-Physical Systems architecture for Industry 4.0-based manufacturing systems. *Manufacturing Letters, 3,* 18–23. https://doi.org/10.1016/j.mfglet.2014.12.001.
17. Colombo, A. W., Karnouskos, S., Kaynak, O., Shi, Y., & Yin, S. (2017). Industrial cyberphysical systems: A backbone of the fourth industrial revolution. *IEEE Industrial Electronics Magazine, 11*(1), 6–16.
18. Leitão, P., Colombo, A. W., & Karnouskos, S. (2016). Industrial automation based on cyber-physical systems technologies: Prototype implementations and challenges. *Computers in Industry, 81,* 11–25. https://doi.org/10.1016/j.compind.2015.08.004.
19. Olivencia Polo, F., Alonso del Rosario, J., & Cerruela García, G. (2010). Supervisory control and automatic failure detection in grid-connected photovoltaic systems. *Trends in Applied Intelligent Systems* (pp. 458–467).
20. Miller, W. T., III, Glanz, F. H., & Kraft, L. G., III. (1990). Cmas: An associative neural network alternative to backpropagation. *Proceedings of the IEEE, 78,* 1561–1567.
21. Basheer, I. A., & Hajmeer, M. (2000). Artificial neural networks: Fundamentals, computing, design, and application. *Journal of Microbiological Methods, 43,* 3–31.
22. Zhang, G. Q., Patuwo, B. E., & Hu, M. Y. (1998). Forecasting with artificial neural networks: The state of the art. *International Journal of Forecasting, 14,* 35–62.
23. Curry, B., Morgan, P., & Beynon, M. (2000). Neural networks and flexible approximations. *IMA Journal of Management Mathematics, 11,* 19–35.

24. Malcolm, B., Bruce, C., & Morgan, P. (1999). Neural networks and finite-order approximations. *IMA Journal of Management Mathematics, 10,* 225–244.
25. Kuo, C. (2011). Cost efficiency estimations and the equity returns for the US public solar energy firms in 1990–2008. *IMA Journal of Management Mathematics, 22,* 307–321.
26. Lapedes, A., & Farber, R. (1987). Nonlinear signal processing using neural networks.

27. Wilson, R. L. (1986). Operations and support cost model for new product concept development. *Computers & Industrial Engineering, 11,* 128–131.
28. Guillén, A. J., Crespo, A., Gómez, J., & Sanz, M. D. (2016). A framework for effective management of condition based maintenance programs in the context of industrial development of E-Maintenance strategies. *Computers in Industry*, *82*, 170–185. https://doi.org/10.1016/j.compind.2016.07.003.
29. Niu, G., Yang, B. S., & Pecht, M. (2010). Development of an optimized condition-based maintenance system by data fusion and reliability-centered maintenance. *Reliability Engineering and System Safety*, *95*(7), 786–796.
30. Zio, E. (2009). Reliability engineering: Old problems and new challenges. *Reliability Engineering & System Safety*, *94*(2), 125–141.
31. MIMOSA (Machinery Information Management Open Standards Alliance) (2011). Open Systems Architecture for Condition Based Maintenance (OSA-CBM), v3.2.19.
32. Crespo, A. (2007). *The maintenance management framework*. United Kingdom: Springer London Ltd.
33. Mitchell, E., Robson, A., & Prabhu, V. B. (2002). The impact of maintenance practices on operational and business performance. *Managerial Auditing Journal, 11*(1), 25–39.
34. Crespo Márquez, A. (2007). *The maintenance management framework: Models and methods for complex systems maintenance*. London: Springer.
35. Saxena, A., Celaya, J., Saha, B., Saha, S., & Goebel, K. (2010). Metrics for Offline Evaluation of Prognostic Performance. *International Journal of Prognostics and Health Management*, *1*, 2153–2648.
36. Vachtsevanos, G., Lewis, F., Roemer, M., Hess, A., & Wu, B. (2006). *Intelligent Fault Diagnosis and Prognosis for Engineering Systems*. NJ, John Wiley and Sons: Hoboken.

Chapter 9
Failure Mode Prediction and Energy Forecasting of PV Plants to Assist Maintenance Task by ANN Based Models

Fernando Olivencia Polo, Jesús Ferrero Bermejo, Juan F. Gómez Fernández and Adolfo Crespo Márquez

Abstract In the field of renewable energy, reliability analysis techniques combining the operating time of the system with the observation of operational and environmental conditions, are gaining importance over time. In this chapter, reliability models are adapted to incorporate monitoring data on operating assets, as well as information on their environmental conditions, in their calculations. To that end, a logical decision tool based on two artificial neural networks models is presented. This tool allows updating assets reliability analysis according to changes in operational and/or environmental conditions. The proposed tool could easily be automated within a supervisory control and data acquisition system, where reference values and corresponding warnings and alarms could be now dynamically generated using the tool. Thanks to this capability, on-line diagnosis and/or potential asset degradation prediction can be certainly improved. Reliability models in the tool presented are developed according to the available amount of failure data and are used for early detection of degradation in energy production due to power inverter and solar trackers functional failures. Another capability of the tool presented in the chapter is to assess the economic risk associated with the system under existing conditions and for a certain period of time. This information can then also be used to trigger preventive maintenance activities.

Keywords Renewable energy · Maintenance · Condition based maintenance · Artificial neural network · Proportional Weibull reliability

---

F. Olivencia Polo · J. Ferrero Bermejo
Magtel Operaciones, Seville, Spain
e-mail: fernando.olivencia@magtel.es

J. Ferrero Bermejo
e-mail: jesus.ferrero@magtel.es

J. F. Gómez Fernández · A. Crespo Márquez (✉)
Intelligent Maintenance System Research Group (SIM), Department of Industrial Management, School of Engineering, University of Seville, Camino de los Descubrimientos s/n, 41092 Seville, Spain
e-mail: adolfo@us.es

J. F. Gómez Fernández
e-mail: juan.gomez@iies.es

A. Crespo Márquez et al. (eds.), *Value Based and Intelligent Asset Management*,
https://doi.org/10.1007/978-3-030-20704-5_9

1 Introduction

Renewable energies present a high dependency on the random condition of climatological phenomena. This variability may have a great impact on existing production commitments fulfilment (as established by the legislation in many countries). Therefore, there is a great interest in environmental conditions prediction and forecasting, that can be accomplished in different ways:

- Including climatological forecasting: using physical models, with solar irradiation-generated power curves, for predicting production in terms of irradiation.
- Excluding climatological forecasting: using statistical models based on historical data (monthly-averaged solar irradiation). For example, the Ministry of Industry of Spain provides a prediction tool without climatological calculations, including a coefficient table considering as variables: months, time of the day and climatological zones.

Besides above mentioned forecasting programs, grid-connected PV systems also require advanced processes monitoring, through a sensor distributed network, to carry out regular functional and performance checks. Furthermore, data obtained needs the proper corresponding time series statistical analysis. In fact, automatic failure detection in photovoltaic systems is a complex process, needing the logging of a great number of electrical variables (such as currents flowing through solar panels and voltages on batteries), together with environmental data (such as irradiance and temperature [1, 2]). As a result, companies face a costly maintenance program, only affordable for large sites and companies with the advantage of certain economies of scale.

In order to easy the implementation and to reduce the complexity in this failure detection process, some emerging proposals are using few variables and more complex statistical analyses [3, 4]. This chapter focuses on applying artificial neural networks to this end.

Artificial neural networks (ANNs) are mathematical tools with intensive utilization in the resolution of many real-world complex problems, especially in classification and prediction ones. ANNs (Artificial Neural Networks) pretend to emulate biological human neural networks learning from the experience and generalizing previous behaviours as characteristics time series. To do this, the simple unit is the neuron whose mission is to process the received data as an activating function that could be the entry of other neuron, combining neurons as a directed graph that can carry out information processing by means of its state response to continuous or initial input [5]. The ANN architecture consists on an input layer, an output layer and generally, one or more hidden layers. Their main characteristic is its ability to process information features in non-linear, high-parallelism, fault and noise environments with learning and generalization capabilities [6, 7]. In comparison to traditional model-based methods, ANNs are data-driven self-adaptive methods well implemented in computers on real time, learning from examples and capturing subtle and hidden functional relationships that are unknown or hard to describe. In addition, ANNs provide strong tolerance before noised data because store information redundantly. Thus, ANNs are well suited for solving problems where explicit knowledge is difficult to specify or define, but where there are enough data [8–10]. In this sense, Lapedes and Farber [11] have shown that backpropagation neural network exceeds by an order of magnitude to the conventional lineal and polynomial methods dealing with chaotic time series of data. Consequently, our interest is using ANNs to analyze data and dismiss predictions errors concerning failure appearance.

The application of these techniques to the renewable energies field, and more specifically to power generation of photovoltaic (PV) systems, has been in continuous development during the last years, including:

- Meteorological data forecasting [12, 13].
- PV systems sizing [14–17].
- PV systems modelling, simulation and control [18].

There are previous works on forecasting PV systems electricity output using ANNs [19, 20], however this chapter describes two new algorithms for early detection of failures in PV systems according to the available amount of failure data, with the intention to be included when describing predictive maintenance task resulting from RCM programs implementation (Reliability-Centred Maintenance).

2 Models Evaluation and Decision Making Process

Reliability centred maintenance (RCM) is the most widespread methodology to study the required maintenance program for an asset in a given operational context [21], quantifying the risks [22] and evaluating the remedial measures to detect, avoid or prevent the functional failures [23]. When considering variable operation and environmental conditions, the study of failures may be complex. Attributable to non-optimal operating conditions, failures often occur in assets suffering of changing environmental (cleanliness, fastening, temperature, etc.) and operational (configurations, preventive maintenance, undue handling, etc.) conditions. Moreover, non-evident defects in the assets (design imperfection, implementation errors, quality of materials, etc.) may also lead to failures [24, 25].

Reliability analysis of Renewal Energy Equipment, in line with the RCM method, is a very complex task depending on operating and environmental conditions. This analysis considers the effects, in the equipment function, of the different failure modes degrading the equipment functionality through deviations from standard operating conditions [26]. Based on real data as historic events, this degradation can be observed or predicted following a failure curve. Due to its own complexity, this analysis is associated to quantitative tools and so it have to be mainly implemented in depth in critical equipment or equipment in which failure consequences are not admissible (due to environment, health and safety, etc.).

An example of this is the “Survival Data Analysis”, focused on a group of individuals and how they react to failure after certain length of time [27–29]. Data and information about these contributing factors could be decisive to obtain, and even to update over time, reliability estimations about the contribution of some events, represented through explanatory variables or covariates, in order to obtain the time until the failure (Survival Time). There are several techniques to solve survival estimations [30, 31], in which typical failure distribution functions are asymmetrical (censored to the right). The influence of these explanatory factors may obey different patterns that could be then used to work out the real risk of an asset. These techniques based on explanatory variables could be parametric when the hazard distributions are known, semi-parametric in the case of unknown hazard distribution but with defined assumptions of hazard proportionality with the time and independence between the constant through time covariates, or non-parametric when these are not necessary to be specified [32–34].

In maintenance, the decision-making is usually characterized by conditions of uncertainty, anticipation in order to handle non-controlled variables is frequently required and this is done by studying their historical evolution individually, or on their relation to other variables. In practice, with limited knowledge, maintenance technicians often feel more confident with their experience, and this would influence their decision that could be conservatively based on levels of satisfaction instead of being optimal [35]. Therefore, it is recommended to improve decision capacity using formalized frameworks which are suitable to the level of information required and to the data which is available. Quantitative tools are preferred to seek greater precision in the choice of strategies, but this is the choice of what is "better", among what is "possible" [36]. Also, the decision process is interactive, not only to predict something, but to replicate reality; it should be upgradeable as improvement continues to obtain and share knowledge.

Parametric methods, as Weibull actuarial and graphical models (EM), are usually employed when people have enough information about failures with a regular pattern, so they can be developed to model failures resulting, most of the times, in a Taylor-made suit per equipment. On the other hand, as previously it has already mentioned before the utilization of semi-parametric methods, as the widely applied Proportional Hazard Model (PHM) of Cox [37], based on a log-lineal-polynomial expression of the covariates under the assumptions of independency among them and constant with the time. While, in non-parametric methods stand out ANN methods thanks to be a self-adaptive and empirical process even with noised and non-lineal information and/or time-dependency in covariates.

Parametric, semi-parametric and nonparametric techniques are employed to estimate the reliability function mainly depending on the knowledge about the failure time distribution (from major to minor respectively). However, concerning to the flexibility against to above mentioned covariates assumptions (independency and time-independency), from EM models to ANN models, the flexibility and efficiency showing relationships among the life cycles and other variables are increased, but also the complexity of implementation and the computational load are increased at the same time [38]. Additionally, in numerous papers [38, 39], the PHM and ANN are compared to fit survival functions showing no significant differences between predictions of Cox regression and ANN models when complexity in models is low. In case of complex models, with many covariates and any interaction terms the differences in terms of advantages are important, showing the following results:

- ANN predictions were better than Cox PHM predictions with high rates of censoring (censoring rate of 60% and higher [40]), reducing significant biases.
- ANN predictions provide better predictions to detect complex nonlinear relationships between independent and dependent variables.
- ANN predictions can incorporate quantified potential prognostic factors that may have been overlooked in the past.

As a result, the maintenance decision making in Renewal Energy Equipment under different operating environments can be supported by ANN fulfilling the requirements of:

- Suitable to level of failure information,
- Implementable in SCADA systems,
- Upgradeable iteratively and with reality,
- Flexible and integrated hierarchically.

According to previous paragraphs, this work main contribution is a logic decision tool doing PV systems electricity forecasting, which, at the same time, may serve as predictive maintenance instrument, that can be linked to proper RCM programs outputs to control critical failure modes. In the sequel, this work focuses on applying artificial neural networks (ANNs) to model PV systems failures.

3 Practical Case Materials

To support this practical research, the ANN models over case studies are now presented. The idea is, not only building the models, but also implementing them in a SCADA system.

PV plants have been in production for more than 25 years. Current decrease in government incentives to renewal energy sources has forced companies to study useful life extension possibilities. Due to this, potential plant re-investments must be also re-evaluated; incorporating future operating and environmental conditions within equipment reliability analysis is considered to be crucial to avoid future production disruptions.

This type of photovoltaic plan was usually built modularly, each 100 kW may represent over 600,000 € of investment, therefore the possibility to replicate the same model for different modules and regions is also considered of great interest. With that in mind, this work tries to develop ANN models that are easy to reproduce, and to update, when the most common parameters found in a photovoltaic plant, determining production, suffer changes.

Our prediction models have innovative features compared to previous works in the literature. The ANN models use, not only environment variables as external temperature or radiation, but also assets' conditions variables as internal temperature

Table 1 Standard configuration of one transformer with inverters and panels

CT	ID	kWn	Ref. module	N° strings	N° panels strings
CT15	A8-1	100	IS-220	528	12
	A8-2	100	IS-220	528	12
	A8-3	100	IS-220	528	12
	A8-4	100	IS-220	528	12
	A8-5	100	IS-220	528	12

for the different operating times. Through this, an early detection of degradation will be possible before failures affect production, and a quantitative measure of risk can be computed. It is important to acknowledge how risk of failures could even reach ten times the purchase equipment cost [41], therefore it has to be classified, and modelled properly, the different non reliability related cost along equipment lifecycle, such as warranties, indemnities, reparations, penalties, etc.

Additionally to be exhaustive in failure predictions per equipment, the analysis of failures has to be accomplished per each critical failure mode because symptoms and causes could be dissimilar among them and the effect of equipment conditions could apply in a different manner.

In our case study, functional analysis and failure mode analysis (Failure Mode Effect and Criticality Analysis—FMECA) was carried out in advance for critical equipment of the photovoltaic plant, understanding that these efforts in failure mode analysis could add enormous value for protecting a production of 6,258,966 €/year in our plant. This effort was completed identifying, at the same time, parameters required to predict failure modes (when that was feasible).

Two common systems are selected to illustrate the model implementation over real data and in a SCADA system: a power inverter and a solar tracker. Both of them are from a 6.1 MW photovoltaic plan opened up in September of 2008 (49,640 operation hours), compound by 37,180 photovoltaic panels in groups of 100 kW for each inverter. The solar trackers orient photovoltaic panels toward the sun to maximize collected solar energy, while the power inverter transforms direct current (DC) to alternating current (AC) form strings of panels, aggregating its own energy (210,000 kWh/year) each five inverters jointly into a transformer (see configuration of the selected transformer in Table 1) through which energy is provided to the distributor at an initial price of 0.4886 €/kWh (513,030 €/year of production), and subsequently reduced due to a legal requirement. Consequently reliability aspects are important, not only to consider the direct costs of failures, but also the indirect loss of profit. Anticipation to avoid this loss of profit will be pursued by the monitoring system.

The standard configuration of one of transformer is described in Table 1.

Possible future failures are predicted using a back-propagation neural network that is trained with inverters' historical data of the last five years. This chapter focuses on failures resulting as a consequence of equipment deterioration and useful life

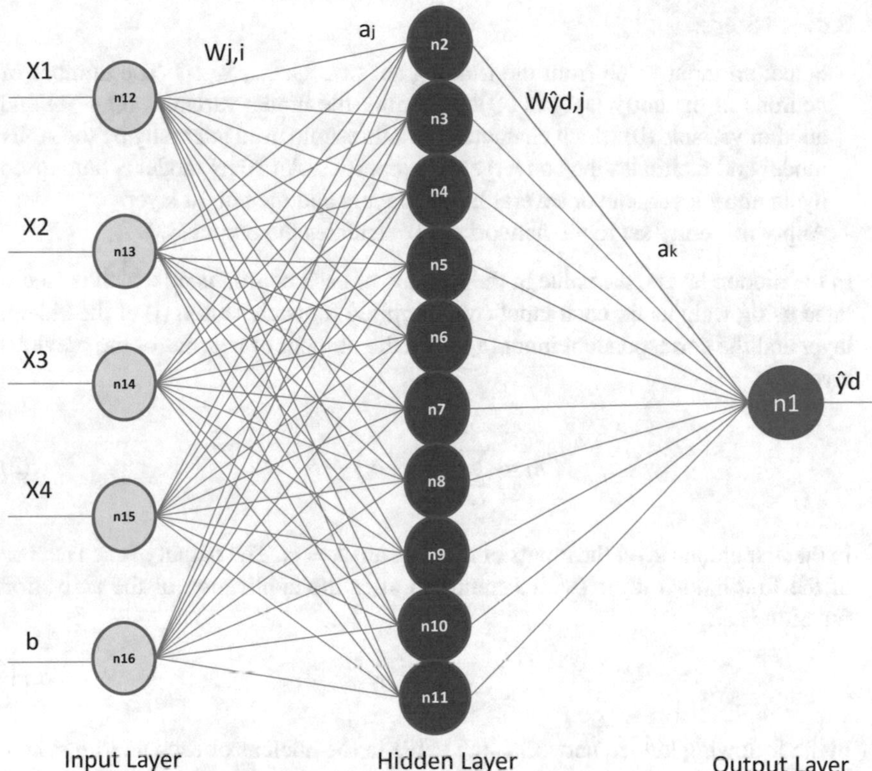

Fig. 1 Developed backpropagation percepton multi-layer ANN

reduction due to operational and geographical (environmental) features that could have a great influence on the equipment. In the following paragraphs of this Section, the chapter first describes the back-propagation training process of the network, and then it concentrates on presenting the overall prediction methodology presented in the chapter, applying it to the case study.

Back-propagation is a popular learning mechanism for solving predictions in multilayer perceptron networks [42], where differentiability is required in the activation function (as in the case of sigmoid function or the hyperbolic tangent function) in the output layer Z, in which its values may vary between 0 and 1, when using normalized variables in the input of the network. The sigmoid or logistic function that is used is presented in (1).

$$Z = f(x) = \frac{e^x}{(1 + e^x)} \tag{1}$$

The back-propagation training consists in the following two steps (see Fig. 1 for a better understanding):

- Forward Steps:
  - Select an input value from the training set (x_1, x_2, …, x_{Q-1}). The number of neurons in the entry layer is (Q), including the inputs variables (Q − 1) and another variable (b) which characterizes a threshold used internally by the ANN model and facilitates the convergence properties. An ANN model is composed by an entry layer, one or several hidden layers and the output layer.
  - Apply this entry set to the network and calculates the output ($\hat{y}_d$).

In the hidden layers, the value in the nucleus of each neuron is n_j, which is calculated using weights for each input ($w_{j,i}$), applied for each neuron (j) of the hidden layer and the correspondent input (a_i) until the number of neurons of the previous layer (Q).

$$n_j = \sum_{i=1}^{Q} (w_{j,i} \cdot a_i) + b \tag{2}$$

In the first hidden layer the inputs of neurons are $a_i = x_i$. The output of each neuron of the first hidden layer (with J neurons) after the application of the activation function is a_j:

$$a_j = f(n_j) \tag{3}$$

In the following hidden networks, the value in the nucleus of each neuron is now based on the past outputs a_j using other weights ($w_{k,j}$), and producing the output a_k applying the activation function on the nucleus value; and at this way for successive hidden layers.
Finally, the output of the ANN (the output layer in the case of one neuron) is $\hat{y}_d$:

$$\hat{y}_d = f(n_{\hat{y}d}) = f\left(\sum_{j=1}^{J} (w_{\hat{y}d,k} \cdot a_k)\right) \tag{4}$$

- Backward Steps:
  - Calculate the errors between the obtained output ($\hat{y}_d$) and the real output (y_d).
  - Adjust the weights in order to decrease the error in reverse way. For this stage, this work has employed a learning coefficient equals (μ) to 1 and so is included in the second term of the sum in Eqs. (5) and (6).

$$w^{new}_{\hat{y}d,k} = w^{old}_{\hat{y}d,k} + a_k \cdot [\hat{y}_d \cdot (1 - \hat{y}_d) \cdot (y_d - \hat{y}_d)] \tag{5}$$

Equation (3) is the weights adjustment for the output neuron of the output layer, where $[\hat{y}_d \cdot (1 - \hat{y}_d) \cdot (y_d - \hat{y}_d)] = \delta_{\hat{y}d}$. An Eq. (4) is the adjustment on any neuron of the hidden layers.

$$w_{j,i}^{new} = w_{j,i}^{old} + w_{j,i}^{old} \cdot a_i \cdot \left[a_j \cdot (1 - a_j) \cdot \left(w_{k,j}^{new} \cdot \delta_j\right)\right] \quad (6)$$

- Repeat forward and backward steps about all the training set until the global error is acceptably low, for this work based on minimizing the root mean square error for the number of observations (n).

$$RMSE = \sqrt{\frac{1}{n}\sum_{d=1}^{n}\left(y_d - \hat{y}_d\right)^2} \quad (7)$$

Because of non-linearity of Z, the learning mechanism of multilayer perceptron networks requires a resolution heuristic algorithm that guarantees the best solution or the global minimum (this is done using the Quasi-Newton resolution method in the free software R or the Levenberg-Marquardt Method in Matlab). To avoid over-adjustment of the network repeating the same employ this time MSE (Mean Square Error) with penalty characterized by λ, mainly employed with a bare quantity of historical data (otherwise λ trend to 0):

$$MSE + \lambda \cdot \sum_{j,i}^{Q,J,K} w_{j,i} = \frac{1}{n} \cdot \sum_{d=1}^{n}\left(y_d - \hat{y}_d\right)^2 + \lambda \cdot \sum_{j,i}^{Q,J,K} w_{j,i}^2 \quad (8)$$

After describing the process of the back-propagation training of the network, let's now concentrate on the Logic Decision Tool based on ANN models that this chapter proposes.

4 Logic Decision Tool and ANN Models

RCM present a generic process for the logic selection of the maintenance actions to correct or prevent the occurrence of failure modes [21], as extension of this for the specific on-condition maintenance actions, the process of decision making is developed addressing before mentioned requirements, see Fig. 2, which includes the following steps:

- The work flow starts with the inspection and failure data collection of external and internal relationships considering the differences in the operational and environmental conditions.
- Then it continues, evaluating if the symptoms of a gradual function loss can be detected effectively.
- Hereafter, the failure modes analysis is developed, determining their effects in the gradual function loss through a set of variables.
- Next, a logic decision tree analysis (LTA) is employed to select among the different prediction models (based on referenced authors):

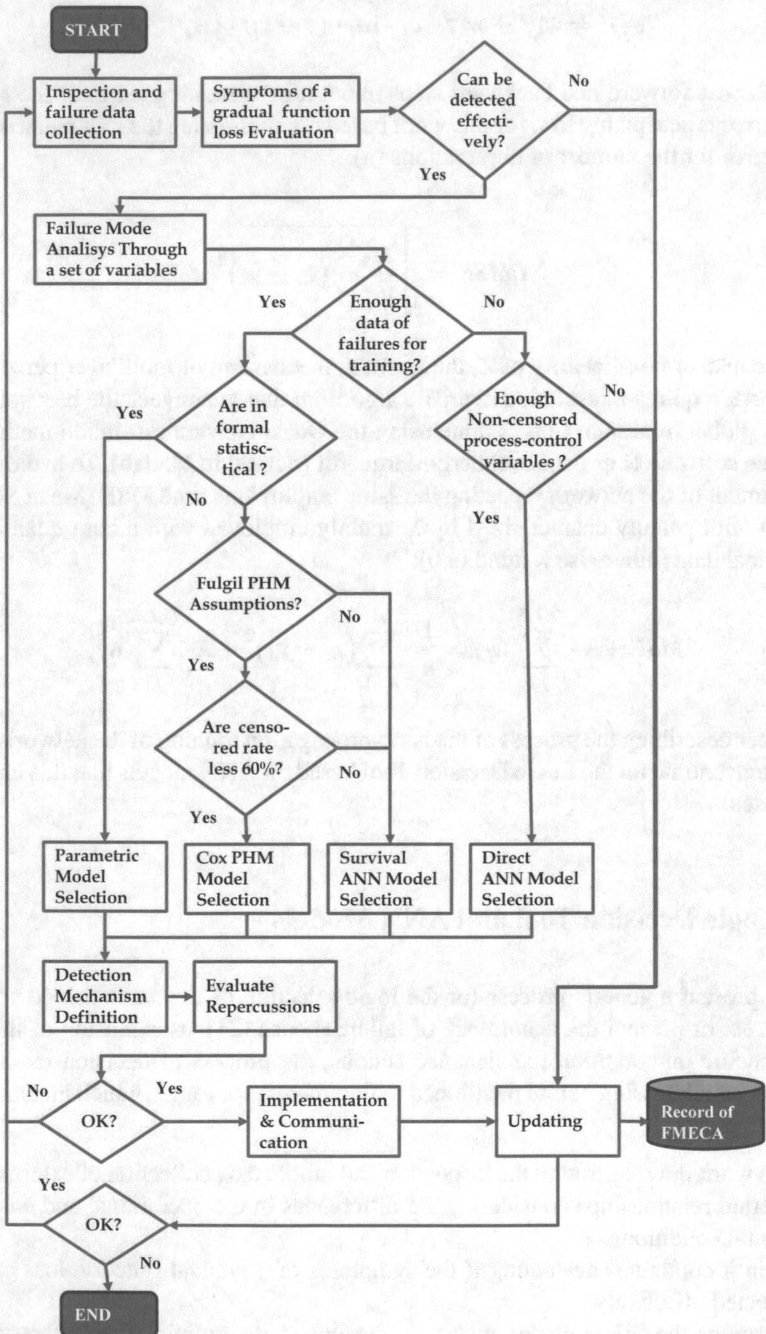

Fig. 2 Work-flow logic decision tree about on-conditions predictions

  - If there are enough formal statistical training to develop or with lineal covariates, parametric models are recommendable.
  - If there are enough data about failures but not as formal statistical training, fulfil the covariates assumptions (that is when the relationship among the hazards of two similar assets with different operating environment factors is clearly proportional), and censoring rate less of 60%, it is suggested PHM.
  - If there are not enough data about failures (formal statistical or not) and not censored, but with enough data of process-control variables (generally noised), it is recommendable to employ direct ANN (even to reproduce complex physic or chemical functions).
  - When there are enough data about failures but not as formal statistical training, and complex interaction with noise and time-dependency among covariates (that is, where the covariates assumptions are not satisfied), or satisfying the covariates assumptions the censoring rate is equals to 60% or higher, then ANN is recommended. Although in this case, ANN has its own limitations, because the data set has to be reorganized to replicate the Survival Function.

- After, the detection mechanism has to be defined.
- Finally, the repercussions of the chosen prediction models have to be evaluated with a cost-benefit analysis, previously to their implementation and communication to the entire organization.
- As a result, the implementation of the approved prediction models is realized.

Updating with in-service data collection has to be maintained as continuous improvement.

Consequently, two new mathematical ANN models are developed in this document showing the aptitudes of ANN to replicate reality self-adaptively in complex and noised operating conditions: Case (A) of direct ANN in absence of failure data, reproducing energy production of the power inverter which is has a physic complex equation; and Case (B) of Survival ANN with enough non-formal failure data and complex covariates interaction, trying to fit the Survival Function of solar trackers.

In both cases, for real time estimation the variables have to be selected from those whose detection is periodically and automatically feasible. All the representative contributions selected have to be compound in a single function which reflects the degradation of the failure model but with two different methodologies. This is done to facilitate the failure discrimination and analysis. The data normalization is undone to the original range of values in the output layer. The ANN architecture has to be developed according to the number of input variables and the final estimation function, see it in Fig. 1 (presented previously), a multi-layer Perceptron using a linear hyperplane as function type base and with a single hidden layer with ten neurons. The activation function of each hidden neuron is a logistic function. The initial weights are randomly selected. The learning backpropagation algorithm used is an error correction supervised minimizing the penalized mean square error through the Quasi-Newton method in the free software R.

The training of the network is realized depending on the architecture and available historical data of variables, where for this work there are the followings:

- Historical Data of variables and production output per hour and days during five years.
- Periods for comparison, to detect the existence of the failure mode when selected for each case.
- In case of a failure mode defect is corrected, the ideal model considers the equipment as new.
- Then, 75% of available data is considered for the network training and the 25% for the network testing.
- In the training the ANN behaviour pattern gives us the network settings such as the number of hidden nodes, which will be validated subsequently with the testing.

4.1 Case (A) Failure Mode Prediction: Lack of Insulation

The selected failure mode of the power inverter is the "lack of insulation" failure, which due to the fact that production losses are significant; this failure mode is considered in SCADA with priority. This failure mode emerges due to corrosion and, the environmental conditions could be determinants in different areas and besides the inverter operating time. The most representative variables of operation, external environment and internal conditions have to be selected and tested to show their effects in the failure mode. The available variables in our SCADA in the case of power inverters are: ambient temperature (°C), the internal temperature of the power inverter (°C), the global horizontal radiation (W/m^2), the operation time of the power inverter (h), and the active energy accumulated of the inverter (kWh).

This case, in our PV plant is characterized by the absence of enough failure data and with non-statistical form. Then, prediction can be realized over process-control variables as the accumulated active energy of the inverter, where physical models for the different components makes the characterization in a transfer function for an accurate estimate of the state of the generator (voltage, current and power) in real time impossible. For all these reasons, researchers have developed several proposals to model these systems [1, 43]. By this reason, the physical model is reproduced with the ANN model.

This case will estimate the accumulated active energy ($\hat{y}d$) of the inverter in absence of the failure mode, in order to compare this with the real accumulated active energy (yd). For this, an ANN will be trained in absence of failures seeking an ideal production model that will be confronted with the real production to distinguish significant changes that denote the failure mode. Deduced by the FMECA analysis of the photovoltaic plant, this ideal production model could be used for detecting at an early stage other failure models even in other equipment, simply comparing it with other internal equipment variables in each case.

Table 2 Data set of variables Case A

Variable	Max.	Ref.	Min.	Unit
Ambient temperature	61	37	1	°C
Internal temperature	57	40	17	°C
Global horizontal radiation	1291	644	10	W/m^2
Operation time		49,640		h
Accumulated active energy	99	60	1	kWh

Table 3 Results of training Case A

Results	Value
MSE training	72.47686
MSE test	83.41932
R^2 training	0.910275
R^2 test	0.8912438

The ANN model has in the input layer with five input neurons, corresponding to the ambient temperature (°C), the internal temperature of the power inverter (°C), the global horizontal radiation (W/m^2), the operation time of the power inverter (h), the accumulated active energy of the inverter (kWh), see Fig. 1 and Table 2, and the threshold neuron. The output layer contains a single output neuron corresponding to the estimated active energy accumulated of the module (kWh). The ANN analysis, done going through the processes of training, predictions and test produces the following results (17,700 measures of two years are processed, 3540 measures per variable from 08:00 am to 17:00 pm, four inputs and one output).

The learning algorithm parameters are as follows: (a) maximum number of cycles = 980, (b) maximum validation failures = 40, (c) min_grad = 1.0e−10, (d) goal = 0, (e) $\mu = 0.005$, (f) μ_dec = 0.1, (g) μ_inc = 10, (h) $\lambda = 0$, (i) min Error = 19.47. The obtained results in this case guarantee a good optimization model, as shown in Table 3.

MSE (Mean Square Error), in the training and testing, validates the ANN signifying the average distance between the obtained prediction and the real production. Besides, R^2 is consistent with this result, explained 89.12% of the predicted model versus actual production. Figure 3 is a representation of deduced predictions, remarking a straight line to indicate the best approximation for error minimization. For validation purposes, the 25% of historical data is used to estimate the generalization error.

Once the ANN model of accumulated active energy is validated, the detection mechanism has to be defined. After training five years of data, the ideal production model pretends to be an approximation to real production in absence of failures, through the transformation of combined experience in abstract conceptualization which is sustained by a non-linear function of a weighted sum of its inputs (modifying the weights of the links that connect the neurons). The ideal production model has to be compared with the real production one, trying to define early prediction. In

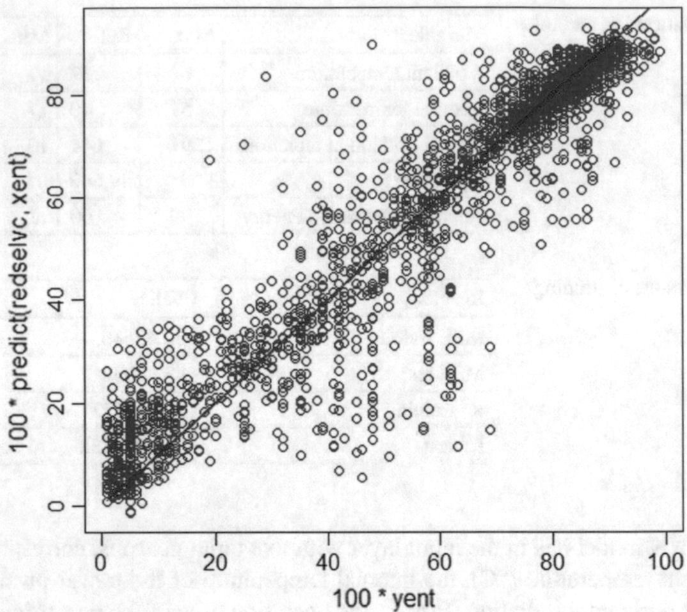

Fig. 3 ANN predictions Case A

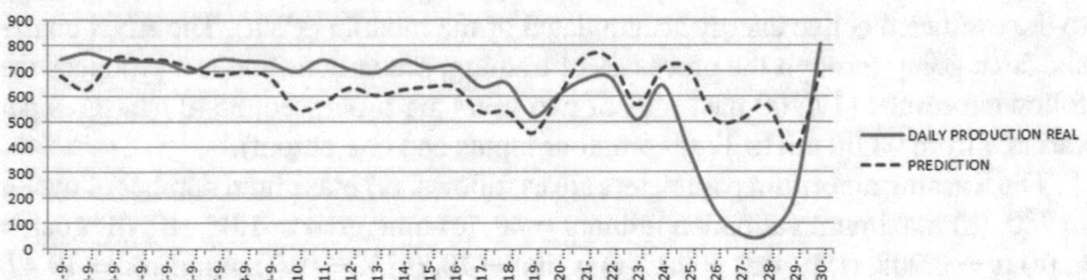

Fig. 4 Detection from ideal and real production comparison

our "lack of isolation" failure mode, an early warning could easily be set up (see Fig. 4) at least 48 h before failure. Notice that the difference between ideal and real production in that case is 40%.

To protect the model against spurious alarms, the 40% difference generating the warning has to be maintained at least during 4 consecutive hours for alarm generation (circumstance modeled in SCADA incrementing a counter by 1 each hour), and requesting to schedule immediately an corrective action.

Hereafter, the case will show the comparison between ideal and real production values (Table 4).

Therefore, based on SCADA systems the model could be implemented easily and replicated for all power inverters in the plant, or in other plants, and the knowledge about this failure mode may increase comparing results among others inverters and

Table 4 Data comparison of ideal and real production from SCADA

Date	Time	Real production	Ideal production	Alarm counter
25-9-12	10	85.03	73.9	0
25-9-12	11	52.82	81.4	0
25-9-12	12	35.88	83.5	1
25-9-12	13	47.99	83.7	2
25-9-12	14	43.59	81.1	3
25-9-12	15	26.37	74.1	4
25-9-12	16	21.06	77.6	5
25-9-12	17	33.69	83	6

redefining the model, or incorporating new or modified variables as the difference among external-internal temperature; or establishing the early alarm to gain more than 48 h.

Thanks to this research the "lack of isolation" failure mode associated indirect cost, as loss of profit, was reduced by 68,591 € per year and plant (575 kW/day with MTBF = 3 per year and for 61 inverters). Furthermore, extrapolating the potential advantage to the life cycle of the plant (5 years) the profits may have reached to thc total production of one inverter during five years.

With the aim of extend the ANN model to other PV plants, the failure mode behavior has been analyzed in two PV plants in different geographical provinces of Spain, Toledo and Zamora, where the operating environments are different. In both of them, the ANN model predicts the lack of insolation of the inverter but in a soften way versus in Cordoba (southern than Toledo and Zamora), due to the difference of meteorological variables, see Fig. 5.

4.2 *Case (B) Failure Mode Prediction: Solar Tracker Blocking*

The selcted failure mode of the solar tracker is the "blocking" failure, which is repetitive in field due to the huge volume of installed units. This failure mode emerges due to corrosion and, the environmental conditions could be determinants in different areas and besides the operating time. This case, in our PV plant is characterized by enough failure data with non-statistical form varying among plants and with high censured rate. Then, prediction can be realized as Survival ANN. The most representative variables of operation and external environment conditions have to be selected and tested to show their effects in the failure mode. The selected variables in our SCADA in the case of solar trackers are relative to diary average: ambient humidity (%), wind speed (m/s), the global horizontal radiation (W/m^2), the operation time of the solar tracker (days) (see Table 5). However, to develop Survival Function

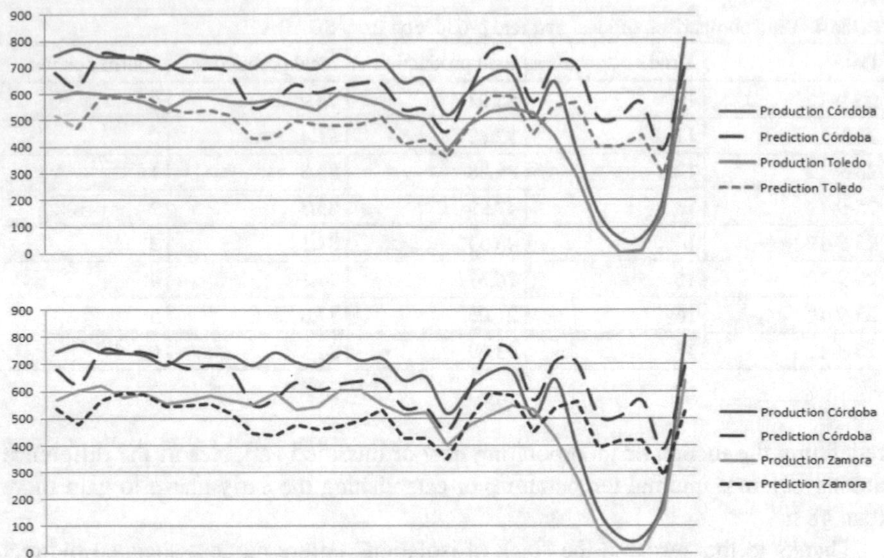

Fig. 5 Detection comparing ideal versus real production with Zamora and Toledo

Table 5 Semi-parametric Weibull parameters for reorganization of survival data

PV1 Fn	TTFi	Pondered α_i	PV2 Fn	TTFi	Pondered α_i
1	105.82	73.34	1	305.58	211.77
2	88.59	61.39	2	119.36	82.71
3	84.06	58.25	3	277.89	192.57
4	128.03	88.73	4	110.34	76.47
5	88.28	61.18	5	99.94	69.26
6	167.21	115.88	6	134.14	92.96
7	188.90	130.91	7	170.53	118.18
8	181.78	125.97	8	375.92	260.51
α	144.15		α	226.58	
β	3.47		β	2.19	

this case has to reorganize the available data, because the training means adjusts iteratively the weight coefficients given the condition in order to approximate the output to the target, which is an input of the ANN. In practice, survival events have to be included and depending on the way to include them, different ANN models are produced [38, 39] in two manners, for example:

- Using ANN instead of the lineal combination of weights coefficients in the Cox PHM, as Farragi and Simon [44], being necessary solve the PHM with Partial Maximum Likelihood Estimation (P-MLE).

- Using an input with the Survival Status over disjoint time intervals where the covariates values are replicated, with a binary variable 0 before the interval of the failure and 1 in the event or after, as Liestol et al. [45] and Brown et al. [46] where each time interval is an input with Survival Status, then a vector of survival status is defined per failure; or
- Employing the Kaplan-Meier (K-M) estimator to define the time intervals as two additional inputs instead of vector, one is the sequence of the time intervals defined by de K-M survival status, and the other is the survival status in each time of the sequence. This is the case of Ravdin and Clark [47] or Biganzoli et al. [48] models which are known as Proportional Kaplan-Meier.

For similarity with the previous case and the intention to utilize the same ANN architecture, now this case has oriented our proposed Survival ANN model based on the ideas of Ravdin and Clark, but with some mathematical modifications:

- With periodic disjoint intervals (for all the failures) of the maximum time to failure, to be suitable to level of failure information, instead of employing Kaplan-Meier estimator intervals.
- With covariates using real data (the average) in each disjoint interval to be upgradeable iteratively and with reality property, instead of repeating the value in each interval.
- With a semi-parametric Weibull estimation of the Survival Status, instead of employing Kaplan-Meier estimator, in order to fitting the curve better and reduce the negative effect of non-monotonically decreasing survival curve.

Thus, the ANN model would have in the input layer with five input neurons, corresponding to ambient humidity (%), wind speed (m/s), global horizontal radiation (W/m^2), the operation time of the solar tracker (days), the modelled semi-parametric survival status, and the threshold neuron. The output layer contains a single output neuron corresponding to the estimated survival function with values from 0 to 1.

The developed semi-parametric Weibull model consists into create time intervals of the maximum time to failure with an increment of the survival function instead of to maintain binary (0 or 1). Therefore, our propose resides in:

1. To estimate in a first step, the survival function with a parametric Weibull over groups of the produced time to failures where the covariates are the same. For example, if the case has two PV plants with 8 failures each one, it has to be realized the Weibull model in two groups, one over the 8 failures of PV plant 1 and other over the 8 failures of PV plant 2. Then, it will obtain a characteristic α and β in each plant, and without using the covariates, only based on time to failures as shows Table 5. Due to this, an estimation of the survival curve shape is obtained.
2. After that, maintaining the β in each plant (which represents the slope of the line), in order to model an estimation of the survival function for each specific failure with a gradual increment from 0 to 1, it is taken the β and the specific time to failure to replace in the Weibull Cumulative Distribution Function (CDF) in each time interval. Consequently, the two additional inputs are developed,

Table 6 Reorganized survival data to train and test the ANN

Failure number (Fn)	1	1	1	1	1	1	1	1	1	1	1
Time interval	10	20	30	40	50	60	70	80	90	100	110
Ambient humidity (%)	95	93	99	91	95	92	84	62	40	53	89
Wind speed (m/s)	11	8.8	6.7	11.3	6.7	11.6	10.9	12.1	8.2	7.3	4.3
G.H. radiation (W/m^2)	35.4	53.4	43.5	31.9	38.7	51.7	80.1	68.1	54.7	68	86
Weibull CDF	0.001	0.011	0.044	0.115	0.232	0.392	0.573	0.741	0.869	0.947	1

one with the time intervals and other with the Weibull CDF with an increment discretized in the time intervals. Although, to match up the CDF curve with a gradual increment from 0 at the beginning of the time to 1 in the exact time of the failure and later, the CDF uses the previous β but α pondered by 0.693, similar to the Median Life (Median Life = α · Ln(2) ^ β = 0.693 but β = 1). As a result for each specific failure, the probability to failure ascends unto reach 1 at the time of the failure and after, using this semi-parametric model, see Eq. (9) with Fn as number of failure in its plant, TTFi as specifics time to failure, ti as the time interval value, and α_i as pondered α. In Table 5, the 16 failures (Fn) and their time to failure (TTF) are presented for each plant (PV1 and PV2) with the initial α and β, and the modified αi with the ponderation.

$$CDF(t) = \begin{cases} 1 - \left(1/e^{\left(\frac{ti}{0.693 \cdot TTFi}\right)^{\beta}}\right) = 1 - \left(1/e^{\left(\frac{ti}{\alpha i}\right)^{\beta}}\right) & \text{if } \; ti < TTFi \\ 1 & \text{if } \; ti >= TTFi \end{cases} \tag{9}$$

Consequently, the data to train and test the ANN are reorganized as in Table 6.

For failure estimation, the output of the ANN model offers an estimation of the CDF or probability of failure, learning from semi-parametric estimation of a Weibull with covariates affection, as roughly proportional to Weibull Survival probability. The ANN analysis, done going through the processes of training, predictions and test produces the following results (3200 measures of two years are processed, 640 measures per variable diary, four inputs and one output) (Table 7).

Now, the learning algorithm parameters are as follows: (a) maximum number of cycles = 1000, (b) maximum validation failures = 40, (c) min_grad = 1.0e−10, (d) goal = 0, (e) μ = 0.005, (f) μ_dec = 0.1, (g) μ_inc = 10, (h) λ = 0, (i) min Error =

Table 7 Data set of variables Case B

Variable	Max.	Ref.	Min.	Unit
Time	400	205	10	h
Relative humidity	100	74.5	27.3	%
Wind average speed	17.2	4.59	0.6	m/s
Global radiation	379.5	106.64	1.4	W/m^2
Survival	1	0.5	0	

Table 8 Results of training Case B in developed model

Results	Value
MSE training	0.01551932
MSE test	0.01641588
R^2 training	0.8681797
R^2 test	0.8540106

Table 9 Results of training Case B with Ravdin and Clark

Results	Value
MSE training	0.08595152
MSE test	0.08493271
R^2 training	0.6371432
R^2 test	0.6520446

0.00001833. The obtained results in this case guarantee a good optimization model, as shown in Table 8.

While, if the Ravdin and Clark model had been employed directly, the results had been with less accuracy (as Table 9 shows).

In this developed model, R^2 explained 85.4% of the survival data. Figure 6 is the representation of deduced predictions, remarking a straight line to indicate the best approximation for error minimization.

As a result, for quick convergence and fitting of the curve, the initial values to train the ANN this case has utilized the semi-parametric estimation of Weibull CDF as an input to obtain the output as close as possible. Then, this case is researching a proportional semi-Weibull ANN model.

These two developed ANN models pretend to explore the capacity of ANN to obtain knowledge about covariates updating it based on experience with new valid data. Although, weighted sum of the inputs of the ANN nodes could not be directly interpreted as the coefficients of the covariates. The aim is to estimate failures with one ANN architecture, either as first approximation to the covariates coefficients, or to be employed as input of other model, or to update the obtained coefficients with other techniques, or to incorporating new inputs, or to compare the quality versus failures in different PV plants or different equipment.

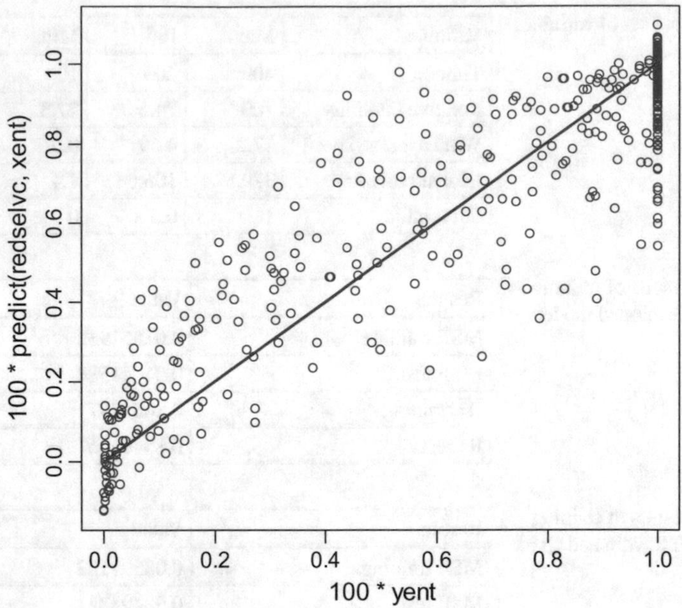

Fig. 6 ANN predictions Case B

5 Conclusions

PV Plants managers want to ensure longer profitability periods with more reliable plants. To ensure profitability along the life cycle of the plant maintenance departments must ensure critical equipment reliability and maximum extension of their life cycle, otherwise failure costs will penalize the expected profit.

Throughout this document, this chapter suggests to apply an ANN model per failure mode and foster a practical implementation in SCADA systems for different plants. This methodology may ease and may improve decision-making processed in condition-based maintenance and risk modelling, enabling reductions of corrective maintenance direct and indirect costs or allowing to show residual life until total equipment failure.

In cases when enough data for significant training is available, a better implementation of our methodology will help to reduce the costs and will improve the knowledge of the life cycle of the plant when suffering non-homogeneous operational and environmental conditions.

ANN capacity of auto-learning among sources of data (sometimes noised or deprived of communication) thanks to reiterative memory is important. In our case study, a vast quantity of data from different remote plants was available, although sometimes this data was affected by problems of sensors readings or communications. Back-propagation perceptron ANN is recommend for automation developments with real-time utilization. Furthermore, advanced ANN models could be applied supporting additional variables.

It is important to know the failure mode behavior in order to pretreatment historical data, eliminating abnormal data that may distort the results.

Values have to be normalized if it is used differentiability activation functions, and with the same scale for all the input values to simplify calculations and analysis. After the normalized values have to be des-normalized before comparison.

Acknowledgements Part of the funding for this research was provided by the SMARTSOLAR project (OPN–INNPACTO-Ref IPT-2011-1282-920000).

References

1. Orioli, A., & Di Gangi, A. (2013). A procedure to calculate the five-parameter model of crystalline silicon photovoltaic modules on the basis of the tabular performance data. *Applied Energy, 102,* 1160–1177.
2. Kostylev, V., & Pavlovski, A. (2011). Solar power forecasting performance—Towards industry standards. In *1st International Workshop on the Integration of Solar Power into Power Systems, Aarhus, Denmark.*
3. Guasch, D., Silvestre, S., & Calatayud, R. (2003). Automatic failure detection in photovoltaic systems. In *Proceedings of 3rd World Conference on Photovoltaic Energy Conversion* (Vols. A–C, pp. 2269–2271).
4. Olivencia Polo, F., Alonso del Rosario, J., & Cerruela García, G. (2010). Supervisory control and automatic failure detection in grid-connected photovoltaic systems. *Trends in Applied Intelligent Systems*, 458–467.
5. Miller, W. T., III, Glanz, F. H., & Kraft, L. G., III. (1990). Cmas: An associative neural network alternative to backpropagation. *Proceedings of the IEEE, 78,* 1561–1567.
6. Basheer, I. A., & Hajmeer, M. (2000). Artificial neural networks: Fundamentals, computing, design, and application. *Journal of Microbiological Methods, 43,* 3–31.
7. Zhang, G. Q., Patuwo, B. E., & Hu, M. Y. (1998). Forecasting with artificial neural networks: The state of the art. *International Journal of Forecasting, 14,* 35–62.
8. Curry, B., Morgan, P., & Beynon, M. (2000). Neural networks and flexible approximations. *IMA Journal of Management Mathematics, 11*, 19–35.
9. Malcolm, B., Bruce, C., & Morgan, P. (1999). Neural networks and finite-order approximations. *IMA Journal of Management Mathematics, 10*, 225–244.
10. Kuo, C. (2011). Cost efficiency estimations and the equity returns for the US public solar energy firms in 1990–2008. *IMA Journal of Management Mathematics, 22,* 307–321.
11. Lapedes, A., & Farber, R. (1987). *Nonlinear signal processing using neural networks*.
12. Mellit, A., Benghanem, M., Bendekhis, M., & IEEE. (2005). Artificial neural network model for prediction solar radiation data: Application for sizing, stand-alone photovoltaic power system. In *2005 IEEE Power Engineering Society General Meeting* (Vols. 1–3, pp. 40–44).
13. Yacef, R., Benghanem, M., & Mellit, A. (2012). Prediction of daily global solar irradiation data using Bayesian neural network: A comparative study. *Renewable Energy, 48,* 146–154.
14. Benghanem, M., & Mellit, A. (2010). Radial basis function network-based prediction of global solar radiation data: Application for sizing of a stand-alone photovoltaic system at Al-Madinah, Saudi Arabia. *Energy, 35,* 3751–3762.
15. Mellit, A., Benghanem, M., Arab, A. H., Guessoum, A., & IEEE. (2003). *Modelling of sizing the photovoltaic system parameters using artificial neural network*.
16. Mellit, A., Benghanem, M., Arab, A. H., Guessoum, A., & Moulai, K. (2004). *Neural network adaptive wavelets for sizing of stand-alone photovoltaic systems*.

17. Mellit, A., Benghanem, M., Arab, A. H., & Guessoum, A. (2005). An adaptive artificial neural network model for sizing stand-alone photovoltaic systems: Application for isolated sites in Algeria. *Renewable Energy, 30,* 1501–1524.
18. Hiyama, T. (1997). Neural network based estimation of maximum power generation from PV module using environmental information—Discussion. *IEEE Transactions on Energy Conversion, 12,* 247.
19. Ashraf, I., & Chandra, A. (2004). Artificial neural network based models for forecasting electricity generation of grid connected solar PV power plant. *International Journal of Global Energy Issues, 21,* 119–130.
20. Mellit, A., & Shaari, S. (2009). Recurrent neural network-based forecasting of the daily electricity generation of a Photovoltaic power system. In *Ecological Vehicle and Renewable Energy (EVER)*, Monaco (pp. 26–29), March 2009.
21. Moubray, J. (1997). *Reliability-centered maintenance*: Industrial Press Inc.
22. Rausand, M., & Høyland, A. (2004). *System reliability theory: Models, statistical methods, and applications* (Vol. 396). Wiley.
23. Campbell, J., & Jardine, A. (2001). *Maintenance excellence: Optimising equipment life-cycle decisions*. In M. Dekker (Ed.). New York, NY.
24. Crespo Márquez, A. (2007). *The maintenance management framework: Models and methods for complex systems maintenance.* Springer Verlag, London.
25. Pham, H., & Wang, H. (1996). Imperfect maintenance. *European Journal of Operational Research, 94,* 425–438.
26. Mobley, K. (2002). *An introduction to predictive maintenance*. Amsterdam: Elsevier.
27. Cox, D. R., & Oakes, D. (1984). *Analysis of survival data*. London: Chapman and Hall.
28. Klein, J., & Moeschberguer, M. (1997). *Survival analysis techniques for censored and truncated data*. New York Inc: Springer.
29. Law, A. M., & Kelton, W. D. (1991). *Simulation modeling and analysis*. New York: McGraw-Hill.
30. Lindsey, J. K. (2001). *The statistical analysis of stochastic processes in time*. Cambridge University Press.
31. Smith, P. J. (2002). *Analysis of failure and survival data*. New York: Chapman-Hall.
32. Hougaard, P. (2000). *Analysis of multivariate survival data*. New York: Springer.
33. Lee, E. T. (1992). *Statistical methods for survival data analysis*. Wiley.
34. Blischke, W. R., & Murthy, D. N. P. (2000). *Reliability modelling, prediction and optimization*. New York: Wiley.
35. Mitchell, E., Robson, A., & Prabhu, V. B. (2002). The impact of maintenance practices on operational and business performance. *Managerial Auditing Journal, 11*(1), 25–39.
36. Stewart, T. (1992). A critical survey on the status of multiple criteria decision-making theory and practice. *OMEGA*—The *International Journal* of *Management Science*, *20*(5/6), 569–586.
37. Cox, D. R. (1972). Regression models and life-tables. *Journal of the Royal Statistical Society Series, 34,* 187–220.
38. Ohno-Machado, L. (2001). Modeling medical prognosis: Survival analysis techniques. *Journal of Biomedical Informatics, 34,* 428–439.
39. Xianga, A., Lapuerta, P., Ryutova, A., Buckley, J., & St. Azena. (2000). Comparison of the performance of neural network methods and Cox regression for censored survival data. *Computational Statistics & Data Analysis, 34*(2), 243–257.
40. Biglarian, A., Bakhshi, E., Baghestani, A. R., Gohari, M. R., Rahgozar, M., & Karimloo, M. (2012). Nonlinear survival regression using artificial neural network. *Journal of Probability and Statistics, 2013*, Article ID 753930 (Hindawi Publishing Corporation).
41. Wilson, R. L. (1986). Operations and support cost model for new product concept development. *Computers & Industrial Engineering, 11,* 128–131.
42. McClelland, J. L., Rumelhart, D. E., & P. R. Group. (1986). Parallel distributed processing. In *Explorations in the microstructure of cognition* (Vol. 2).
43. Blanes, J. M., Toledo, F. J., Montero, S., & Garrigos, A. (2013). In-site real-time photovoltaic I-V curves and maximum power point estimator. *IEEE Transactions on Power Electronics, 28,* 1234–1240.

44. Faraggi, D., & Simon, R. (1995). A neural network model for survival data. *Statistics in Medicine, 14,* 73–82.
45. Liestol, K., Andersen, P. K., & Andersen, U. (1994). Survival analysis and neural nets. *Statistics in Medicine, 13,* 1189–1200.
46. Brown, S. F., Branford, A., & Moran, W. (1997). On the use of artificial neural networks for the analysis of survival data. *IEEE Transaction of Neural Networks, 8,* 1071–1077.
47. Ravdin, P. M., & Clark, G. M. (1992). A practical application of neural network analysis for predicting outcome of individual breast cancer patients. *Breast Cancer Research and Treatment, 22,* 285–293.
48. Biganzoli, E., Boracchi, P., Mariani, L., & Marubini, E. (1998). Feed forward neural networks for the analysis of censored survival data: A partial logistic regression approach. *Statistics in Medicine, 17,* 1169–1186.

Chapter 10
Condition-Based Maintenance for Systems with Aging and Cumulative Damage Based on Proportional Hazards Model

Bin Liu, Zhenglin Liang, Ajith Kumar Parlikad, Min Xie and Way Kuo

Abstract This chapter develops a condition-based maintenance (CBM) policy for systems subject to aging and cumulative damage. The cumulative damage is modelled by a continuous degradation process. Different from previous studies which assume that the system fails when the degradation level exceeds a specific threshold, this chapter argues that the degradation itself does not directly lead to system failure, but increases the failure risk of the system. The Proportional Hazards Model is employed to characterize the joint effect of aging and cumulative damage. CBM models are developed for two cases: one assumes that the distribution parameters of the degradation process are known in advance, while the other assumes that the parameters arc unknown and need to be estimated during system operation. In the first case, an optimal maintenance policy is obtained by minimizing the long-run cost rate. For the case with unknown parameters, period inspection is adopted to monitor the degradation level of the system and update the distribution parameters. A case study of Asphalt Plug Joint in UK bridge system is employed to illustratc the maintenance policy.

Keywords Condition-based maintenance · Aging and degradation · Proportional hazards model · Unknown distribution parameters · Cumulative damage

---

B. Liu · M. Xie · W. Kuo
Department of Systems Engineering and Engineering Management, City University of Hong Kong, Kowloon, Hong Kong
e-mail: binliu9-c@my.cityu.edu.hk

M. Xie
e-mail: minxie@cityu.edu.hk

W. Kuo
e-mail: way@cityu.edu.hk

Z. Liang · A. K. Parlikad (✉)
Department of Engineering, Institute for Manufacturing, University of Cambridge, Cambridge CB3 0FS, UK
e-mail: aknp2@cam.ac.uk

Z. Liang
e-mail: zhenglin_liang@sina.com

A. Crespo Márquez et al. (eds.), *Value Based and Intelligent Asset Management*,
https://doi.org/10.1007/978-3-030-20704-5_10

1 Introduction

With the development of sensor technologies, system condition now can be monitored at a much lower cost, which prompts the application of condition-based maintenance. CBM takes advantage of the online monitoring information to make maintenance decisions. For a system subject to CBM, maintenance actions are carried out only when "necessary", based on the collected condition information [1]. Compared with the traditional time-based maintenance, CBM has shown its priority in preventing unexpected failure and reducing economic losses [2].

CBM is conducted based on the observation that systems usually suffer a degradation process before failure, and the degradation process can be observed by degradation indicators such as temperature, voltage and vibration. In literature, many researchers used multi-state deteriorating model to describe the degradation process and formulated the maintenance strategy as a Markov or semi-Markov decision process [3, 4]. Although Markov model is widely used in degradation modeling, one disadvantage is that the classification of system state is very arbitrary [5].

Recently, more emphasis is paid on continuous degradation processes. In the framework of continuous degradation, the degradation process is usually described by a general path model or a stochastic-process-based model such as Wiener process, Gamma process and inverse Gaussian process [6–8]. Caballé et al. [9] proposed a CBM for systems with continuous degradation and external shocks. Peng and van Houtum [10] developed a joint CBM and lot sizing policy for systems subject to continuous degradation.

An implicit assumption of the previous research is that a system fails when its degradation level exceeds a pre-specified failure threshold. However, in reality, the failure threshold is difficult to determine and usually it is a random variable depending on the environment condition and product's characteristics. In this chapter, the cumulative damage is modeled as a continuous degradation process. We argue that degradation process does not necessarily lead to system failure, but increases the likelihood of failure. Both internal aging and cumulative damage contribute to system failure. Examples of the joint effect of aging and cumulative damage on system failure can be found in systems such as high-voltage power transformers and bridge systems [11, 12]. For a new transformer, its insulation strength can protect it from severe events such as transient overvoltage and lightning strikes. When a transformer ages, its internal condition degrades, which makes it more vulnerable to unusual environment condition and increases the risk of failure. For a bridge, failures are sometimes triggered by external events such as hurricane, flood and overload. If a bridge undergoes severe deterioration, it may hit the point where tiny external influences can lead to failure. The degradation itself does not directly lead to system failure, but it increases the probability of failure when exposed to external events.

A convenient and prevalent way to integrate the aging and degradation effect into system failure is by proportional hazards model (PHM) [13]. PHM incorporates a baseline hazard function which accounts for the aging effect with a link function that takes the inspection information into account to improve the prediction of failure [14].

Applications of PHM can be found in various fields such as finance, manufacturing system and energy generators [15].

In literature, several studies have been conducted on maintenance policy in the PHM framework. Banjevic et al. [16] developed a control-limit maintenance policy for system subject to periodic inspection. Ghasemi et al. [17] proposed a CBM policy for systems with imperfect information, where the condition the system cannot be directly monitored. Wu and Ryan [18] investigated the value of condition monitoring in the PHM setting, where a continuous-time Markov chain was used to describe the system condition. Wu and Ryan [11] further extended the model by considering Semi-Markov covariate process and continuous monitoring. Tian & Liao [19] proposed a CBM policy for multi-component systems using PHM. Lam & Banjevic [20] investigated the issue of inspection scheduling for CBM. In all of these previous studies, the degradation process is characterized via Markov or semi-Markov model. In addition, the distribution parameters in the PHM are assumed as known in advance.

This chapter aims to develop CBM policies for systems subject to aging and cumulative damage. PHM is used to model the joint effect of aging and cumulative in the framework of failure rate. The effect of cumulative damage is modeled as the stochastic covariate in the PHM framework. The system is subject to periodic inspection, which is assumed to be perfect. At inspection, maintenance actions are carried out based on the observed condition information. Optimal maintenance policies are obtained by minimizing the long-run cost rate. Specifically, two CBM models are developed by assuming respectively known distribution parameters and unknown distribution parameters. In the case where the distribution parameters are unknown, the parameters have to be estimated and updated at each inspection, and maintenance decisions are made subsequently.

This chapter differs from other existing works in that: (a) It incorporates the influence of both aging and cumulative damage in modeling the failure rate. (b) It argues that degradation itself does not result in system failure, but increases the risk of failure. (c) It utilizes the observed condition information to update distribution parameters for making appropriate maintenance decisions. The research presented in this chapter was first published by the authors in Reliability Engineering and System Safety journal.

The remainder of this chapter is organized as follows. Section 2 presents the degradation-integrated failure model, where PHM is used to describe the impact of aging and cumulative damage. Section 3 formulates two maintenance models. One assumes known distribution parameters while the other assumes unknown distribution parameters. Application of the maintenance models to Asphalt Plug Joints in UK bridge system is presented in Sect. 4. Finally, concluding remarks and future research suggestions are given in Sect. 5.

2 Degradation-Integrated Failure Model

This chapter considers a single-unit system subject to aging and cumulative damage. The cumulative damage is modeled as a continuous degradation process. For systems such as bridges, which are subject to traffic load hours by hours, a continuous degradation process is reasonable to characterize the cumulative damage over time. In this chapter, we use “cumulative damage” and “degradation process” interchangeably. Different from previous studies which assume that soft failure occurs when the degradation level hits a pre-specified threshold, we here consider sudden failure, which depends on both the aging and cumulative damage. For most infrastructure systems, failures usually happen due to external shocks or serious events, and degradation makes it more vulnerable when exposed to shocks. As previously described, the degradation process itself does not directly lead to system failure, but it increases the failure rate of the system. PHM is used to characterize the influences of degradation level on system failure rate. The degradation level of the system is represented as the value of covariate in the PHM framework. Based on PHM, the failure rate at time t is given as

$$h(t; X_t) = h_0(t)\varphi(X_t) \tag{1}$$

where $h_0(t)$ is the baseline failure rate at time t, which is a non-decreasing function of t. X_t is the degradation level at time t, and $\varphi(\cdot)$ is a positive function projecting the degradation level to the failure rate function. Let $X = \{X_t, t \geq 0\}$ be a continuous stochastic process that depicts the degradation process. Various stochastic processes can be used to describe the degradation process, among which a wide used candidate is the general path [21]. Assume that $X_t = g(t; \theta, \alpha, \varepsilon(t))$, where $g(\cdot)$ is a parametric function that characterizes the evolution of the degradation process, θ is a random variable that accounts for unit-to-unit variability, α is a random parameter that captures the initial degradation level among the components' population, $\varepsilon(t)$ is an independent and identically distributed (iid) random error term [22]. The selection of $g(\cdot)$ depends on system characteristics and can take a variety of forms such as linear, exponential or logarithmic. In this chapter, for simplicity, we assume that $g(\cdot)$ is a linear function. The degradation process can be denoted as $X_t = \alpha + \theta t + \varepsilon(t)$ [23, 24], where the error term $\varepsilon(t)$ follows a Gaussian distribution with mean zero and variance σ^2, α and θ follow Gaussian distribution, with mean $\mu_0' = \mu_0 - \sigma^2/2$ and variance σ_0^2, and mean μ_1 and variance σ_1^2. In Eq. (1), the baseline failure rate function, $h_0(t)$, accounts for the aging effect, which can be explained as the normal failure rate when no cumulative damage is imposed. The influence of cumulative damage is incorporated in the degradation level X_t. Obviously X_t follows a Gaussian distribution,

$$X_t \sim N\left(\mu_0 + \mu_1 t - \sigma^2/2, \sigma_0^2 + \sigma_1^2 t^2 + \sigma^2\right) \tag{2}$$

It is assumed that $\mu_0 + \mu_1 t - \sigma^2/2 \gg \sigma_0^2 + \sigma_1^2 t^2 + \sigma^2$, such that the probability of X_t being negative can be neglected and X_t stochastically increases with t almost surely. Given the degradation process x, the conditional reliability can be obtained as

$$R(t; x) = P(T > t | x_s, 0 \leq s \leq t) = \exp\left(-\int_0^t h_0(s)\varphi(x_s)ds\right) \quad (3)$$

where T is the time to failure and x_s is the realization of X_s at time s. The probability density function (pdf) is given as

$$f_T(t; x) = \lim_{\Delta t \to 0^+} \frac{P(t \leq T < t + \Delta t | T > t)}{\Delta t} = \frac{h_0(t)\varphi(x_t)}{\exp\left(\int_0^t h_0(s)\varphi(x_s)ds\right)} \quad (4)$$

The expected lifetime of the system can be obtained as

$$E[T] = E\left[E_{|X_s, 0<s<T}[T]\right] = \int_0^\infty \int_0^\infty t \cdot f(t|x_s) f_{X_s}(x_s) dx_s dt \quad (5)$$

where f_{X_s} is the pdf of degradation level by time s. If the projecting function $\varphi(\cdot)$ is exponential, $h(t; X_t) = h_0(t)\exp(\beta X_t)$, where β is the coefficient, then we have $\log h(t; X_t) = \log h_0(t) + \beta X_t$, which implies that the failure rate function follows a lognormal distribution,

$$\log h(t) \sim N\left(\beta\left(\mu_0 + \mu_1 t - \sigma^2/2\right) + \log h_0(t), \beta^2\left(\sigma_0^2 + \sigma_1^2 t^2 + \sigma^2\right)\right) \quad (6)$$

The lognormal distribution fits numerous reliability data and reflects the failure due to crack propagation [25].

3 Maintenance Model Formulation

This section aims to establish maintenance models for systems with known and unknown distribution parameters respectively. The system is assumed as non-repairable, thus the inspection/replacement policy is adopted [26]. The system failure is self-announcing, but the degradation level is not evident, which can only be detected at inspection. Periodic inspection is carried out to detect the degradation level, with the cost C_i. Two maintenance actions are available upon the system: preventive replacement and corrective replacement. Preventive replacement can be overhaul of the system, while corrective replacement refers to physical replacement of the system [26]. Both preventive replacement and corrective replacement restore the system to the "as good as new" state. At inspection, if the degradation level or the age of the system exceeds certain threshold, preventive replacement will be implemented, with the cost C_p. If the system fails unexpectedly, corrective replacement is performed immediately, with the cost C_r. The corrective replacement cost includes the replacement cost of the system and also the cost comprising various costs with respect to failure-induced problems. Intuitively, C_r is more complex and more cost intensive ($C_p < C_r$).

Assume that the system is inspected every τ time units, where τ is a given parameter associated with the system characteristics. Given that the system functions at the kth inspection, the probability that the system survives through $(k+1)\tau$ is

$$P(T > (k+1)\tau | T > k\tau, x_s, k\tau \le s \le (k+1)\tau) = \exp\left(-\int_{k\tau}^{(k+1)\tau} h_0(s)\varphi(x_s)ds\right) \quad (7)$$

Proposition 1 *Given the degradation level* X_s, *for* $k\tau \le s \le (k+1)\tau$, *the cumulative hazard rate function between two consecutive inspections increases in* k. *In addition, the inequality*

$$E\left[\int_0^t h_0(s)\varphi(X_s)ds\right] \ge \int_0^t h_0(s)\varphi(E[X_s])ds$$

holds for the cumulative hazard rate of the system.

The detailed proofs of the propositions in this chapter are provided in Appendix. Proposition 1 implies that the expected cumulative failure rate is at least larger than the cumulative failure rate given at mean degradation level. Based on Proposition 1, we can readily obtain that the conditional reliability of the system surviving through the next inspection, $P(T > (k+1)\tau | T > k\tau,\ x_s,\ k\tau \le s \le (k+1)\tau)$, decreases with the inspection index k.

3.1 Maintenance Model with Known Distribution Parameters

In this section, we assume that the parameters of the failure rate function are known in advance. Since both the age and the degradation level influence the failure rate, maintenance operations are carried out based on the hazard rate at inspection, which explicitly incorporates the effects of aging and degradation. If the hazard rate at inspection exceeds a specific threshold ζ, preventive replacement is implemented. Otherwise, the system is left as it be. Long-run cost rate is used as the criterion to evaluate the maintenance policy. Our objective is to minimize the long-run cost rate by seeking an optimal ζ. According to the renewal theorem, the long-run cost rate is given as

$$\psi(\zeta) = \lim_{t\to\infty}\frac{C(t)}{t} = \frac{E[C(S)]}{E[S]} = \frac{C_p P_a + C_r P_b + C_i E[N_I]}{E[\min\{T, N_I\tau\}]} \quad (8)$$

where S is the length of a renewal cycle, P_a is the probability that a renewal cycle ends with preventive replacement, P_b is the probability that a renewal cycle ends with corrective replacement, and N_I is the number of inspection.

At time t, the probability that the failure rate of the system exceeds the threshold ζ can be obtained as

$$\begin{aligned} P(h(t, X_t) > \zeta) &= P\left(\varphi(X_t) > \frac{\zeta}{h_0(t)}\right) = P\left(X_t > \varphi^{-1}\left(\frac{\zeta}{h_0(t)}\right)\right) \\ &= \Phi\left(\frac{(\mu_0 + \mu_1 t - \sigma^2/2 - \varphi^{-1}(\zeta/h_0(t)))}{\sqrt{\sigma_0^2 + \sigma_1^2 t^2 + \sigma^2}}\right) \end{aligned} \tag{9}$$

Since a renewal cycle occurs either after a preventive replacement or corrective replacement, it is appealing to analyze the renewal cycle separately. The probability that preventive replacement is performed at the kth inspection is expressed as

$$\begin{aligned} P_a(k) = \exp\left(\begin{array}{l} -\int_0^{(k-1)\tau} h_0(t) \int_0^{\ln(\zeta/h_0(t))} \varphi(x(t)) f_{X(t)} dx dt \\ -\int_{(k-1)\tau}^{k\tau} h_0(t) \int_0^{\infty} \varphi(x(t)) f_{X(t)} dx dt \end{array}\right). \\ (P(h(k\tau) > \zeta) - P(h((k-1)\tau) > \zeta)) \end{aligned} \tag{10}$$

The probability that failure occurs within the interval $((k-1)\tau, k\tau)$ can be obtained as

$$\begin{aligned} P_b(k) &= \left(1 - \exp\left(-\int_{(k-1)\tau}^{k\tau} h_0(t) \int_0^{\infty} \varphi(x(t)) f_{X(t)} dx dt\right)\right). \\ &\exp\left(-\int_0^{(k-1)\tau} h_0(t) \int_0^{\ln(\zeta/h_0(t))} \varphi(x(t)) f_{X(t)} dx dt\right). \\ &\Phi\left(\frac{(\mu_0 + \mu_1 t - \sigma^2/2 - \varphi^{-1}(\zeta/h_0(t)))}{\sqrt{\sigma_0^2 + \sigma_1^2 t^2 + \sigma^2}}\right) \end{aligned} \tag{11}$$

Detailed derivations of Eqs. (10) and (11) are provided in Appendix.

If preventive replacement is carried out at the kth inspection, the cost and length of a renewal cycle can be obtained as $C_a = C_p + kC_i$ and $S_a = k\tau$. If a failure occurs in the interval $((k-1)\tau, k\tau)$, the cost and in a renewal cycle is expressed as $C_b = C_r + (k-1)C_i$ and the expected length is calculated as $E[S_b] = k\tau - \int_{(k-1)\tau}^{k\tau} (k\tau - t) f(t) dt$. The long-run cost rate can be achieved by combining the renewal cycles ending with preventive replacement and corrective replacement. After some calculations, the long-run cost rate is given as

$$\psi(\zeta) = \frac{E[C]}{E[S]} = \frac{C_p \sum_{k=1}^{\infty} P_a(k) + C_r \sum_{k=1}^{\infty} P_b(k) + C_i \sum_{k=1}^{\infty} (kP_a(k) + (k-1)P_b(k))}{\sum_{k=1}^{\infty} k\tau P_a(k) + \sum_{k=1}^{\infty} \left(k\tau - \int_{(k-1)\tau}^{k\tau} (k\tau - t) f(t)dt\right) P_b(k)} \tag{12}$$

The optimal maintenance threshold ζ can be obtained by minimizing Eq. (12), i.e., $\zeta^* = \arg\min\{\psi(\zeta) : 0 < \zeta\}$.

3.2 Maintenance Model with Unknown Distribution Parameters

In this section, we assume that the distribution parameters of the degradation process are unknown and have to be estimated with the inspected information. Denote X_k as the observed degradation level at the kth inspection. If the parameters α and θ are known, we can have the joint distribution of the observations $X_1, \ldots, X_k$ as

$$f(X_1, \ldots, X_k|\alpha, \theta) = \left(\frac{1}{\sqrt{2\pi\sigma^2}}\right)^k \cdot \exp\left(-\sum_{i=1}^{k}\left(\frac{X_i - \alpha - \theta i\tau}{2\sigma^2}\right)\right) \tag{13}$$

However, the exact values of α and θ are unknown, due to the unit-to-unit variation. We assume that the prior distribution of α and θ are known, which can be obtained from the reliability characteristics of the population of the components. In accordance with previous sections, let the prior distributions of α and θ be Gaussian distributions with mean $\mu_0^{'}$ and variance σ_0^2 and mean μ_1 and variance σ_1^2.

Given the inspected information $X_1, \ldots, X_k$, the posterior distributions of α and θ are bivariate normal distribution with parameters [23]

$$\mu_\alpha = \frac{\left(\sum_{i=1}^k X_i\sigma_0^2 + \mu_0^{'}\sigma^2\right)\left(\sum_{i=1}^k (i\tau)^2\sigma_1^2 + \sigma^2\right) - \left(\sum_{i=1}^k i\tau\sigma_0^2\right)\left(\sum_{i=1}^k X_i i\tau\sigma_1^2 + \mu_1\sigma^2\right)}{\left(k\sigma_0^2 + \sigma^2\right)\left(\sum_{i=1}^k (i\tau)^2\sigma_1^2 + \sigma^2\right) - \left(\sum_{i=1}^k i\tau\sigma_1^2\right)\left(\sum_{i=1}^k i\tau\sigma_0^2\right)}$$

$$\mu_\theta = \frac{\left(k\sigma_0^2 + \sigma^2\right)\left(\sum_{i=1}^k X_i i\tau\sigma_1^2 + \mu_1\sigma^2\right) - \left(\sum_{i=1}^k i\tau\sigma_1^2\right)\left(\sum_{i=1}^k X_i\sigma_0^2 + \mu_0^{'}\sigma^2\right)}{\left(k\sigma_0^2 + \sigma^2\right)\left(\sum_{i=1}^k (i\tau)^2\sigma_1^2 + \sigma^2\right) - \left(\sum_{i=1}^k i\tau\sigma_1^2\right)\left(\sum_{i=1}^k i\tau\sigma_0^2\right)}$$

$$\sigma_\alpha^2 = \sigma^2\sigma_0^2 \frac{\sum_{i=1}^k (i\tau)^2\sigma_1^2 + \sigma^2}{\left(k\sigma_0^2 + \sigma^2\right)\left(\sum_{i=1}^k (i\tau)^2\sigma_1^2 + \sigma^2\right) - \left(\sum_{i=1}^k i\tau\right)^2\sigma_0^2\sigma_1^2}$$

$$\sigma_\theta^2 = \sigma^2\sigma_1^2 \frac{k\sigma_0^2 + \sigma^2}{\left(k\sigma_0^2 + \sigma^2\right)\left(\sum_{i=1}^k (i\tau)^2\sigma_1^2 + \sigma^2\right) - \left(\sum_{i=1}^k i\tau\right)^2\sigma_0^2\sigma_1^2}$$

$$\rho = \frac{-\sigma_0\sigma_1\sum_{i=1}^k i\tau}{\sqrt{\left(k\sigma_0^2 + \sigma^2\right)\left(\sum_{i=1}^k (i\tau)^2\sigma_1^2 + \sigma^2\right)}}$$

where ρ is the correlation. The above equations imply that the degradation process is nonstationary evolution process whose parameters are updated according to the observed health condition of the system. The joint pdf of $\tilde{\alpha}$ and $\tilde{\theta}$ is given as

$$f(\tilde{\alpha}, \tilde{\theta}) = \frac{1}{2\pi\sigma_\alpha\sigma_\theta\sqrt{1-\rho^2}}\exp\left(-\frac{z}{2(1-\rho^2)}\right) \tag{14}$$

where

$$z = \frac{(\tilde{\alpha}-\mu_\alpha)^2}{\sigma_\alpha^2} - \frac{2\rho(\tilde{\alpha}-\mu_\alpha)\left(\tilde{\theta}-\mu_\theta\right)}{\sigma_\alpha\sigma_\theta} + \frac{\left(\tilde{\theta}-\mu_\theta\right)^2}{\sigma_\theta^2}$$

Proposition 2 *The correlation* ρ *and* σ_θ^2 *decrease with the inspection interval* τ, *while* σ_α^2 *increases with* τ.

Proposition 2 can be used to reduce the variance of estimates by varying the inspection length. If θ exerts dominant impact on the degradation process, which can be evaluated via the prior distribution, then inspection interval is suggested to be extended so as to reduce the uncertainty of the estimates. If the degradation process is largely influenced by α, then the inspection interval should be short so as to improve the accuracy of the estimates.

Corollary 1 *Under continuous monitoring, the variances of* α *and* θ *are constant.*

Corollary 1 can be achieved by letting τ approach to 0 and k approach to ∞. Corollary 1 can be explained as the consequence of continuous monitoring and perfect inspection, which significantly reduces the associated uncertainty. After estimating the distribution parameters at the kth inspection, the distribution of the degradation level at time $t + k\tau$ can be predicted, which follows a Gaussian distribution with mean [23]

$$\tilde{\mu}(t+k\tau) = \mu_\alpha + \mu_\theta(t+k\tau) - \frac{\sigma^2}{2} \tag{15}$$

and variance

$$\tilde{\sigma}^2(t+k\tau) = \sigma_\alpha^2 + (t+k\tau)^2\sigma_\theta^2 + \sigma^2 + 2\rho(t+k\tau)\sigma_\alpha\sigma_\theta \tag{16}$$

Since the parameters are updated whenever an inspection is carried out, maintenance decision based on a stationary failure rate may lead to a suboptimal solution. Instead, we focus on a dynamic maintenance policy, which captures the predictive information of the degradation process. We use the "failure probability till next inspection" (FPI) as the indicator to make maintenance decisions. In this way, the maintenance procedure goes as follows: at each inspection, the distribution parameters are updated based on the inspected information, if the FPI of the system exceeds certain threshold, preventive replacement is performed. Otherwise, the system is left unattained.

Since the maintenance decision is made one inspection after another, we focus on the expected cost till the subsequent inspection. Given that the system functions through the previous k inspections, and the estimates of the distribution parameters, $\tilde{\alpha}$ and $\tilde{\theta}$ are available, the FPI of the system can be denoted as

$$P\Big(T < (k+1)\tau | T > k\tau, \tilde{\alpha}, \tilde{\theta}\Big) = 1 - \exp\left(-\int_{k\tau}^{(k+1)\tau}\int_0^{\infty} h_0(s)\varphi\Big(x(s)|\tilde{\alpha}, \tilde{\theta}\Big) f_{x(s)|\tilde{\alpha},\tilde{\theta}} dx(s)ds\right) \tag{17}$$

If the system fails before the kth inspection, corrective replacement is performed at once, and a renewal cycle follows subsequently. If the FPI is larger than ω_k at the kth inspection, i.e., $\int_{k\tau}^{(k+1)\tau}\int_0^{\infty} h_0(s)\varphi(x(s)) f_{X(s)}(x(s))ds > \log(1/(1-\omega_k))$, preventive replacement is carried out. Within a renewal cycle, if the system does not fail before the kth inspection, then at the kth inspection, the expected cost within the period $(k\tau, (k+1)\tau)$ is

$$\begin{aligned} E\big[C_{pe}(k)\big] = C_i + C_r \cdot \left(1 - \exp\left(-\int_{k\tau}^{(k+1)\tau}\int_0^{\infty} h_0(s)\varphi(x(s)) f_{X(s)}(x(s))ds\right)\right) \\ + C_p P\left\{\int_{k\tau}^{(k+1)\tau}\int_0^{\infty} h_0(s)\varphi\Big(x(s)|\tilde{\alpha}, \tilde{\theta}\Big) f_{X(s)|\tilde{\alpha},\tilde{\theta}} ds > \log(1/(1-\omega_k))\right\} \end{aligned} \tag{18}$$

The expected length within the period $(k\tau, (k+1)\tau)$ is

$$\begin{aligned} E\big[T_{pe}(k)\big] = \left(\left(\int_{k\tau}^{(k+1)\tau} (t-k\tau) f_T(t)dt\right) + \tau \exp\left(-\int_{k\tau}^{(k+1)\tau}\int_0^{\infty} h_0(s)\varphi(x(s)) f_{X(s)} ds\right)\right) \cdot \\ P\left\{\int_{k\tau}^{(k+1)\tau}\int_0^{\infty} h_0(s)\varphi\Big(x(s)|\tilde{\alpha}, \tilde{\theta}\Big) f_{X(s)|\tilde{\alpha},\tilde{\theta}} ds \le \log(1/(1-\omega_k))\right\} \end{aligned} \tag{19}$$

Equation (19) is obtained based on the event that no preventive replacement is carried out at the kth inspection. Effectiveness of the maintenance policy is highly dependent on the observation data; a closed-form expression of the long-run cost rate is difficult to obtain. For simplicity, we make the period-by-period maintenance decision by comparing the FIR with the ratio of C_p and C_r. The decision π is then presented as

$$\pi = \begin{cases} 1, \text{ if FPI} > C_p/C_r \\ 0, \text{ otherwise} \end{cases} \tag{20}$$

where 1 denotes preventive replacement and 0 implies doing nothing. Note that for safety-critical systems where a high reliability is required, more constraints are imposed on ω_k.

Maintenance actions have to satisfy the reliability constraint while minimizing the maintenance cost.

4 Case Study

To illustrate the practical value of the designed approach, it is applied to support the maintenance decision of bridge joints in UK. Bridge joints are used to accommodate the necessary movements of bridge decks, withstand the traffic load, and protect bearing from induced moisture and chloride ion. In this example, Asphalt Plug Joint (APJ) is studied and analyzed in particular. APJ is one of the most common bridge joints due to its waterproof and noise reduction properties. It also exhibit the property of low cost and easiness to install, repair and replace. APJ is constituted by a metal plate, which spans between bridge decks to accommodate longitudinal expansion and contraction (up to 40 mm). The plate is then covered by asphaltic plug making a smooth riding surface and preventing the debris and water. Figure 1 shows the structure of a bridge and the location of APJ.

APJs have an expected lifetime between 5 and 15 years based on the operating environment. According to the local maintenance experts, apart from the regular aging process, the deterioration of APJ is influenced by environmental factors such as accumulated debris, corrosion and traffic load. Additionally, it is also influenced by the induced damaged from other bridge components, such as the water leakage on the underside of the deck, the performance of bearing and superstructure movement. In this example, we mimic the overall impact of the factor as a time-dependent covariate factor. When an APJ is functioning improperly, it will cause problems on the bridge deck and bearing. To mitigate the risk of APJ failure, general inspection is regulated with a two-year interval to assess the condition of APJ joints. The inspection cost is £250. The replacement cost is £6341. The failure cost includes replacement cost, traffic management cost and add-on cost, which is £15,751 in total. The local

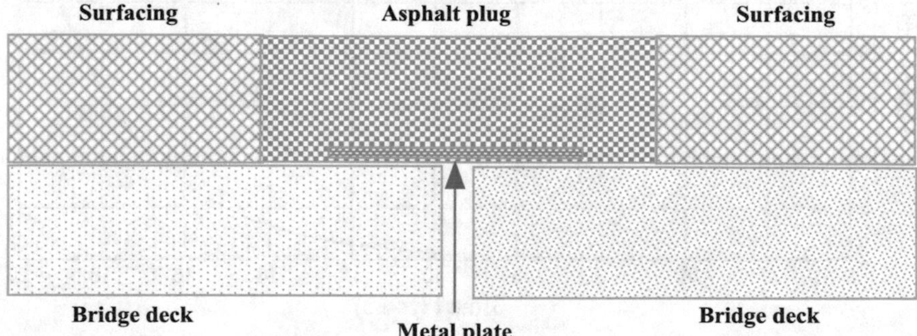

Fig. 1 Sketch of asphalt plug joint

maintenance team is keen to find the optimal threshold to replacement APJ so that the operation and maintenance cost can be minimized.

4.1 CBM with Known Distribution Parameters

According to the practical experience from the experts in UK Council, the baseline hazard rate function follows Weibull distribution, $h_0(t) = bt^{b-1}/a^b$, where the scale parameter is set as $a = 40$ (year) and shape parameter $b = 3$. Weibull distribution was widely used in modelling the crack proposition [27, 28] The link function is assumed as exponential, e.g., $\varphi(X_t) = \exp(X_t)$.

The parameters of the degradation process are set as $\mu_0 = 0.5, \mu_1 = 0.2, \sigma^2 = 0.01, \sigma_0^2 = 0.005$, and $\sigma_1^2 = 0.005$. According to Eq. (6), the failure rate function follows a lognormal distribution, which is plotted in Fig. 2. As can be observed, the failure rate increases rapidly after 10 years, which implies a high risk of failure and intervention actions should be implemented in time. In addition, we plot the variation of system reliability and pdf in Figs. 3 and 4.

Sensitivity analysis is conducted to investigate the influences of parameters on the lifetime distribution of the system. Figure 5 shows how the failure rate and system reliability vary with different u_α. Obviously, a larger u_α leads to a higher failure rate; system reliability function shifts to left when u_α increases. In addition, Fig. 6 plots the influences of different u_θ on the failure rate and system reliability. Compared with u_α, different u_θ lead to larger differences of the failure rate and system reliability. The results imply that degradation rate exerts significant effect on system lifetime

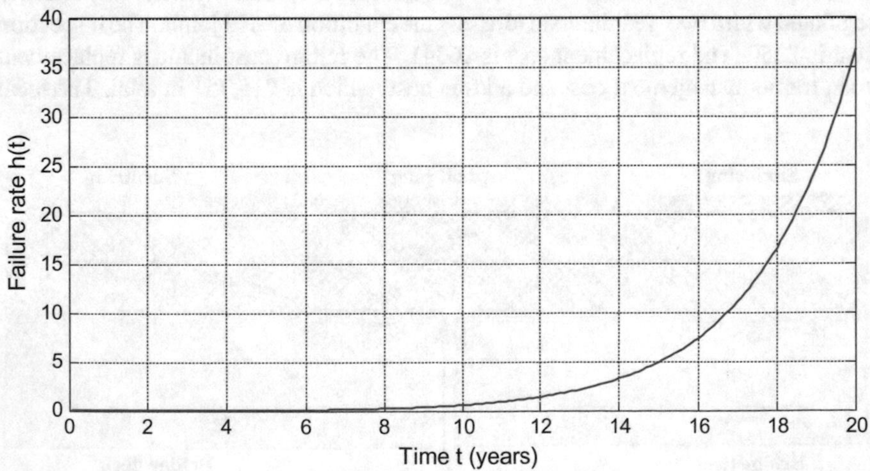

Fig. 2 Plot of failure rate

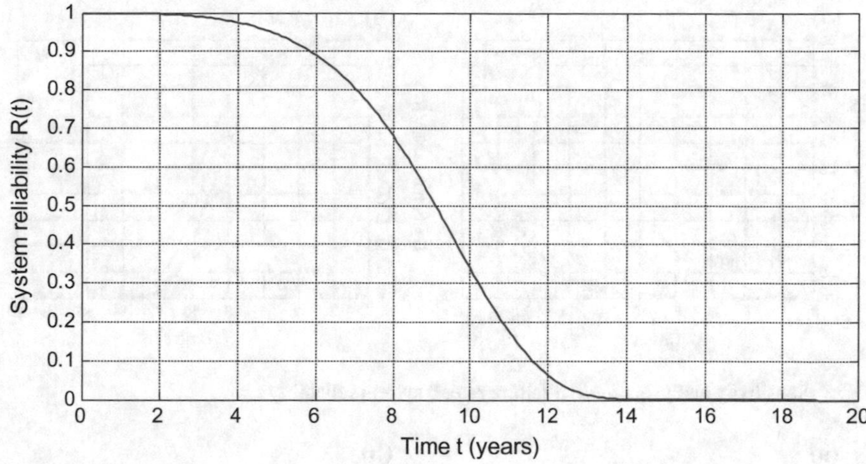

Fig. 3 Plot of system reliability

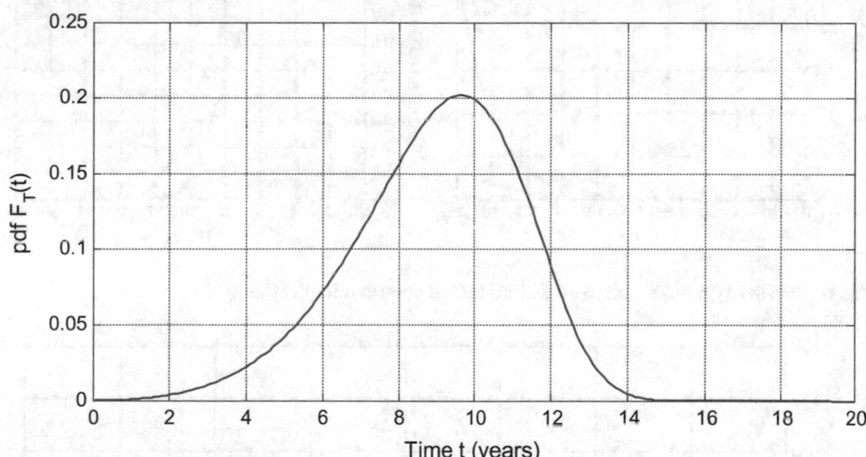

Fig. 4 Plot of pdf

distribution and the managers or engineers are suggested to invest more resource to accurately determine the value of degradation rate.

In current operation, the APJs are inspected every two years, $\tau = 2$. According to Eq. (12), the optimal maintenance policy is obtained as $\zeta = 0.038$, which implies that preventive replacement is carried out when the failure rate at inspection exceeds 0.038. The minimal long-run cost rate is achieved as $\psi^*(\zeta) = 1078$. Figure 7 shows how the long-run cost rate varies with respect to ζ.

Based on Eq. (1), we can obtain the optimal degradation threshold for preventive replacement with respect to system age, as shown in Fig. 8. Obviously, the optimal degradation threshold for preventive replacement decreases with system age. The

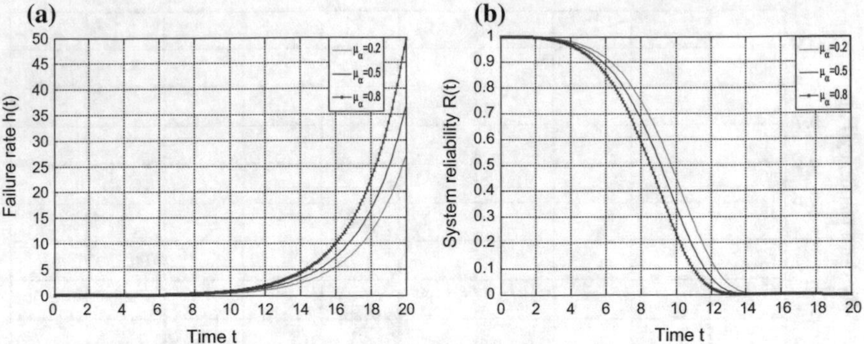

Fig. 5 Sensitivity analysis on u_α, **a** failure rate, **b** system reliability

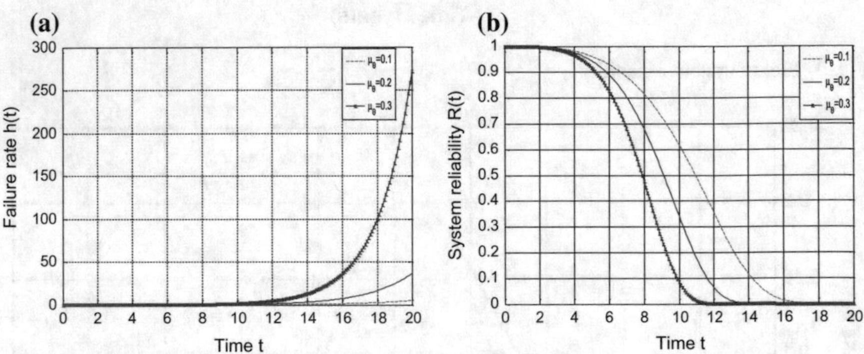

Fig. 6 Sensitivity analysis on u_θ, **a** failure rate, **b** system reliability

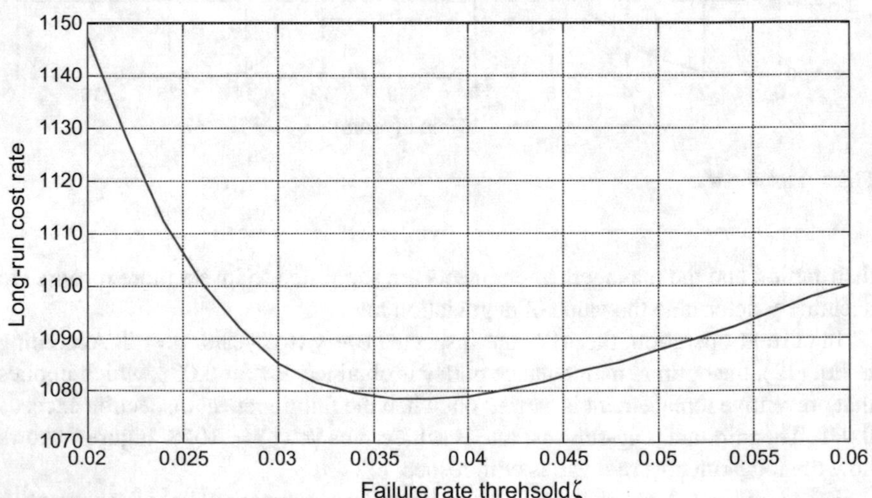

Fig. 7 Long-run cost rate with respect to ζ

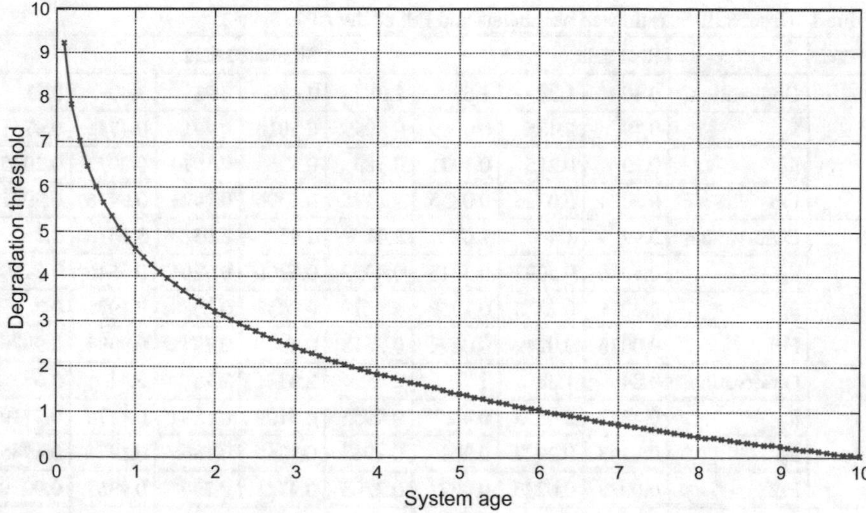

Fig. 8 Degradation threshold of preventive replacement with respect to system age

result presented in Fig. 8 is useful in practice. Engineers or managers can simply make maintenance decisions by comparing the observed degradation level with the threshold, which significantly facilitates the implementation of maintenance operations.

4.2 CBM with Unknown Distribution Parameters

When the parameters of the degradation process are unknown, inspection is performed to observe the system state and update the estimation of the parameters. For the bridge system, we only have eight-year inspection data, where the system is inspected every two years. For illustration purpose, we also simulate the system state data for the later eight years. Four APJs are under investigation. The parameters of the prior distribution are obtained by historical experience and expert judgement, which are given as $\mu_0 = 0.5$, $\mu_1 = 0.2$, $\sigma^2 = 0.01$, $\sigma_0^2 = 0.005$, and $\sigma_1^2 = 0.005$. The parameters are estimated according to Proposition 2 and FPI of the system is calculated based on Eq. (17).

Table 1 shows the observed system state along with the estimated parameters and FPI. It can be seen that the observed system state increases with the inspection index. In addition, the FPI increases rapidly with the inspection index. This is due to the fact that the link function $\varphi(\cdot)$ is exponential, which leads to exponential increasing of the failure rate. The estimated parameters $\tilde{u}_\alpha$ and $\tilde{u}_\theta$ are close to the prior, which implies the effectiveness of the prior distribution.

Table 1 Observations, estimated parameters and FPI of the APJs

Item		Real data				Simulated data			
1	Observation	0.86	1.38	1.63	1.98	2.47	2.91	3.40	3.71
	$\tilde{u}_\alpha$	0.49	0.485	0.4989	0.5099	0.5016	0.491	0.4743	0.4751
	$\tilde{u}_\theta$	0.19	0.215	0.1971	0.189	0.1933	0.1974	0.2027	0.2025
	FPI	0.0024	0.0236	0.0855	0.2175	0.4506	0.7423	0.9476	0.9971
2	Observation	0.99	1.4	1.62	2.41	2.45	3.05	3.14	3.8
	$\tilde{u}_\alpha$	0.5086	0.5092	0.5315	0.4952	0.5268	0.5204	0.5546	0.5450
	$\tilde{u}_\theta$	0.2271	0.2238	0.1952	0.2219	0.2057	0.2038	0.1975	0.2
	FPI	0.0026	0.0249	0.0874	0.2618	0.4975	0.7931	0.9494	0.9976
3	Observation	0.84	1.32	1.71	2	2.31	2.75	3.41	3.68
	$\tilde{u}_\alpha$	0.4871	0.4835	0.4825	0.4955	0.5109	0.5107	0.4775	0.4719
	$\tilde{u}_\theta$	0.1843	0.2027	0.204	0.1943	0.1865	0.1865	0.197	0.1985
	FPI	0.0023	0.0226	0.087	0.2203	0.4331	0.706	0.9357	0.9959
4	Observation	0.81	1.22	1.85	2.09	2.21	2.85	3.5	3.85
	$\tilde{u}_\alpha$	0.4829	0.4815	0.4552	0.466	0.5063	0.5308	0.4756	0.4479
	$\tilde{u}_\theta$	0.1757	0.1823	0.2162	0.2082	0.1875	0.1779	0.1953	0.2027
	FPI	0.0023	0.0211	0.0902	0.2342	0.4348	0.6784	0.9315	0.9967

Based on the proposed maintenance policy, the system is replaced preventively when the FRI is larger than $C_p/C_r = 0.4026$. With the calculated FPI, we can conclude that, if the APJs are not failed, they should be preventively replaced at the fourth inspection, so as to achieve maximal economic benefits.

5 Conclusions

This chapter investigates the condition based maintenance policy for systems with aging and cumulative damage. The joint effect of aging and cumulative damage is described by proportional hazards model. Maintenance models are developed with consideration of known and unknown distribution parameters respectively. The results in this chapter show that the degradation rate exerts a significant impact on system lifetime distribution. Engineers or managers are suggested to pay more attention to improving the accuracy of the degradation rate estimation. The proposed condition based model can be widely applied for infrastructure systems which are subject to cumulative damage and exhibit a long life cycle.

Extensions of this research can be conducted by generalizing the one-dimensional cumulative damage into multi-dimensional. Then multiple sensors should be equipped to inspect the damage (degradation) indicators. Parameter estimation of the distribution parameters could be complicated as interactions may exist among the multi-dimensional cumulative damages. In addition, the form of link function can be explored with a variety of candidates. Exponential function is used for simplicity in this chapter; in reality, various link functions can be tested if the associated data are available.

Acknowledgements The work described in this chapter was partially supported by a theme-based project grant (T32-101/15-R) of University Grants Council, and a Key Project (71532008) supported by National Natural Science Foundation of China.

Appendix

Proof of Proposition 1 Denote $w(t)$ as the cumulative hazard rate function of the system, i.e.,

$$w(t; x_s) = \int_0^t h_0(s)\varphi(x_s)ds$$

In the following, for notational simplicity, we will suppress x_s of the $w(t; x_s)$. the derivative of $w(t)$ with respect to t can be obtained as

$$\frac{dw(t)}{dt} = h_0(t)\varphi(x_t) > 0$$

and

$$\frac{d^2w(t)}{d^2t} = \frac{dh_0(t)}{dt}\varphi(x_t) + h_0(t)\frac{d\varphi(x_t)}{dx_t} \cdot \frac{dx_t}{dt} > 0$$

Here we unofficially use dx_t/dt to denote the derivative of x_t. The inequality holds since $h_0(t)$ and $\varphi(x_t)$ are is non-decreasing functions in t and x_t, and X_t is stochastically increasing in t. we can conclude that $w(t)$ is a convex function in t. Based on the Jensen's inequality, we have $w(t_1) + w(t_3) > 2w(t_2)$, for any $0 < t_1 < t_2 < t_3$,

On the other hand, the cumulative hazard rate between two consecutive inspections can be rewritten as

$$\int_{k\tau}^{(k+1)\tau} h_0(s)\varphi(x_s)ds = \int_0^{(k+1)\tau} h_0(s)\varphi(x_s)ds - \int_0^{k\tau} h_0(s)\varphi(x_s)ds = w((k+1)\tau) - w(k\tau)$$

Readily we can obtain

$$\int_{k\tau}^{(k+1)\tau} h_0(s)\varphi(x_s)ds = w((k+1)\tau) - w(k\tau) > w(k\tau) - w((k-1)\tau) = \int_{(k-1)\tau}^{k\tau} h_0(s)\varphi(x_s)ds$$

On the other hand, Jensen's inequality states that $E[g(x)] > g(E[x])$, for any convex function $g(x)$, which completes the proof. □

Derivation of Equations (10) and (11) Denote U_a as the event that given no failure occurs, the system is preventively replaced at the kth inspection, $U_a = 1\{h(t) \le \zeta, t \in [0, (k-1)\tau] \cap h(k\tau) > \zeta\}$, and V_a as the event that no failure occurs before $k\tau$, $V_a = 1\{\text{no failure occurs before } k\tau\}$. We can have

$$
\begin{aligned}
&P_a(k) = P(V_a \cap U_a) = P(V_a|U_a) \cdot P(U_a) \\
&= P(V_a|U_a) \cdot P(h(t) \le \zeta, t \in [0, (k-1)\tau] \cap h(k\tau) > \zeta) \\
&= \exp\left(-\int_0^{(k-1)\tau} h_0(t) \int_0^{\ln(\zeta/h_0(t))} \varphi(x) f(x) dx dt - \int_{(k-1)\tau}^{k\tau} h_0(t) \int_0^{\infty} \varphi(x) f(x) dx dt\right) \cdot \\
&(P(h(k\tau) > \zeta) - P(h((k-1)\tau) > \zeta)) \\
&= \exp\left(-\int_0^{(k-1)\tau} h_0(t) \int_0^{\ln(\zeta/h_0(t))} \varphi(x) f(x) dx dt - \int_{(k-1)\tau}^{k\tau} h_0(t) \int_0^{\infty} \varphi(x) f(x) dx dt\right) \cdot \\
&\left(\Phi\left(\frac{\left(\mu_0 + \mu_1 k\tau - \sigma^2/2 - \varphi^{-1}(\zeta/h_0(k\tau))\right)}{\sqrt{\sigma_0^2 + \sigma_1^2 (k\tau)^2 + \sigma^2}}\right) - \Phi\left(\frac{\left(\mu_0 + \mu_1 (k-1)\tau - \sigma^2/2 - \varphi^{-1}(\zeta/h_0((k-1)\tau))\right)}{\sqrt{\sigma_0^2 + \sigma_1^2 ((k-1)\tau)^2 + \sigma^2}}\right)\right)
\end{aligned}
$$

Denote U_b as the event that no preventive replacement is carried out before the kth inspection, $U_b = 1\{h(t) \le \zeta, t \in [0, (k-1)\tau]\}$, and V_b as the event that given no preventive replacement, failure occurs within the interval $((k-1)\tau, k\tau)$, $V_b = 1\{\text{failure occurs within } ((k-1)\tau, k\tau)\}$. The probability that failure occurs in the period $((k-1)\tau, k\tau)$ is given as

$$
\begin{aligned}
P_b(k) &= P(V_b \cap U_b) = P\left(V_b|U_b\right) \cdot P\left(U_b\right) \\
&= P\left(V_b|U_b\right) \cdot P(h(t) \le \zeta, t \in [0, (k-1)\tau]) \\
&= \begin{pmatrix} \exp\left(-\int_0^{(k-1)\tau} h_0(t) \int_0^{\ln(\zeta/h_0(t))} \varphi(x) f(x) dx dt\right) \\ -\exp\left(-\int_0^{(k-1)\tau} h_0(t) \int_0^{\ln(\zeta/h_0(t))} \varphi(x) f(x) dx dt - \int_{(k-1)\tau}^{k\tau} h_0(t) \int_0^{\infty} \varphi(x) f(x) dx dt\right) \end{pmatrix} \cdot \\
&\Phi\left(\frac{\left(\mu_0 + \mu_1 t - \sigma^2/2 - \varphi^{-1}(\zeta/h_0(t))\right)}{\sqrt{\sigma_0^2 + \sigma_1^2 t^2 + \sigma^2}}\right) \\
&= \left(1 - \exp\left(-\int_{(k-1)\tau}^{k\tau} h_0(t) \int_0^{\infty} \varphi(x) f(x) dx dt\right)\right) \cdot \exp\left(-\int_0^{(k-1)\tau} h_0(t) \int_0^{\ln(\zeta/h_0(t))} \varphi(x) f(x) dx dt\right) \cdot \\
&\Phi\left(\frac{\left(\mu_0 + \mu_1 t - \sigma^2/2 - \varphi^{-1}(\zeta/h_0(t))\right)}{\sqrt{\sigma_0^2 + \sigma_1^2 t^2 + \sigma^2}}\right)
\end{aligned}
$$

□

Proof of Proposition 2 Let

$$\rho_2(\tau^2) = \rho^2 = \frac{\left(\sum_{i=1}^{k} i\tau\right)^2 \sigma_0^2 \sigma_1^2}{(k\sigma_0^2 + \sigma^2)\left(\sum_{i=1}^{k} (i\tau)^2 \sigma_1^2 + \sigma^2\right)}$$

Then the derivative of ρ_2 with respect to τ^2 is given as

$$\rho_2' = \frac{\left(\sum_{i=1}^k i\right)^2 \sigma_0^2\sigma_1^2\left(k\sigma_0^2+\sigma^2\right)\left(\sum_{i=1}^k (i\tau)^2\sigma_1^2+\sigma^2\right) - \left(\sum_{i=1}^k i\tau\right)^2 \sigma_0^2\sigma_1^2\left(k\sigma_0^2+\sigma^2\right)\left(\sum_{i=1}^k (i)^2\sigma_1^2\right)}{\left(k\sigma_0^2+\sigma^2\right)^2\left(\sum_{i=1}^k (i\tau)^2\sigma_1^2+\sigma^2\right)^2}$$

$$= \frac{\left(\sum_{i=1}^k i\right)^2 \sigma_0^2\sigma_1^2\sigma^2\left(k\sigma_0^2+\sigma^2\right)}{\left(k\sigma_0^2+\sigma^2\right)^2\left(\sum_{i=1}^k (i\tau)^2\sigma_1^2+\sigma^2\right)^2} > 0$$

Since $\rho = -\sqrt{\rho_2}$, it can be concluded that ρ decreases with τ.
On the other hand, σ_α^2 can be rewritten as

$$\sigma_\alpha^2 = \sigma^2\sigma_0^2\frac{1/\left(k\sigma_0^2+\sigma^2\right)}{1-\rho^2}$$

which implied that σ_α^2 increases with τ.

In addition, σ_θ^2 can be rewritten as $\sigma_\theta^2 = \sigma^2\sigma_1^2\frac{k\sigma_0^2+\sigma^2}{\left((k\sigma_0^2+\sigma^2)\sigma_1^2\sum_{i=1}^k (i)^2-\left(\sum_{i=1}^k i\right)^2\sigma_0^2\sigma_1^2\right)\tau^2+(k\sigma_0^2+\sigma^2)\sigma^2}$.

Since $k\sum_{i=1}^k (i)^2 > \left(\sum_{i=1}^k i\right)^2$ for any k, we can easily conclude that σ_θ^2 decreases with τ. □

Proof of Corollary 1 Denote $\tau k = M$, where M is a constant with bounds, $M < \infty$. Let $\sum_{i=1}^{k} (i\tau)^2 = \frac{\tau^2 k(k+1)(2k+1)}{6}$ and $\sum_{i=1}^{k} i\tau = \frac{\tau k(k+1)}{2}$, we can have

$$\frac{1}{\sigma_\alpha^2} = \frac{1}{\sigma^2\sigma_0^2}\left[\left(k\sigma_0^2+\sigma^2\right) - \frac{3k^2(k+1)^2\sigma_0^2\sigma_1^2}{2k(k+1)(2k+1)\sigma_1^2+12\sigma^2\tau^{-2}}\right]$$

$$\frac{1}{\sigma_\theta^2} = \frac{1}{\sigma^2\sigma_1^2}\left\{\frac{\tau^2\sigma_1^2 k(k+1)}{2}\left[\frac{2k+1}{3} - \frac{k(k+1)\sigma_0^2}{2\left(k\sigma_0^2+\sigma^2\right)}\right]+\sigma^2\right\}$$

When $\tau \to 0$ and $k \to \infty$, we have $\frac{1}{\sigma_\alpha^2} \to \infty$, which implies $\lim_{\tau\to 0}\sigma_\alpha^2 = 0$.
Similarly, we can obtain $\lim_{\tau\to 0}\sigma_\theta^2 = 0$, which completes the proof. □

References

1. Liu, B., Xu, Z., Xie, M., & Kuo, W. (2014). A value-based preventive maintenance policy for multi-component system with continuously degrading components. *Reliability Engineering & System Safety, 132,* 83–89.
2. Wu, S., Chen, Y., Wu, Q., & Wang, Z. (2016). Linking component importance to optimisation of preventive maintenance policy. *Reliability Engineering & System Safety, 146,* 26–32.
3. Maillart, L. M. (2006). Maintenance policies for systems with condition monitoring and obvious failures. *IIE Transactions, 38*(6), 463–475.

4. Srinivasan, R., & Parlikad, A. K. (2013). Value of condition monitoring in infrastructure maintenance. *Computers & Industrial Engineering, 66*(2), 233–241.
5. Chen, N., Ye, Z. S., Xiang, Y., & Zhang, L. (2015). Condition-based maintenance using the inverse Gaussian degradation model. *European Journal of Operational Research, 243*(1), 190–199.
6. Ye, Z. S., & Xie, M. (2015). Stochastic modelling and analysis of degradation for highly reliable products. *Applied Stochastic Models in Business and Industry, 31*(1), 16–32.
7. Ye, Z. S., Chen, N., & Shen, Y. (2015). A new class of Wiener process models for degradation analysis. *Reliability Engineering & System Safety, 139,* 58–67.
8. Liu, B., Xie, M., & Kuo, W. (2016). Reliability modeling and preventive maintenance of load-sharing systems with degrading components. *IIE Transactions, 48*(8), 699–709.
9. Caballé, N. C., Castro, I. T., Pérez, C. J., & Lanza-Gutiérrez, J. M. (2015). A condition-based maintenance of a dependent degradation-threshold-shock model in a system with multiple degradation processes. *Reliability Engineering & System Safety, 134,* 98–109.
10. Peng, H., & van Houtum, G. J. (2016). Joint optimization of condition-based maintenance and production lot-sizing. *European Journal of Operational Research, 253*(1), 94–107.
11. Wu, X., & Ryan, S. M. (2011). Optimal replacement in the proportional hazards model with semi-markovian covariate process and continuous monitoring. *IEEE Transactions on Reliability, 60*(3), 580–589.
12. Wardhana, K., & Hadipriono, F. C. (2003). Analysis of recent bridge failures in the United States. *Journal of Performance of Constructed Facilities, 17*(3), 144–150.
13. Lin, D. Y., & Wei, L. J. (1989). The robust inference for the Cox proportional hazards model. *Journal of the American statistical Association, 84*(408), 1074–1078.
14. Pham, H. T., Yang, B. S., & Nguyen, T. T. (2012). Machine performance degradation assessment and remaining useful life prediction using proportional hazard model and support vector machine. *Mechanical Systems and Signal Processing, 32,* 320–330.
15. Jardine, A. K., & Tsang, A. H. (2013). *Maintenance, replacement, and reliability: Theory and applications*. Boca Raton, FL: CRC Press.
16. Banjevic, D., Jardine, A. K. S., Makis, V., & Ennis, M. (2001). A control-limit policy and software for condition-based maintenance optimization. *INFOR: Information Systems and Operational Research, 39*(1), 32–50.
17. Ghasemi, A., Yacout, S., & Ouali, M. S. (2007). Optimal condition based maintenance with imperfect information and the proportional hazards model. *International Journal of Production Research, 45*(4), 989–1012.
18. Wu, X., & Ryan, S. M. (2010). Value of condition monitoring for optimal replacement in the proportional hazards model with continuous degradation. *IIE Transactions, 42*(8), 553–563.
19. Tian, Z., & Liao, H. (2011). Condition based maintenance optimization for multi-component systems using proportional hazards model. *Reliability Engineering & System Safety, 96*(5), 581–589.
20. Lam, J. Y. J., & Banjevic, D. (2015). A myopic policy for optimal inspection scheduling for condition based maintenance. *Reliability Engineering & System Safety, 144,* 1–11.
21. Lu, C. J., & Meeker, W. O. (1993). Using degradation measures to estimate a time-to-failure distribution. *Technometrics, 35*(2), 161–174.
22. Elwany, A. H., Gebraeel, N. Z., & Maillart, L. M. (2011). Structured replacement policies for components with complex degradation processes and dedicated sensors. *Operations Research, 59*(3), 684–695.
23. Gebraeel, N. Z., Lawley, M. A., Li, R., & Ryan, J. K. (2005). Residual-life distributions from component degradation signals: A Bayesian approach. *IIE Transactions, 37*(6), 543–557.
24. Haghighi, F., & Bae, S. J. (2015). Reliability estimation from linear degradation and failure time data with competing risks under a step-stress accelerated degradation test. *IEEE Transactions on Reliability, 64*(3), 960–971.
25. Provan, J. W. (1987). Probabilistic approaches to the material-related reliability of fracture-sensitive structures. In *Probabilistic fracture mechanics and reliability* (pp. 1–45). Springer Netherlands.

26. Huynh, K. T., Barros, A., Berenguer, C., & Castro, I. T. (2011). A periodic inspection and replacement policy for systems subject to competing failure modes due to degradation and traumatic events. *Reliability Engineering & System Safety, 96*(4), 497–508.
27. Bažant, Z. P. (2004). Probability distribution of energetic-statistical size effect in quasibrittle fracture. *Probabilistic Engineering Mechanics, 19*(4), 307–319.
28. Cook, R. F., & Clarke, D. R. (1988). Fracture stability, R-curves and strength variability. *Acta Metallurgica, 36*(3), 555–562.

Chapter 11
Implementation of a Condition Monitoring System on an Electric Arc Furnace Through a Risk-Based Methodology

Cristian Colace, Luca Fumagalli, Simone Pala, Marco Macchi, Nelson R. Matarazzo and Maurizio Rondi

Abstract This chapter presents the deployment of a condition monitoring system on an electric arc furnace in a steel making company, ranging from the development of the system until its implementation and the results achieved by its use in the plant. A step-wise risk-based methodology is introduced and it is adopted to deploy the condition monitoring system. The electric arc furnace is a relevant asset for safety issues; due to the characteristics of the furnace—running continuously at high temperatures and in harsh environmental conditions—many components cannot be visually inspected, thus a maintenance system, with real-time monitoring capabilities, represents a proper solution to keep under control the asset health state. Besides the monitoring activity, appropriate risk information must also be shown to maintenance personnel to effectively improve maintenance activity. In this concern, the condition monitoring system, herein presented, can be considered an E-maintenance tool, integrated within an existing industrial ICT infrastructure, and representing one practical application of E-maintenance concept within industry.

Keywords Condition based maintenance · Production system safety · Hazard and operability analysis · E-maintenance

L. Fumagalli (✉) · S. Pala · M. Macchi
Department of Management, Economics and Industrial Engineering, Politecnico di Milano, Piazza Leonardo da Vinci, 32, Milan 20133, Italy
e-mail: luca1.fumagalli@polimi.it

M. Macchi
e-mail: marco.macchi@polimi.it

C. Colace · N. R. Matarazzo · M. Rondi
Tenaris Dalmine S.p.A., Piazza Caduti 6 Luglio 1944, 1, Dalmine, BG 24044, Italy
e-mail: mrondi@tenaris.com

A. Crespo Márquez et al. (eds.), *Value Based and Intelligent Asset Management*,
https://doi.org/10.1007/978-3-030-20704-5_11

1 Introduction

Safety is claimed as an important issue in process design and in the operations in the chemical process industry [1]. This applies also to steel making industry, where the risks for workers must be carefully considered [2]: to this end, a correct maintenance activity plays a key role.

Different approaches may be used to address the issue of safety. Supporting approaches and tools, namely CBM (Condition Based Maintenance) [3, 4] and Intelligent Maintenance Systems (IMSs) [5] are worth being considered as a lever to foster the achievement of good results in asset and maintenance management. Indeed, CBM is considered as a promising maintenance policy [6].

Intelligent maintenance systems are often linked in scientific literature with the concept of E-maintenance. Tsang [7] suggests the role of E-maintenance in the ICT support of maintenance management and processes; moreover, E-maintenance can also be seen as the ability to monitor plant floor assets and integrate the gathered information to upper level enterprise applications [8] and enable proactive decision process execution [9]. See further discussion about E-maintenance concepts in Iung et al. [10].

Crespo-Marquez and Iung [11] further suggests that in future the E-maintenance will introduce a revolutionary change in the present approach; indeed, recent research novelties can contribute to the E-maintenance framework, such as prognostic tools [12, 13] and mixed reality tools [14, 15].

Notwithstanding the large discussion on E-maintenance concept in scientific literature, an industrial adoption of E-maintenance platforms seems still not so diffused. Interesting cases prove the capabilities of E-maintenance platforms through industrial demonstrations (see for instance [16, 17]), but a few works present E-maintenance solutions that are actually run in real industrial contexts. Industrial perspective of E-maintenance was further investigated in recent works [18, 19], where, initially, value assessment and, more recently, value-driven engineering of E-maintenance platforms were addressed as an important matter, aiming at a discussion with industrial practitioners on costs and benefits of E-maintenance solutions.

This research aims at deploying a practical application in the steel making industry of a condition monitoring system developed in accordance to the E-maintenance concept. The chapter shows how data coming from plant floor level are transformed into information, being elaborated and displayed at an upper level in order to support risk-informed decisions of maintenance planners. The chapter grounds on some first information provided by Colace et al. [20], aiming at deepening the research facets of the proposed system and contributing to the scientific literature by:

1. providing practical step-wise risk-based methodological approach leading from risk analysis to deployment of a condition monitoring system;
2. providing empirical evidences of useful possibility of implementation of the E-maintenance theoretical concepts in industry.

The proposed approach designs the Condition Monitoring (CM) system starting from the information produced by a Process Hazard Analysis (PHA); this can be considered a novelty compared with traditional methods for Condition Monitoring design, mainly based on pure technical and technological considerations [21]; well summarizes the typical techniques and the approaches followed to build CM systems.

The system has been developed directly in an industrial environment, at Tenaris-Dalmine plant, located in Dalmine (Italy), with the objective of its practical adoption in the day-by-day operations, according to the direction of improvement of tools to be used in the plant [22].

The tool is a kind of virtual inspector (shortly speaking, VI), able to supervise in real-time the health of the electric furnace (i.e. one of the key production equipment of the plant). This VI uses information sourced by the control system of the electric furnace and develops additional knowledge of the production system behaviour, by health monitoring activity.

The electric furnace is a complex machine and, in fact, the developed VI allows to keep under control only certain phenomena. More specifically, the condition monitoring system, which has been named Burnersys, aims at monitoring the variables of the burning system of the furnace.

Some steps have been carried out to develop and implement the system and represent the practical step-wise risk-based approach herein proposed.

The steps are presented in Sect. 2 and then discussed through the following structure of the Chapter. Section 3 introduces the industrial context; Sect. 4 summarizes the use of a PHA method, as methodological approach for risk-based analysis, understandable by production process specialists and process supervisors. Section 5 describes what signals from the field have been taken in account and provides the criteria followed to select the signals. Section 6 presents the development of the algorithm for health monitoring. Section 7 presents the implementation of the algorithm within the existing IT architecture which is adopted in TenarisDalmine plant, in order to properly connect the VI with the input signals and realize the user interface that allows the operators to read the output information provided by the VI. Eventually, Sect. 8 provides conclusions of the research, highlighting the results achieved by using the VI during several months of furnace operation.

2 Step-Wise Risk-Based Approach for Condition Monitoring System Design

The step-wise risk-based approach proposed by this Chapter encompasses:

1. Process Hazard Analysis (PHA) to map the asset behaviour;
2. Identification of the key variables to monitor the asset;
3. Definition of the condition monitoring algorithm, encompassing the definition of reference values representing good asset health conditions;
4. Implementation and integration of the condition monitoring system, namely Burnersys, in the ICT architecture of the plant.

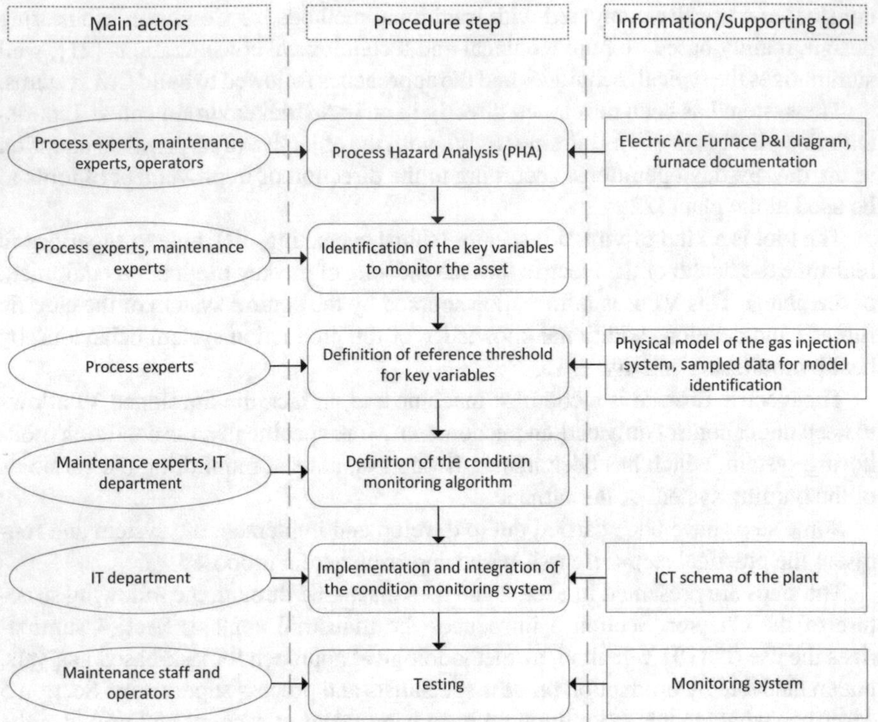

Fig. 1 Schema of the step-wise risk-based approach

The work has been developed according to these four steps that are also summarized by Fig. 1, indicating also tools and information to use to support the deployment of the CM solution. The steps are split in the practical activities done to carry out the procedure that are then detailed in the following sections.

2.1 *Process Hazard Analysis*

Different techniques can be applied to perform this step. Several methods can be used to study technical failures (FMEA, HAZOP, fault trees, event trees, reliability diagrams, and so on). The choice depends mainly on the study aims [23, 24]. Specifically, the Hazard and Operability Analysis (HAZOP) has been selected to model the behaviour of the system in normal and abnormal circumstances; it allows a formalized adoption and enables a rigorous and reproducible application [25]. Other PHA techniques such as the following can be applied [26, 27]: Checklist, What-If, Failure Mode Effect Analysis (FMEA) and Fault Tree Analysis (FTA). Checklist has not been applied in the presented case because, according to [26], it is primarily used for process described by standards and industry practices; moreover, since it relies on the experience it may not identify all hazardous situations. What-If analysis is a brainstorming approach that leads to investigate on the possible process deviations or failures; as in the previous case, the technique needs to be applied by highly experienced staff in order to avoid missing important items. Moreover, according again to [26], a combination of Checklist and What-If analysis can compensate the weakness of each other. Then, Failure Mode and Effect Analysis (FMEA) may be not efficient for identifying an exhaustive list of combinations of equipment failures that lead to accidents [26]. Fault Tree Analysis (FTA) is an optimal technique to identify combinations of basic equipment and human failures, indeed according to [26], FTA is well suited for analysing highly redundant system but for asset in which single failures can lead to accidents HAZOP and FMEA can be considered better than FTA. Moreover [26], indicates HAZOP, along with What-If and What-If/Checklist, are as better able in handling batch processes system than FTA or FMEA; metal melting through the electric arc furnace is a batch process.

The information provided by HAZOP have not been used solely to identify the risk, but it provides a background to be exploited in the next steps for the design of the Condition Monitoring (CM) algorithm. Indeed, a CM system collects data and through the information provided by HAZOP can map the actual state of the system. In this way, it is possible for the CM to indicate whether the system is behaving in a normal or abnormal condition; hence, the HAZOP provides also an information background on how to solve identified problem during the plant runtime. More precisely, this information can be integrated in the CM and showed to the plant operator if an abnormal condition is detected. HAZOP analysis is further discussed in Sect. 4.

2.2 *Identification of the Key Variables*

This phase is devoted to identify the main variables, the so called key variables, that best and fully describe the behaviour of the system. The objective is to define a small group of variables that are highly representative of the system, or some of its features.

2.3 *Definition of the Condition Monitoring Algorithm*

The monitoring system needs to follow a specific algorithm in order to assess in which state the furnace is working. Therefore, in this step it is defined the algorithm based on the information coming from the previous steps: the formalization of the working condition from the step 1 and the key variables from the step 2.

2.4 Implementation in the ICT Architecture of the Plant

The algorithm defined at step 3 is then coded into the ICT system and the output must be shown to the operator in the control room. The information collected through the previous step 1 by the HAZOP analysis are used also here; in fact, they are useful to provide the operator with information not only limited to the alarm and the warning, but also to indicate possible causes and counteractions or suggestions.

3 Industrial Context

3.1 The Plant

Tenaris is the leading global manufacturer and supplier of tubular products and services used in the drilling, completion and production of oil and gas, in process and power plants and in specialized industrial and automotive applications.

An important part of Tenaris's manufacturing capacity is located in Italy at the state-of-the-art Dalmine Seamless Pipe Mill in Dalmine, Bergamo. TenarisDalmine is the headquarter of Tenaris Italia, which has 3000 employes, 5 premises and an yearly production capacity of 950,000 tons of finite goods. Tenaris's process for producing seamless steel pipe is based on the Electric Arc Furnace (EAF), Ladle Furnace, Vacuum Degassing and Continuous Casting process. The Electric Arc Furnace is charged by selected scrap and varying percentages of pig iron.

3.2 Furnace Description

The electric furnace (Fig. 2) is constituted by a lower sheet metal keel (lower-part) coated with refractory bricks, which serves to contain the liquid steel, and cooled by a cage with a structural function that supports panels cooled by a water circuit [28]. In TenarisDalmine plant, there are two electric furnaces with the following characteristics: capacity 105 t, diameter 6.1 m, weight of the structure (considering also refractory bricks) 230 t.

The characteristics of each furnace allow to obtain 98 t of liquid steel, with an average power request of 67 MW per cast and a maximum power peak of 89 MW (performance of 2013). The main contribution of energy needed for melting the scrap is supplied from the electricity for about two thirds and for one thirds by chemical energy supplied by the lances and the burners located on board. The chemical process is controlled by a system that monitors gases exiting the furnace. In particular, the upper part of the furnace encompasses, among others, three oxygen injection points (see Fig. 3), particularly relevant for this research: the condition monitoring system herein presented is deployed to monitor the parameters of the burner system. Some

Fig. 2 Graphical representation of the furnace

Fig. 3 Graphical representation of the furnace with carbon injectors and oxygen lances

specific data are provided with focus on the oxygen injection pipeline. The system thus focuses on supporting maintenance activity for the oxygen injection system of electric arc furnace.

The typical rate of oxygen injection through the KT oxygen lances is 2000 Nm3/h in order to enhance the decarburization rate during the very short refining period. More recent modifications considered the installation of an additional KT Carbon

Injector, to be placed in close proximity to the third KT Oxygen Lance (see further detail in Ferro et al. [28]).

4 Hazop Analysis

Hazard studies provide a systematic methodology for identification, evaluation and mitigation of potential process hazards that can cause severe human, environmental and economic losses. To this end, different methods are practiced at various stages during the plant life cycle [29]. Among the available techniques, hazard and operability (HAZOP) analysis is one of the most used one, and according to the reasons explained in Sect. 2.1 it has been applied to the presented case. The HAZOP technique uses a systematic process to identify possible deviations from normal operations, and it ensures that adequate countermeasures are taken to help preventing accidents. To this end, NSW Government Department of Planning [30] provides useful guidelines. Indeed, HAZOP studies had evolved from the Imperial Chemical Industries' "Critical Examination" technique formulated in the mid-1960s and one decade later HAZOP was formally published as a disciplined procedure to identify deviations from the design intent [31]. Lawley [32] defined and delineated the principles needed to carry out operability studies and hazard analysis, and Weber et al. [33], and Labowsky et al. [34] show an application case of HAZOP analysis to a real case. Furthermore, this approach has been proven to be useful when dealing with setup of monitoring and condition based maintenance system [35] even if still low attention is paid to this important topic. For instance, in the literature review provided by Dunjó et al. [31], even if maintenance topic is considered, condition based maintenance is not explicitly addressed.

Thus, the first step of the proposed step-wise procedure is the application of HAZOP in this novel context, that is, in order to identify the variable of interests and to classify the effects that any deviation from the reference value of such variable may cause. Therefore, HAZOP is initially introduced as preliminary analysis for the deployment of a monitoring tool within this research. In this regard, the aim is to exploit the capability of the HAZOP technique in the design phase of a condition monitoring system.

Following the typical HAZOP practice, HAZOP adopts special adjectives (such as more, less, no, etc.) in combination with process variables (speed, flow, pressure, etc.) to consider in a systematic way all the possible deviations. The HAZOP analysis, herein presented, focuses on the oxygen injection system of electric arc furnace.

The Burner KT5—oxygen injection pipeline is considered for showing the analysis done in this first step of the methodology. In Table 1, the variables and the guide words used in the analysis are summarized; Tables 2, 3, 4, 5, and 6 show the HAZOP analysis output for each deviation.

It is worth highlighting that HAZOP has provided useful information for development of the VI tool Burnersys. Such information represents engineering knowledge that has been directly included within the tool under the shape of knowledge

Table 1 Summary of the variables and guide words used in the HAZOP analysis

		Guide word		
		High	Low	No
Variable	Pressure	High pressure Item n. 1.1 (see Table 2)	Low pressure Item n. 1.3 (see Table 4)	Lower limit is atmospheric pressure (it can be assumed as low pressure)
	Flow	High flow Item n. 1.2 (see Table 3)	Low flow Item n. 1.4 (see Table 5)	No flow (It can be assumed as low flow)
	Unity failure		Leakage/broken unity Item n. 1.5 (see Table 6)	Leakage/broken unity Item n. 1.5 (see Table 6)

Table 2 HAZOP table for deviation: high pressure

Item n. 1.1—Deviation: high pressure			
Causes	Effects	Counteractions	Suggestions
• The pipeline is obstructed by slag • Injection nozzle is crushed • Malfunction in pressure sensor • Malfunction in flow control valve • Errors in PLC data communication • Crushed pressure sensor • Crushed flexible pipe • Wrong mounted nozzle • Malfunction in non-return-valve	• Increase in pressure level may cause a damage in flexible pipe, possible fire triggering with explosion or damage to other plants • Malfunction in the burner with consequent missed melting of the scrap • An error in pressure measure may cause the control system to turn off the burner • Possible mixing of different fluid in the pipeline • During supersonic injection phase of the oxygen the burner may cause some melted metal to be projected and/or have a poor oxidation of metal	• Check nozzle status during planned stops • Check annually pressure level of the pipelines • Increase flexible pipe reliability • Change nozzle • Annually check flow control valve status • In case of explosion immediately stop the system by pushing emergency button	• Evaluate the possibility to add indicators on operator monitor to help them in finding possible malfunctions • Evaluate the possibility to implement a safety stop under particular conditions • Evaluate the possibility to divide working area in different compartments

Table 3 HAZOP table for deviation: high flow

Item n. 1.2—Deviation: high flow			
Causes	Effects	Counteractions	Suggestions
• Malfunction in flow control valve • Errors in PLC data communications	• Burner malfunctions and and missed melting of the scrap • Possible fire triggering with explosion or damage to other plants	• Check nozzle status during planned stops • Check annually pressure level of the pipelines • Increase flexible pipe reliability • Annually check flow control valve status • In case of explosion immediately stop the system by pushing emergency button	• Evaluate the possibility to add indicators on operator monitor to help them in finding possible malfunctions • Evaluate the possibility to implement a safety stop under particular conditions • Evaluate the possibility to divide working area in different compartments

Table 4 HAZOP table for deviation: low pressure

Item n. 1.3—Deviation: low pressure			
Causes	Effects	Counteractions	Suggestions
• Leakage from the pipeline • Crushed or bad-mounted injector nozzle • ON/OFF valve not completely open • Malfunction in pressure sensor • Malfunction in flow control valve • Errors in PLC data communications	• Possible fire triggering with explosion or damage to other plants • Malfunction in the burner with consequent missed melting of the scrap • An error in pressure measure may cause the control system to turn off the burner • During supersonic injection phase of the oxygen the burner may cause some melted metal to be projected and/or have a poor oxidation of metal	• Check nozzle status during planned stops • Check annually pressure level of the pipelines • Increase flexible pipe reliability • Change nozzle • Annually check flow control valve status • In case of explosion immediately stop the system by pushing emergency button	• Evaluate the possibility to divide working area in different compartments • Evaluate if it possible to change the position of cooling pipe in a more safe position • Evaluate the possibility to add indicators on operator monitor to help them in finding possible malfunctions • Evaluate the possibility to implement a safety stop under particular conditions

Table 5 HAZOP table for deviation: low flow

Item n. 1.4—Deviation: low flow			
Causes	Effects	Counteractions	Suggestions
Malfunction in flow control valve ON/OFF valve not completely open Errors in PLC data communications The pipeline is obstructed by slag Obstructed or bad-mounted injector nozzle Malfunction in non-return-valve	Malfunction in the burner with consequent missed melting of the scrap Possible fire triggering with explosion or damage to other plants	Check nozzle status during planned stops Check annually pressure level of the pipelines Increase flexible pipe reliability Change nozzle Annually check flow control valve status In case of explosion immediately stop the system by pushing emergency button	Evaluate the possibility to add indicators on operator monitor to help them in finding possible malfunctions Evaluate the possibility to implement a safety stop under particular conditions Evaluate the possibility to divide working area in different compartments

Table 6 HAZOP table for deviation: leakage

Item n. 1.5—Deviation: leakage			
Causes	Effects	Counteractions	Suggestions
• Direct exposition to electric arc irradiation • High pressure (see deviation n. 1.1) • Injector not-well–maintained • Bad mounting of injector • Impact with an external item	• Possible fire triggering with explosion or damage to other plants • Malfunction in the burner with consequent missed melting of the scrap	• Check annually pressure level of the pipelines • Increase flexible pipe reliability • In case of explosion immediately stop the system by pushing emergency button	• Evaluate the possibility to add indicators on operator monitor to help them in finding possible malfunctions • Evaluate the possibility to implement a safety stop under particular conditions • Evaluate the possibility to divide working area in different compartments

rich data to be displayed to the operator. In fact, visual information sourced from HAZOP tables is displayed by the VI Burnersys, allowing a better understanding of the behaviour of the monitored equipment (i.e. the oxygen injection system of the EAF).

5 Key Variable Identification

In order to explain the logic behind the identification of the key variables, it is worth properly defining the logic function of a CM system within the entire measurement chain. To this end, MIMOSA reference is introduced [36]. Machinery Information Management Open Standards Alliance (MIMOSA) is a standard body operating in manufacturing. It developed the OSA-CBM specification, a standard architecture for managing information in a condition-based maintenance system. In order to unify an approach to condition based maintenance (CBM), a series of functional blocks are defined (see also ISO-13374). The blocks are: Data Acquisition, Data Manipulation, State Detection, Health Assessment, Prognostic Assessment, Advisory Generation.

CBM, in the schema proposed by OSA-CBM, is an automated (or semi-automated) set of activities. Thus, this has properly fitted with the need of preparing the algorithms to deploy Burnersys tool. Starting from the Data Acquisition, tools as sensors are present at this level and then Data Manipulation is entrusted to proper algorithms implemented on calculators that follow the sensors in the measurement chain. This analysis pertain to the first level of the plant automation architecture. They will be addressed in Sect. 7. Herein this section, instead, it is worth highlighting how variables to be gathered have been selected, according to the tasks required by the State Detection function.

The data to be analysed must be chosen among the relevant variables to be monitored as highlighted by means of HAZOP analysis. Then, the key focus is on the tasks of State Detection. The principle to be followed in this step of analysis is to identify a variable whose deviation is directly related with a change of the health state of the asset. In this first phase, it is not yet important to select variables whose behaviour describe a health state, but only the variables whose change indicate something is happening to the asset. The variables considered by the HAZOP analysis (summarized in the Table 1) are the ones identified as key variables; the State Detection function monitors their values and compares them to references thresholds to find possible deviations. Table 7 summarizes the possible deviations and defines the rules to detect deviations.

Table 7 shows the identified key variables, they are the pressure and flow (which is aligned to the outcome of HAZOP analysis). Analysing their behaviour according to the proposed rules it is possible to detect an abnormal working condition. Therefore, the State Detection function can be built following the comparison proposed by the rules, those compares two values: one is the measure of the variables from the field, and the second one is the reference value, or in other words, the expected value of the variable.

Table 7 State Detection rules based on variable identified through HAZOP analysis

Variable	Rules	Deviation
Pressure	Measured pressure > Reference pressure	Deviation item n. 1.1
	Measured pressure < Reference pressure	Deviation item n. 1.3
	Measured pressure = Reference pressure	No deviation detected
Flow	Measured flow > Reference flow	Deviation item n. 1.2
	Measured flow < Reference flow	Deviation item n. 1.4
	Measured flow = Reference flow	No deviation detected
Pressure & Flow	Measured flow < Reference flow and Measured pressure < Reference pressure	Deviation item n. 1.5

By hypothesis, each pipeline is considered "as good as new", thus the pressure drops are assumed constants and known (traditionally modelled with Bernoulli's law): a single, unique pressure value corresponds thus to each flow rate value. This conceptually allows to think about a possible reference value, used for comparison with deviations. Once the reference value is identified, then, the deviations identified by HAZOP are considered in order to model different states of the equipment.

According to the fluid dynamics of the pipelines, in case of leakage, equivalent to a decrease of pressure drop, one can expect to measure low values of pressure that due to the control system generates a low flow rate. Conversely, in case of a pipe obstruction, an increase in pressure drop is expected. This allows the detection of two alternative phenomena, thus the state of the system:

- flow rate = set-point (reference value);
- flow rate < set-point (reference value).

The latter case, however infrequent, already shows a mismatch between the set point value and the effective flow rate, so it is not required to analyse the pressure drop to detect an abnormal behaviour.

Then, considering the need to analyse the pressure values to properly monitor the state, it has been investigated what is the best method that allows to correlate the two variables (i.e. flow and pressure) in a precise manner, in order to quickly identify and analyse a deviation, considering:

- the availability of certain parameters already monitored for production process control;
- the impossibility to measure some variables in some, not accessible points of the asset.

6 State Detection and Health Assessment Algorithm

The state detection and health assessment algorithm has a specific logic for the condition monitoring system: it is thought to be a source of information for generating appropriate alerting messages when a change in the health state is diagnosed.

The algorithm is based on the knowledge provided by HAZOP and outputs information related with the risks associated with the deviation of each single variables. In particular, the algorithm analyses the deviation of each single variable (i.e. deviation towards LOW value or HIGH value) to realize the State Detection function; then, since the algorithm has been built according to a path of analysis based on HAZOP, for each deviation, it indicates the possible causes, thus realizing the Health Assessment function; more precisely, the Health Assessment is triggered by the discovered deviation and leads, according to the deviation, and guided by the knowledge available from Tables 2, 3, 4, 5, and 6, to identify the possible causes. Eventually, the algorithm introduces a means also to analyse the combination of deviations of the two variables (i.e. flow and pressure).

6.1 Review of Possible Approaches

Norm ISO 13379 interestingly provides definition of methods for health assessment. The norm classified the methods in two categories. The former refers to knowledge-based methods which rely on the use of fault models, behaviour models, physical models or case based reasoning. The latter refers to numerical methods/techniques (such as neural networks, pattern recognition, statistical, or other numerical approaches). Jardine Andrew [21] details the classification identifying: (i) Physical model based approaches (i.e. knowledge-based methods), (ii) Artificial Intelligence (AI) approaches, and (iii) Statistical approaches (i.e. numerical methods). Indeed, following [21] classification some methods were considered as candidates for being the basis for state detection and health assessment algorithm.

Herein, the model based approach is specifically analysed, as the Physical approach with the Bernoulli's law is the main background for the case. For what concern AI approaches, a short review of possible solutions to be applied to the case is also provided. Last but not least, for statistical approaches, control charts and regression models are considered as they are adopted for the State Detection function.

As well known, flow rate and pressure are related by the Bernoulli's law (1), in this case valid for compressible fluids. This method aims at developing a physical model and it can have some advantages if compared to statistical or AI methods.

Bernoulli's law (1) approach has been already used by Memoli et al. [37] in EAF related research to determine the outlet flow of oxygen once known the inlet pressure.

$$\frac{\gamma}{\gamma - 1}\frac{p}{\rho} + \frac{v^2}{2} = constant \tag{1}$$

Nevertheless, this approach goes under some assumptions: short conduct with small curvature and divergence, perfect oxygen gas, negligible stresses due to viscosity and negligible energy loss and heat exchange (adiabatic flow). However, such assumptions may be difficult to respect considering our industrial case, subject to possible changes in the geometry of the pipeline and to harsh operations, not always respecting ideal conditions.

This approach was thus considered too costly and difficult for implementation in the industrial case: this method is very sensitive to the variations in the geometry of the pipelines, frequent in the furnace burners in TenarisDalmine, which would finally impact on the values identified as reference. This would require model updates in order to fit to the changes made on the pipeline, resulting in a not affordable extra-work.

The second method considers the use of systems of AI techniques like Artificial Neural Networks (ANNs). To this end [38–42], show examples of application of artificial intelligence for condition monitoring, fault detection and predictive maintenance. The main advantage offered by ANN is their flexibility and adaptability to different decision making problems. However, the application of ANN has some drawbacks. The training of the network might need a substantial amount of time and the size of the training data has to be very large [43]. This poses questions regarding ANN feasibility when data are related to rare events, to sparse observations, or to expensive experiments. In addition, ANNs somehow hide the logic behind the calculation on data and, considering also the difficulty to implement ANNs through proper software libraries in TenarsDalmine ICT system, this solution was excluded. Other approaches, such as radial basis functions networks (RBFNs) [44], support vector machines (SVMs) [45] and Gaussian Processes [46] are suitable alternatives to ANNs, for non-linear modelling, function approximation and regression problems. However, all of them entail limitations similar to ANNs, and for this reason they were not considered for our case study.

The third method involves the construction of control charts for each pipeline, considering the flow-pressure characteristics curves that, if implemented with a link to the control architecture, would also allow the control of the oxygen injection system.

The control chart can be built both on pressure and on flow stored data. They are based on the principle that the measured samples should lay around a reference value called centre line; other main elements of the control chart are upper and lower control limits, that indicate the threshold at which the process is considered to have an "unexpected" behaviour. The definition of such thresholds and the rules to be applied to trigger advised for detected deviations may follow different approaches: for instance, EWMA [47] or CUSUM [48] control charts.

The adoption of control charts seems a not-expensive solution, but it was considered of low flexibility, and sensitive to changes, with need for updates of the parameters, especially considering the centre line reference value, when some even small changes in the system happens. Changes would also need training period to set-up the control limits of the charts, and this may impact on the flexibility to rapidly adapt the algorithm to the changes of the pipeline.

Another option within the statistical approaches, effectively chosen for the VI Burnersys deployment, is the method of least squares or polynomial regression. Often a variable y can be expressed as a polynomial of a second variable x as in formula (2)

$$y_i = A + Bx_i + Cx_i^2 \tag{2}$$

In our case, y and x are, respectively, pressure and flow rate. Once known a series of values (x_i, y_i), with $i = 1, \ldots N$, for every x_i the value y_i can be obtained by the formula (2), once estimated values of A, B, and C. After tests on plant data, the method has been verified to be applicable to the case of Burnersys. Besides good results obtained applying this approach, another positive issue related with this is the flexibility introduced by this method. It relies on the easiness of possible future re-configuration and new setting of the tool if changes in the system happen. Indeed, the method of regression found good applicability to our case also considering the easy use and the possibility to check its goodness of modelling thanks to a clear estimator as R^2. Moreover, the possibility to easy integrate the mathematics behind the method within the company architecture were also positively evaluated.

Overall, an overview on our case, through the Data acquisition, Data manipulation, State Detection and Health Assessment functions, has been provided. In the next subsection further details on the State Detection and Health Assessment algorithm will be presented. Instead, for what concern MIMOSA Prognostic Assessment, no algorithms are considered within Burnersys to carry out such task, that would deal with the evaluation of the Remaining Useful Life (RUL) (see Jardine Andrew [21] for further details). On the other hand, the Advisory Generation plays a key role to provide the information to the maintenance planner. The information can regard the priority of the maintenance task under analysis, highlighting the possible impacts of continuing operating the equipment in the actual health state. The Advisory Generation phase will be addressed by showing in the remainder how the higher level of the ICT architecture is tailored to accomplish with this task.

6.2 Development of the Algorithm for Burnersys

Values of variables (i.e. flow and pressure) were collected (flow in Nm^3/h and pressure in bar) through a data acquisition campaign of field data, resulting in the coefficients of formula (2): $A = 672 \times 10^{-9}$, $B = 451 \times 10^{-5}$, and $C = -0.392$; and a coefficient of determination [formula (3)] $R^2 = 0.999375$.

$$R^2 = 1 - \frac{\sigma_{err}^2}{\sigma_{tot}^2} \tag{3}$$

σ_{err}^2 is the sample variance of the estimated residuals and σ_{tot}^2 is the sample variance of the dependent variable.

Considering that values of R^2 close to 1 mean that the regressors predict well the value of the dependent variable in the sample, the obtained result confirms that the method is effectively able to model the scenario reflected for the asset. Hence, the model shows a good quality in terms of fitted data, further confirming the choice of this method as a good one.

Further on, this method revealed to be inexpensive and very flexible, and it is deemed to be easily adaptable to any kind of future modification to the oxygen supply pipelines. Since the industrial case imposed relevant requirements on flexibility, such a characteristic of the method was deemed essential for further implementation.

The algorithm for Burnersys is then developed levering on the regression model and the knowledge gathered from the HAZOP. In particular, the so built regression model allows to identify variables to be used as reference values for revealing the state of the asset. Starting from the reference values of the variables, representing the process in good working condition, it is possible to develop an algorithm in order to identify deviations and then to handle the alarms to the operator or to activate safety mechanisms of the furnace. The algorithm is based on an iterative approach. During each iteration it collects flow rate and pressure data from the field and by using the formula expressed in (2) it computes the measured flow rate to obtain the expected pressure of oxygen injector. Keeping into account an appropriate tolerance threshold, the measured pressure is compared with the expected pressure and the measured flow rate is compared with the set point flow rate.

The tolerance threshold is defined according to the measuring errors that occur at sensor level and that are typical for the sensors used in the Tenaris case. Moreover, the tolerance thresholds also consider the variance of the reference value of the variables in normal operating condition.

The results of the comparisons between the key variables and the thresholds are the input of the next steps of the algorithm.

A flow chart has been built to represent the algorithm and its role as troubleshooting scheme (see Fig. 4).

If an alarm is detected, the Advisory Generation task is triggered: it provides an output video alarm on the operator screen with a check list of operations to perform in order to solve the problem. The alarm is displayed according to the flow chart of Fig. 4 that links to the specific items provided by HAZOP tables. The operator terminal shows a list of possible actions to do in order to solve the cause of the malfunctioning (see Sect. 4), listed from the most likely to happen till the less one. Hereafter an example of such a list is presented:

- Stop the furnace and check nozzle status;
- Check the pressure level of the pipelines by other portable systems (error in the pressure measurement system is possible);
- Stop the furnace and change nozzle;
- Check the camera to see possible problems occurring in the furnace (risk of imminent explosion is possible).

Fig. 4 Flow chart of the algorithm

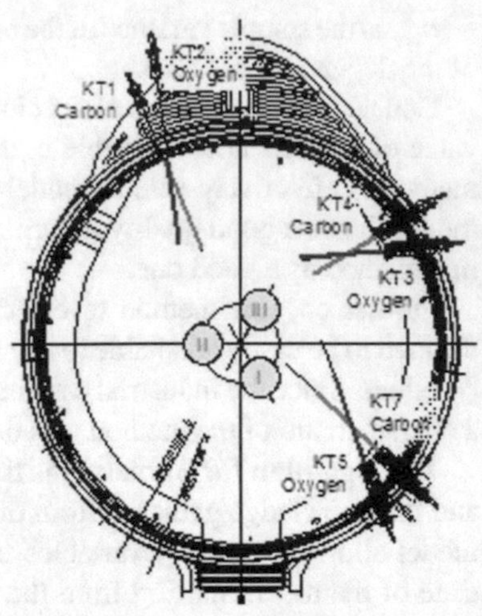

The possible lists to be displayed by the VI tool has been created thanks to the HAZOP analysis (see Table 1, column "Counteractions") that hence concurred to create a complete condition monitoring system for advising the maintenance operator with the proper actions to do. In addition, this allows the identification of highly impacting risks from the HAZOP tables that may result in an automatic disconnection of supply of oxygen under certain conditions with impact in terms of safety of EAF operation.

7 Implementation of the Algorithm Within the Existing ICT Architecture

In order to understand the ICT architecture of TenarisDalmine systems, represented in Fig. 5, the reader can refer to standard IEC 62264:2003 [49] and the hierarchical levels therein defined.

The oxygen injection system works according to the set point values of the flow rate, set by a process expert through an user interface (Fig. 5, level 2 of the architecture), then used by the field system (i.e. Chemical control PLC) to ensure that the flow rate follows the set-point value independently from other factors (Fig. 5, level 1).

The oxygen injection system is equipped with the necessary sensors (level 0) in order to ensure the proper functioning of the developed algorithms.

Each pipeline is equipped with both flow meter and pressure sensor. Then the state detection and health assessment algorithm, encompassing definition and analysis of the deviations from good working condition, is implemented on the control levels above the sensors (i.e. Levels 1 and 2).

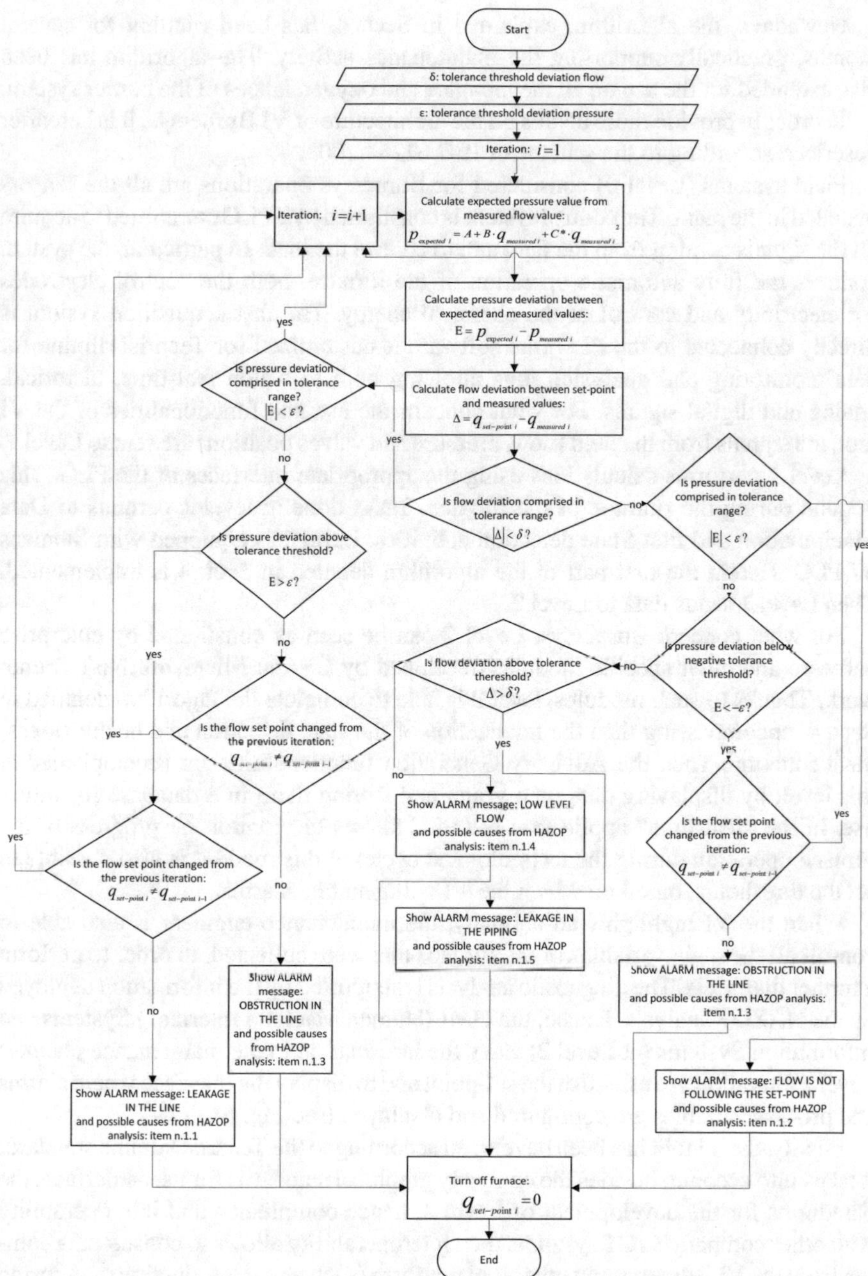

Fig. 5 ICT architecture supporting the virtual inspector Burnersys

Nowadays, the algorithm, explained in Sect. 4, has been running for several months, practically supporting the maintenance activity. The algorithm has been also extended for the use on all the methane and oxygen lances of the burner system.

In order to provide more details on the architecture of VI Burnersys, it is hereafter described according to the scheme of IEC 62264:2003.

Field systems (Level 0) considered for Burnersys operations are all the sensors installed in the plant. The control system is constituted by a PLC committed to acquire all the signals coming from the field and to control the load. In particular, the system controls the fully automatic operation of the furnace, both the control electrodes for electricity and control of the chemical energy. The fast acquisition system is directly connected to the PLC (the software is customized for TenarisDalmine for data monitoring and analysis), thus allowing both to watch real-time, historical, analog and digital signals. For what concern the use and functionalities of the VI tool, the signals from the field (flow, pressure and valves position) are sent to Level 1.

Level 1 performs calculations using the appropriate interfaces of the PLC. This is done during the runtime of the furnace. Tasks done at level 1 pertains to Data Manipulation and first State detection activities. Level 1 is equipped with Siemens S7 PLC. Herein the first part of the algorithm detailed in Sect. 4 is implemented. Then Level 1 sends data to Level 2.

For what concern Burnersys, Level 2 can be seen as constituted by enterprise software and other specific modules developed by C# and Microsoft .NET framework. Thanks to such modules, level 2 is able to complete the algorithm detailed in Sect. 4, encompassing then the finalization of the state detection and health assessment function. Then the Advisory Generation functionalities are accomplished at this level, by displaying data on a graph and storing them in a database for future use. In the "real-time" application the PLC allows to monitor the progress of the furnace operation during the technological cycle; in this mode it is also possible to set the thresholds, based on which the VI will generate alarms.

When the VI highlights an anomaly, the maintenance engineer is also able to consult all the main variables of the furnace that were collected, in order to perform a further diagnosis. The diagnostic analysis is supported by the information deployed by the HAZOP analysis. Hence, the HMI (Human Machine Interface) Systems and Information Systems (at Level 2) carry the information to the maintenance planner: Level 2 thus allows to establish the set-point and to display the interface where alarms and process variables are configured and displayed (see Fig. 6).

Finally, the VI tool has been developed according to the TenarisDalmine standard: it takes into account, besides the company graphical standards for user-interface, the paradigms for the development of Level 2, hence compliance and interoperability with other company's ICT systems; then interoperability allows to consult data coming from the VI Burnersys on other tool platform (such as e.g. production monitoring system) connected with Level 2. This integration with company's ICT systems represents one of the main constraint that had to be addressed to guarantee the effectiveness of the tool and its direct use in the company, besides the proper functionalities of the implemented algorithms. This represents a further positive aspect related with the E-maintenance concept: the developed condition monitoring system is capable

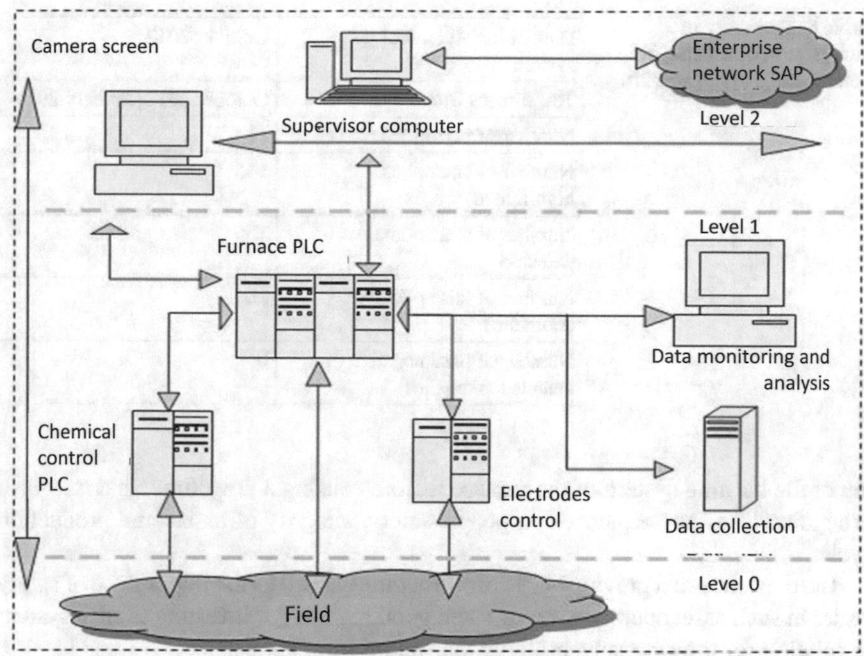

Fig. 6 Interface of advisory generation: example of obstruction ("Otturazione") identified in the KT3 methane lance—in Italian as implemented in the company

of a full integration with the company ICT systems enabling its use in the day-by-day operations (according to [8]'s vision of E-maintenance). Moreover, it is worth remarking that the developed system can be considered as a process monitoring tool, integrating some features typical of a SCADA (Supervisory Control and Data Acquisition) enlarged to health monitoring. Since it is respecting IEC 62264:2003 and company's standard, such a specific SCADA is seamlessly interoperating with the company's ICT architecture.

8 Conclusions

Production process in steel making industry involves many variables and operators cope with the tasks of monitoring, control and diagnose the health state of the assets. Often, this results in high difficulties to effectively analyze the current assets' behavior, to detect and diagnose process anomalies. This finally impacts on the capability to quickly take appropriate control actions.

The experience gained by the TenarisDalmine maintenance staff gave the foundations for initiating a novel approach, introducing the herein presented Condition Monitoring system. The system enables to detect incipient failures through monitor-

Table 8 Summary of Burnersys performance from October 2012 to January 2014

Date of installation of the system	October 2012
Test time of Burnersys	October 2012–January 2014
Number of performed casts	11,650
Number of anomalies highlighted	145
Number of real anomalies detected	135
Number of false positive anomalies	10
Number of false negative (not detected anomalies)	0

ing of the burning system of the furnace, before causing a downtime, that may also bring disastrous consequences on people safety, integrity of assets and production cycle.

The existence of a previous ICT infrastructure, as well as the importance of safety issues in such asset operation, represented good reasons for investing in such system as solution for the company, bringing innovation for maintenance.

Nowadays, the system is running and the following Table 8 summarizes the main information about its adoption in a period of several months.

Positive advantages have been achieved, allowing detecting possible anomalies. False positive anomalies are around 7%. This represents for the company a good result; room for improvement is possible, considering that the system has been recently installed. Moreover, because safety is under concern, false positive advices do not represent a problem; indeed, what represents a good result for the company is that the system has not shown any false negative advice. This is an important improvement, meaning that during the period of use the system allowed to prevent all the possible risks derived from problem on the burner system.

The research herein presented pursued two main novelties, considering both industrial and scientific point of views.

The first is the practical implementation of an E-maintenance tool within a real industrial environment. The presented system can be considered as a proof of practical results of implementation within an industrial environment of the research results envisioned by E-Maintenance; indeed, the dissemination in scientific literature is still lagging behind the richness of interesting works on E-maintenance concept, and the present work aimed at contributing to the research discussion in this concern.

The second novelty is the formalization of CM system (i.e. CM system design) based on operational risk, as a result of the application of a risk-based methodology, grounding on HAZOP, for the design and deployment of the CM system as development of an initial idea postulated by Medina Oliva et al. [35] and Leger et al. [50]. The presented chapter aimed at explaining how to use HAZOP to design CM systems, by explaining the methodology adopted starting from a traditional HAZOP analysis; indeed, the proposed approach is flexible enough to use as a starting point any variant of HAZOP approach available in literature.

Further developments can focus on the enhancement of the stepwise methodology proposed herein, taking into account the damage propagation model as shown in Saxena et al. [51] and possible modeling of propagation of variable deviations along the monitored asset as shown in Bregon et al. [52]. This novel use of HAZOP for CM system deployments seems promising also for a generalized approach to be used in other industrial sectors.

Acknowledgements The work has been developed within the scope of a project work of MeGMI, Master Executive in Gestione della Manutenzione Industriale—Executive Master on Industrial Maintenance Management, delivered by MIP—School of Management—Politecnico di Milano and SdM—School of Management Università degli Studi di Bergamo (www.mip.polimi.it/megmi).

References

1. Zhao, J., Cui, L., Zhao, L., Qiu, T., & Chen, B. (2009). Learning HAZOP expert system by case-based reasoning and ontology. *Computers & Chemical Engineering, 33,* 371–378.
2. Barreto, S. M., Swerdlow, A. J., Smith, P. G., & Higgins, C. D. (1997). A nested case-control study of fatal work relate injuries among Brazilian steel workers. *Occupational and Environmental Medicine, 54,* 599–604.
3. Deloux, E., Castanier, B., & Bérenguer, C. (2008). Maintenance policy for a deteriorating system evolving in a stressful environment. *Proceedings of the Institution of Mechanical Engineers, Part O: Journal of Risk and Reliability, 222,* 613–622.
4. Huynh, K. T., Barros, A., & Bérenguer, C. (2012). Adaptive condition-based maintenance decision framework for deteriorating systems operating under variable environment and uncertain condition monitoring. *Proceedings of the Institution of Mechanical Engineers, Part O: Journal of Risk and Reliability, 226*(6), 602–623.
5. Djurdjanovic, D., Lee, J., & Ni, J. (2003). Watchdog agent an infotronics-based prognostics approach for product performance degradation assessment and prediction. *Advanced Engineering Informatics, 17,* 109–125.
6. Aberdeen Group. (2006). The Asset Management Benchmark Report: Moving Toward Zero Breakdown, http://www.aberdeen.com.
7. Tsang, A. H. C. (2002). Strategic dimensions of maintenance management. *Journal of Quality in Maintenance Engineering, 8*(1), 7–39.
8. Lee, J., Ni, J., Djurdjanovic, D., Qiu, H., & Liao, H. (2006). Intelligent prognostics tools and e-maintenance. Special issue on e-maintenance. *Computers in Industry, 57*(6), 476–489.
9. Muller, A., Crespo Marquez, A., & Iung, B. (2008). On the concept of e-maintenance review and current research. *Reliability Engineering and System Safety, 93*(8), 1165–1187.
10. Iung, B., Levrat, E., Crespo Marquez, A., & Erbe, H. (2009). Conceptual framework for e-maintenance: Illustration by e-maintenance technologies and platforms. *Annual Reviews in Control, 33*(2), 220–229.
11. Crespo-Marquez, A., & Iung, B. (2008). A review of e-maintenance capabilities and challenges. *Journal on Systemics, Cybernetics and Informatics, 6,* 162–166.
12. Abichou, B., Voisin, A., & Iung, B. (2012). Choquet integral parameters inference for health indicators fusion within multi-levels industrial systems: Application to components in series. *IFAC Proceedings Volumes, 45*(31), 193–198.
13. Lorton, A., Fouladirad, M., & Grall, A. (2013). Computation of remaining useful life on a physic-based model and impact of a prognosis on the maintenance process. *Proceedings of the Institution of Mechanical Engineers, Part O: Journal of Risk and Reliability, 227,* 434–449.

14. Espíndola, D., Fumagalli, L., Garetti, M., Botelho, S., & Pereira, C. (2011). Adaption of OSA-CBM Architecture for Human-Computer Interaction through Mixed Interface, INDIN 11 conference, Caparica, Lisbon, Portugal, 26–29 July 2011, 485–490.
15. Espíndola, D., Fumagalli, L., Garetti, M., Pereira, C. E., Botelho, S., & Ventura Henriques, R. (2013). A model-based approach for data integration to improve maintenance management by mixed reality. *Computers in Industry, 64*(4), 376–391.
16. Mascolo, J., Nilsson, P., Iung, B., Levrat, E., Voisin, A., Krommenacker, N., et al. (2010). Industrial demonstrations of e-maintenance solutions. In *E-maintenance* (pp. 391–474). London: Springer.
17. López-Campos, M. A., Márquez, A. C., & Fernández, J. F. G. (2013, June). Modelling using UML and BPMN the integration of open reliability, maintenance and condition monitoring management systems: An application in an electric transformer system. *Computers in Industry, 64*(5), 524–542.
18. Macchi, M., Barberá Martínez, L., Crespo Márquez, A., Holgado Granados, M., & Fumagalli, L. (2012). Value assessment of an e-maintenance platform. *IFAC Proceedings Volumes, 45*(31), 145–150.
19. Macchi, M., Crespo Márquez, A., Holgado, M., Fumagalli, L., & Barberá Martínez, L. (2014). Value driven engineering of e-maintenance platforms. Special issue on Advanced Maintenance Engineering, Services and Technology. *Journal of Manufacturing Technology Management, 25*(4), 569–598.
20. Colace, C., Fumagalli, L., Pala, S., Macchi, M., Matarazzo, N. R., & Rondi, M. (2013). An intelligent maintenance system to improve safety of operations of an electric furnace in the steel making industry. *Chemical Engineering, 33,* 397–402.
21. Jardine Andrew, K. S., Daming, Lin, & Dragan, Banjevic. (2006). A review on machinery diagnostics and prognostics implementing condition-based maintenance. *Mechanical Systems and Signal Processing, 20*(7), 1483–1510.
22. Rondi, M., & Memoli, F. (2003). Increase of productivity in Dalmine steel plant through the application of innovative electrical and chemical technologies. *La Metallurgia Italiana, 95*(10), 53–60.
23. Lèger, A., Weber, P., Levrat, E., Duval, C., Farret, R., & Iung, B. (2009). Methodological developments for probabilistic risk analyses of socio-technical systems. *Proceedings of the Institution of Mechanical Engineers, Part O: Journal of Risk and Reliability, 223*, 313–332.
24. Villemeur, A. (1992). *Reliability, availability, maintainability and safety assessment. Volume 1: Methods and techniques*. New York: Wiley.
25. Flaus, J. M. (2008). A model-based approach for systematic risk analysis. *Proceedings of the Institution of Mechanical Engineers, Part O: Journal of Risk and Reliability, 222,* 79–93.
26. Center for Chemical Process Safety. (2008). Guidelines for Hazard evaluation procedures, 3rd ed., Wiley, ISBN: 978-0471978152.
27. Hyatt, N. (2003). *Guidelines for process hazards analysis (PHA, HAZOP), hazards identification, and risk analysis*. CRC Press, ISBN: 978-0849319099.
28. Ferro, L., Giugliano, P., Galbiati, P., Memoli, F., Giavani, C., & Maiolo, J. (2007). The electric arc furnace of Tenaris Dalmine: From the application of the new technologies of digital electrode regulation and multipoint injection to the dynamic control of the process. *16th IAS Steelmaking Conference, 6–8 November, Rosario, Argentina*, pp. 59–72.
29. Swann, C. D., & Preston, M. L. (1995). Twenty-five years of HAZOP. *Journal of Loss Prevention in Process Industries, 8*(6), 349–353.
30. New South Wales Government Department of Planning. (2008). HAZOP Guidelines, Hazardous Industry Planning Advisory Paper No 8.
31. Dunjó, J., Fthenakis, V., Vílchez, J. A., & Arnaldos, J. (2009). Hazard and operability (HAZOP) analysis. A literature review. *Journal of Hazardous Materials, 173*(1–3), 19–32.
32. Lawley, H. G. (1974). Operability studies and hazard analysis. *Chemical Engineering Progress, 70*(4), 45–56.
33. Weber, P., Medina-Oliva, G., Simon, C., & Iung, B. (2012). Overview on Bayesian networks applications for dependability, risk analysis and maintenance areas. *Engineering Applications of Artificial Intelligence, 25*(4), 671–682.

34. Labowsky, J., Svandova, Z., Markos, J., & Jelemensky, L. (2003). Model-based HAZOP study of a real MTBE plant. *Journal of Loss Prevention in the Process Industries, 20,* 230–237.
35. Medina Oliva, G., Iung, B., Barberá, L., Viveros, P., Ruin, Y. (2012). Root cause analysis to identify physical causes. *Proceedings of PSAM 2011 & ESREL, Helsinki, Finland, 25–29 June*.
36. Machinery Information Management Open Systems Alliance (MIMOSA). (2008). OSA-CBM UML Specification 3.2.1 Release November 2008, viewed 30 September 2009. http://www.mimosa.org/.
37. Memoli, F., Mapelli, C., Ravanelli, P., & Corbella, M. (2004). Simulation of oxygen penetration and decarburisation in EAF using supersonic injection system. *ISIJ International, 44*(8), 1342–1349.
38. Fumagalli, L., Ierace, S., Dovere, E., Macchi, M., Cavalieri, S., & Garetti, M. (2011). Agile diagnostic tool based on electrical signature analysis. *IFAC Proceedings Volumes, 44*(1), 14067–14072.
39. Uraikul, V., Chan, C. W., & Tontiwachwuthikul, P. (2007). Artificial intelligence for monitoring and supervisory control of process systems. *Engineering Applications of Artificial Intelligence, 20,* 115–131.
40. Korbicz, J. (2006). Robust fault detection using analytical and soft computing methods. *Bulletin of the Polish Academy of Sciences, Technical Sciences, 54*(1), 75–88.
41. Ierace, S., Marinaro, P., Tatavitto, P., & Troiano, L. (2010). Profiling the power usage of industrial machinery by ANN. *SoCPaR, 2010,* 413–418.
42. Ierace, S., Pinto, R., Troiano, L., & Cavalieri, S. (2010). Neural network as an efficient prognostics tool: A case study in a textile company. *IFAC Proceedings Volumes, 43*(3), 122–127.
43. Hagan, M. T., Demuth, H. B., & Beale, M. H. (1996). *Neural network design*. Boston, MA: PWS Publishing.
44. Lowe, D., & Broomhead, D. (1988). Multivariable functional interpolation and adaptive networks. *Complex systems, 2,* 321–355.
45. Cortes, C., & Vapnik, V. (1995). Support-vector networks. *Machine Learning, 3*(20), 273–297.
46. Rasmussen, C. E. (2006). *Gaussian processes for machine learning*. MIT Press. ISBN 026218253X.
47. Crowder, S. V., & Hamilton, M. D. (1992). An EWMA for monitoring a process standard deviation. *Journal of Quality Technology, 24,* 12–21.
48. Gan, F. F. (2011). An optimal design of CUSUM quality control charts. *Journal of Quality Technology, 23,* 279–286.
49. IEC 62264:2003 Enterprise-control system integration.
50. Leger, J. B., Iung, B., Ferro De Beca, A., & Pinoteau, J. (1999). An innovative approach for new distributed maintenance system: Application to hydro power plants of the REMAFEX project. *Computers in Industry, 38,* 131–148.
51. Saxena, A., Goebel, K., Simon, D., & Eklund, N. (2008). Damage propagation modeling for aircraft engine run-to-failure simulation. *1st International Conference on Prognostics and Health Management (PHM08), October 6–9, Denver, CO*, pp. 1–9.
52. Bregon, A., Daigle, M., Roychoudhury, I., Biswas, G., Koutsoukos, X., & Pulido, B. (2014). An event-based distributed diagnosis framework using structural model decomposition. *Artificial Intelligence, 210,* 1–35.

Chapter 12
A Dynamic Opportunistic Maintenance Model to Maximize Energy-Based Availability While Reducing the Life Cycle Cost of Wind Farms

Asier Erguido Ruiz, Adolfo Crespo Márquez, Eduardo Castellano and Juan F. Gómez Fernández

Abstract Operations and maintenance costs of the wind power generation systems can be reduced through the implementation of opportunistic maintenance policies at suitable indenture and maintenance levels. These maintenance policies take advantage of the economic dependence among the wind turbines and their systems, performing preventive maintenance tasks in running systems when some other maintenance tasks have to be undertaken in the wind farm. The existing opportunistic maintenance models for the wind energy sector follow a static decision making process, regardless of the operational and environmental context. At the same time, on some occasions policies do not refer to practical indenture and maintenance levels. In this chapter, a maintenance policy based on variable reliability thresholds is presented. This dynamic nature of the reliability thresholds, which vary according to the weather conditions, provides flexibility to the decision making process. Within the presented model, multi-level maintenance, capacity constraints and multiple failure modes per system have been considered. A comparative study, based on real operation, maintenance and weather data, demonstrates that the dynamic opportunistic maintenance policy significantly outperforms traditional corrective and static opportunistic maintenance strategies, both in terms of the overall wind farm energy production and the Life Cycle Cost.

A. Erguido Ruiz
Ikerlan Technology Research Centre, Operations and Maintenance Technologies Area, Pº J. M. Arizmendiarrieta, 2, Gipuzkoa, Spain
e-mail: aerguido@ikerlan.es

A. Erguido Ruiz · A. Crespo Márquez (✉) · J. F. Gómez Fernández
Department of Industrial Management, School of Engineering, University of Seville, Camino de los Descubrimientos s/n, 41092 Seville, Spain
e-mail: adolfo@us.es

J. F. Gómez Fernández
e-mail: juan.gomez@iies.es

E. Castellano
MIK Research Centre, Mondragon University, 20500 Gipuzkoa, Spain
e-mail: ecastellano@mondragon.edu

A. Crespo Márquez et al. (eds.), *Value Based and Intelligent Asset Management*,
https://doi.org/10.1007/978-3-030-20704-5_12

Keywords Opportunistic maintenance model · Dynamic reliability thresholds · Life cycle cost · Wind energy · Weather conditions

1 Introduction

The growing importance of renewable energy in terms of installed capacity and technological advances has been remarkable during the last years. This growth has been particularly notorious within the wind energy sector, which occupies a leading position among renewable energies [1]. Furthermore, the sector has not only suffered a great development for the last two decades but it is expected to continue its expansion during the following years, being firmly reinforced by the main World Powers energy plans [2].

Along with this progress new challenges have arisen, especially in terms of new technologies' reliability [3] and logistics associated to wind farms' (WF) maintenance [4] (see nomenclature in Table 1). Moreover, due to the trend of WFs' location shift towards offshore sites [5, 6], to deal with these challenges is getting even more difficult. As a result, operations and maintenance costs can rise to a 32% or a 12–30% of the total life cycle cost (LCC) in offshore or in onshore WFs respectively [5, 7].

Within this context, asset management acquires high relevance in the sector since it is crucial to search optimal maintenance strategies that allow to improve wind turbines' (WT) reliability and to reduce maintenance cost while raising availability [8, 9]. In spite of its importance, asset management strategies are not optimised in practice nowadays, being corrective maintenance (CM) and time-based minor preventive maintenance (PM) such as routing checks for minimizing degradation effects [10], the most applied maintenance strategies [11].

In addition to these two maintenance policies, Condition Based Maintenance (CBM), enhanced by the several Condition Monitoring Systems (CMS) available for wind energy installations [12, 13], is the third maintenance method currently applied to the wind power systems [11]. In fact, based on their ability to prevent failures [14], CBM strategies have been proved to be cost effective [15, 16] and have been widely researched [17, 18].

Nevertheless, maintenance strategies based uniquely on WTs' health condition monitoring do not take into account that the WTs are multi-component systems composed by a number of subsystems, with dependencies among them that directly affect to the adequacy of the maintenance strategies [19, 20]. According to Nicolai and Dekker [21] these dependencies can be classified as: economic, when performing maintenance activities in different systems simultaneously have different economic consequences than implementing them individually [22]; structural, when a maintenance activity in a system implies performing further actions in other systems [23]; and stochastic, when the risk of failure of two different systems is not independent [24].

Table 1 Nomenclature and acronyms definition

Nomenclature			
LCC	Life cycle cost	w_{ik}	Weight that determines reactivity of SRT_{ikj} and DRT_{ik} to wind speed
WF	Wind farm	K	Number of FM considered for each system
WT	Wind turbine	J	Levels of PM types considered for each FM
FM	Failure mode	GRP	Generalized renewal process
O&M	Operations and maintenance	VA_{hikt}	Virtual age associated to FM *k* in system *i* in WT *h* in period *t*
CM	Corrective maintenance	α_{ik}	Weibull scale parameter of FM *k* of system *i*
PM	Preventive maintenance	β_{ik}	Weibull shape parameter of FM *k* of system *i*
CBM	Condition based maintenance	q_{ikj}^{pr}	Restoration factor of *j* PM level on system *i* for FM *k*
CMS	Condition monitoring system	q_{ik}^{c}	Restoration factor of CM on system *i* for FM *k*
TTF	Time to failure	NT	Number of MTs
MT	Maintenance team	C	Capacity of each MT (in hours)
v_t	Average wind speed in period t	c^{na}	Cost of no availability or opportunity cost
v^i	Cut in wind speed	c^p	Penalty cost due to unplanned maintenance
v^o	Cut out wind speed	c_{ik}^{c}	Cost of tools and materials needed for performing CM of FM *k* in system *i*
v^r	Wind speed at which rated power is obtained	c_{ikj}^{pr}	Cost of tools and materials needed for performing PM *j* of FM k in system *i*
GP_t	Generated power in period *t*	c^{team}	Cost of MT
RP	Rated power of the WT	c^{et}	Extra time cost
$R_{ik}(VA)$	Reliability of system *i* and FM k at virtual age VA	NT^{max}	Maximum number of MTs
SRT_{ikj}	Fixed reliability threshold for applying perfect or imperfect PM *j* on system *i* and FM k	NT^c	Number of MTs working on CM
SRT_{ikjt}	System reliability threshold in period *t* for applying perfect or imperfect PM *j* system *i* and FM k	c^{disp}	Cost of maintenance dispatch

(continued)

Table 1 (continued)

Nomenclature			
DRT_{ik}	Fixed dispatch reliability threshold	m_{ik}^{c}	Maintainability of CM for FM *k* in system *i*
DRT_{ikt}	Dispatch reliability threshold in period *t*	m_{ik}^{pr}	Maintainability of PM for FM *k* in system *i*
V	Wind speed threshold for determining reliability thresholds variation	OT	Total operating time
P	Periods of time considered for wind speed forecasting	D^{t}	Required WF time-based availability
		D^{o}	Required WF energy-based availability
		T	Maximum iteration periods

When any of the mentioned dependencies exist among subsystems, the optimal maintenance strategies are not those that consider the subsystems separately in the maintenance decision process [25]:

> Obviously, the optimal maintenance action for a given subsystem at any time point depends on the states of all subsystems in the system: the failure of one subsystem results in the possible opportunity to undertake maintenance on other subsystems (opportunistic maintenance).

In such circumstances, both group maintenance policies and opportunistic maintenance policies are the most suitable maintenance policies, and thus, the most studied ones [19]. On the one hand, group maintenance strategies establish different groups of systems that will undergo maintenance activities attending to the number of failures suffered by the systems, their age or their operation time [26, 27]. However, group maintenance policy is especially cost effective when disassemble and reassembly costs are high [25], which is not particularly the case in the wind energy sector. On the other hand, opportunistic maintenance policy takes advantage of short term circumstances, performing maintenance in non-failed systems when a failure has already happened in another one; making the maintenance decision according to different thresholds regarding systems' age, reliability or health condition. Several models have demonstrated the suitability of opportunistic maintenance policy in diverse sectors following varied strategies, such as age limits [28, 29], combined failure distribution of the systems [30] or accumulated operated periods of the systems [19].

Although opportunistic maintenance policies have not been traditionally implemented in the wind power systems [31], more recently some authors have studied and demonstrated their suitability in the sector [11, 22, 31–36]; mainly due to the positive economic dependence among WTs [33]. Furthermore, they allow handling some of the main conflicting objectives concerning the decision making process of the wind energy sector: the maximization of revenue, power and reliability and the minimization of operations and maintenance costs [37].

1.1 Previous Research

According to the reviewed opportunistic maintenance models for the wind energy sector (see Table 2), Besnard et al. [32] demonstrate that it is possible to reduce the maintenance cost and the opportunity cost due to failures by taking advantage of both low wind speed periods and the dispatches for CM in order to perform some prearranged PM.

Tian et al. [31] focus their research on developing an opportunistic maintenance policy based on the condition monitoring data. Aided by this data, the authors identify the useful remaining life of the systems and they calculate the WT's reliability. So, if the reliability of the WT does not surpass a determined threshold, systems within the WT are replaced.

Ding and Tian [20] propose an opportunistic maintenance model where both imperfect and perfect maintenance levels are considered. This is, systems will not always return to a status as good as new after a repair (the reader is addressed to [38] for further information). In this research, the authors set two different age thresholds for each system, which are based on their Mean Time To Failure (MTTF), in order to make the maintenance decision. The same authors extended their research in Ding and Tian [11] considering different age thresholds for systems belonging to failed and running WTs.

The main focus of the research performed by Sarker and Faiz [35] is to find an opportunistic maintenance strategy that optimises maintenance cost following a multilevel preventive maintenance policy. With this purpose the authors establish several age thresholds for the systems, which determine the PM activities to be performed.

Atashgar and Abdollahzadeh [34] go a step further as they find multi-objective optimal maintenance strategies for minimizing both maintenance cost and loss of production in WFs with redundant WTs. Within this research the WTs are grouped into blocks and opportunistic repairs are performed to WTs of the same block, according to the reliability thresholds associated to perfect or imperfect maintenance.

In Abdollahzadeh et al. [33], reliability thresholds that determine optimal maintenance activities are set for each component. Both in [33, 34], maintenance teams (MT) can be preventively dispatched to the WF, instead of having to wait for a failure to happen.

Finally, Zhu et al. [36] study three different maintenance strategies for an offshore WF: periodic routine, reactive maintenance and opportunistic maintenance. In this research each system can only have a failure mode (FM), which can be either hard or soft according to the consequences. Moreover, in this research the impact of the CMS accuracy on opportunistic maintenance is also analysed.

Table 2 Comparative analysis of the reviewed opportunistic maintenance models for the wind energy sector

(a)

Reference	*Main Focus*	Restoration Effect: Effect of maintenance activities ①②③④ (1. Not Considered 2. According to the system 3. According to the maintenance task 4. According to the system and maintenance task)	Restoration Effect: Method ①② (1. Not Considered 2. Virtual Age model)	Opportunistic maintenance policy: Nature ①② (1. Static 2. Dynamic)	Opportunistic maintenance policy: Maintenance levels ①②③④ (1. Not Considered 2. One level 3. Two Level (Perfect/Imperfect repairs) 4. Several levels)	Opportunistic maintenance policy: Failed/Running turbines ①②③ (1. Only for failed WTs 2. Equal for Failed/Running (F/R) WTs 3. Different for F/R WTs)
Besnard et al. [32]	Weather Conditions	●○○○	●○	●○	●○○○	○●○
Tian et al. [31]	CMS	●○○○	●○	●○	○●○○	○●○
Ding and Tian [20]	Two-level PM	○○●○	○●	●○	○○●○	○●○
Ding and Tian [11]	Two-level PM in F/R WTs	○○●○	○●	●○	○○●○	○○●
Abdollahzadeh et al. [33]	Bi-objective model	○○○●	○●	●○	○○●○	○○●
Atashgar and Abdollahzadeh [34]	Redundancies	○○●○	○●	●○	○○●○	○○●
Sarker and Faiz [35]	Multi-level PM	○○●○	○●	●○	○○○●	○●○
Zhu et al. [36]	Imperfect prediction signal	●○○○	●○	●○	○●○○	●○○

(b)

Reference	Optimisation process: Objective Function ①② (1. Cost 2. Cost & Prod)	Maintenance task consideration: Severity of Failures ①② (1. Not Considered 2. Considered)	Maintenance task consideration: Maintenance Time ①② (1. Not Considered 2. Considered)	Characteristics of WF, WTs and Systems: WF Location ①②③ (1. Onshore 2. Offshore 3. Not specified)	Characteristics of WF, WTs and Systems: Models of WTs considered ①② (1. One 2. Several)	Characteristics of WF, WTs and Systems: Number of systems ①② (1. None (WT as a whole) 2. Several)
Besnard et al. [32]	○●	●○	○●	○●○	●○	●○
Tian et al. [31]	●○	●○	○●	○○●	●○	○●
Ding and Tian [20]	●○	●○	●○	○○●	●○	○●
Ding and Tian [11]	●○	●○	●○	○○●	●○	○●
Abdollahzadeh et al. [33]	○●	●○	○●	○○●	○●	○●
Atashgar and Abdollahzadeh [34]	○●	●○	○●	○○●	○●	○●
Sarker and Faiz [35]	●○	●○	●○	○●○	●○	○●
Zhu et al. [36]	●○	○●	○●	○●○	●○	○●

1.2 Proposed Approach

It is remarkable that the reviewed researches present static opportunistic maintenance policies that base the decision making process on fixed age, reliability or health thresholds. However, WTs operate under non-stationary conditions that highly condition the repair jobs [39–41]. So, ideally, the maintenance models should be able to adequately fit the decision making process to the conditions under which the WTs are operating at each time.

In order to deal with this challenge, a dynamic opportunistic maintenance model is presented in this chapter. This model adjusts the decision making process according to the weather conditions, pursuing a double objective: (1) the optimisation of the total LCC due to maintenance and (2) the improvement of the WF energy-based availability.

With this purpose, the decision making process is determined by variable reliability thresholds, that change according to the wind speed conditions; fostering the performance of maintenance activities during low wind speed periods and hindering them during high wind speed periods. This dynamic nature, in addition to lead to a better performance of the maintenance strategies both in terms of energy-based availability and LCC, will also allow handling some of the most conflicting factors that appear in the wind power industry [37]:

1. Maximization of revenue and power while maximization of reliability [37].
2. MTs' safety while maintenance performance [10].
3. Minimization of the opportunity cost [42–44].
4. Improvement of reliability within high velocity wind periods [41].

In order to ensure a realistic approach of the model, several constraints have been included in the study, such as capacity limitations due to the available MTs and systems' maintainability. Furthermore, several FMs are considered per system, bearing their different impact on cost and availability. In fact, although the different FMs within each system directly condition the WT's performance and hence, the resources deployed in the maintenance strategy [45], in the reviewed opportunistic maintenance models only a FM is considered per system.

Finally, in order to prove the suitability of the model for establishing the adequate maintenance strategy, an agent-based simulation has been developed, taking advantage of the ability of simulation techniques to handle the stochastic nature of the sector [33]. The simulation results have been obtained from real operational and reliability data about WFs located in the north of Spain, provided by a leading company in the sector. Likewise, in order to search as realistic scenarios as possible, the simulation has also been fed with real wind data, according to the WF location.

The chapter is organized as follows: in Sect. 2 the problem is defined and the presented dynamic opportunistic maintenance policy is explained and analytically derived; in Sect. 3 the simulation process is explained; in Sect. 4 computational results are shown and discussed for a specific case study, comparing both traditional and presented approaches; finally, in Sect. 5, concluding remarks are presented and future research lines are established.

2 Problem Definition and Model Description

2.1 Problem Definition

The WF consists of H WTs of similar characteristics that have N critical systems connected in series. Each system might fail in k different FMs, classified according to their severity ($k = 1, 2, \ldots, K$). Consequently, after a failure, corresponding k CM will be performed. Likewise, systems can also undergo different PM levels associated to the different FMs, prior to their occurrence ($j = 1, 2, \ldots, J$).

Generally, the repairs return the systems to an operational condition worse than the new one but better than just before the maintenance task is performed. This concept leads to a classification of maintenance as perfect or imperfect, according to the ability of each maintenance activity to restore the system [38]. Accordingly, in this work both perfect and imperfect repairs have been considered, being $j = J$ a perfect repair and $j = 1$ the most imperfect repair.

Among the several methods that treat the restoration effect of maintenance (see [38]), in this work the Generalized Renewal Process (GRP) proposed by Yañez et al. [46] is used. This method has been specifically utilised in the presented model due to its flexibility for modelling both the behaviour of the systems before failures and the quality of repairs during the different life stages of the systems. To do so, GRP considers a q_{ik} rejuvenation parameter [0, 1] associated to the efficiency of the restoration effect of the maintenance activity j on the system i ($q = 0$ for the most imperfect maintenance and $q = 1$ for perfect maintenance). Consequently, after a maintenance activity j, Eq. 1 is followed to update the virtual age of the system i (the reader is addressed to [46] for further information). Then, in order to identify systems' reliability after an imperfect repair, failure probability distribution conditioned to the survival of the new virtual age is calculated in Eq. 2 (adopted from [46]):

$$VA_i^{new} = VA_i^{old}(1 - q_{ij}) \tag{1}$$

$$F\left(t \middle| VA_i^{new}\right) = P\left[T_{ij} \le t \middle| T_{ij} > VA_i^{new}\right] = \frac{F(t) - F\left(VA_i^{new}\right)}{1 - F\left(VA_i^{new}\right)} \tag{2}$$

In the wind energy sector, Weibull and Exponential distributions have been respectively utilised for modelling the reliability of mechanical and electrical systems [47]. Since Exponential distribution is also contemplated by Weibull distribution as a particular case, Eq. 2 is particularized ad hoc for Weibull distribution in Eq. 3, according to the scale (α_{ik}) and shape parameters (β_{ik}) that define the Weibull distribution for each FM k of system i (see [33]):

$$R\left(t \middle| VA_{hik}^{new}\right) = 1 - F\left(t \middle| VA_{hik}^{new}\right) = \exp\left[\left(\frac{VA_{hik}^{new}}{\alpha_{ik}}\right)^{\beta_{ik}} - \left(\frac{t}{\alpha_{ik}}\right)^{\beta_{ik}}\right] \tag{3}$$

Finally, every failure occurrence or PM involves some fixed and variable maintenance costs that have to be considered in the model. Every maintenance activity implies dispatching a MT to the WF, which involves a relevant cost (c^{disp}). Likewise, each maintenance task will require a material cost $\left(c_{ik}^{c},\ c_{ik}^{pr}\right)$ and a time to repair $\left(m_{ik}^{c},\ m_{ik}^{pr}\right)$. However, since in the wind energy sector there are some failures caused by sensors' false alarms, each system has been provided with a FM $k = 1$ that has an impact in terms of availability, but not in terms of material, human resources nor dispatch cost. The time to repair is the source of the opportunity and penalty costs (c^{na}, c^{p}) and the human resources' need, which can be either internal or external. According to internal resources a number of MTs are hired (NT^{max}) and are considered to be constant through the whole analysis. The MTs have a certain annual cost (c^{team}) and capacity (C). Finally, whereas when no own resources are available to perform required CM activities extra time (ET) is externally hired at an extra cost (c^{et}), PM will only be performed with own resources.

2.2 *Dynamic Opportunistic Maintenance Policy*

Two different maintenance decisions are considered within the presented dynamic opportunistic maintenance policy according to respective dynamic reliability thresholds: the MT dispatch to the WF, based on DRT_{ik}, and the PM decision, based on SRT_{ikj}. Accordingly, on the one hand, if the reliability of any FM (R_{ik} (VA)) does not reach the DRT_{ik}, the decision of preventively dispatching a MT to the WF will be made, ensuring a minimum reliability for every system and FM. On the other hand, once the decision of dispatching a MT to the WF has been made—either preventively or correctively-, the imperfect PM decision is made according to SRT_{ikj} (see Figs. 1 and 2) .

As stated, whereas in the reviewed researches these thresholds are static, in the presented model they are dynamic. Both DRT_{ik} and SRT_{ikj} will vary with regards to wind speed conditions according to Eqs. 4–6; being increased in low wind speed periods and decreased in high speed periods. Thus, PM will be fostered during low wind periods and hindered during high wind speeds periods (see Fig. 1). As shown in Eqs. 4–6 the variation of the thresholds -and hence, the maintenance strategy- is determined by the following factors:

1. Wind speed threshold (V): during wind speed periods above this value, the reliability thresholds will decrease, hindering the PM activities; and on the contrary during low wind speed periods. Wind speed during the next p periods of time is forecasted and compared to V in order to determine if reliability thresholds should decrease or increase (Eq. 6).
2. Generated power (GP_t) and reactivity weight (w_{ik}): determine to which extent the reliability thresholds should be reactive to wind conditions. The gradient of both DRT_{ik} and SRT_{ikj} is proportional to the difference between the generated power at each time period (GP_t) and the rated power of WTs (RP).

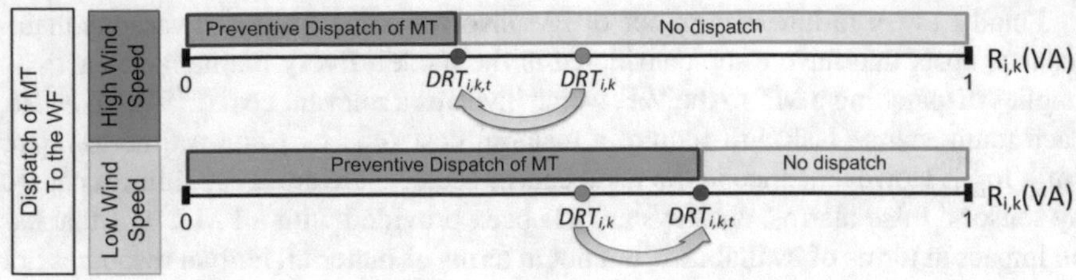

(a) Decision-making structure for the dispatch of the maintenance teams

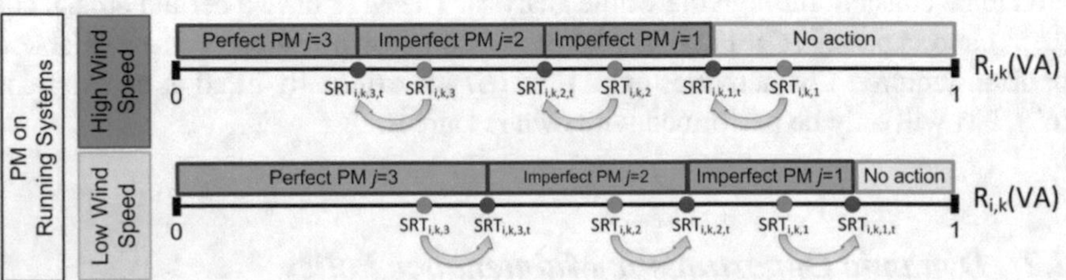

(b) Decision-making structure for performing PM. Example with 3 PM levels per FM

Fig. 1 Decision-making structure for the dynamic opportunistic maintenance model

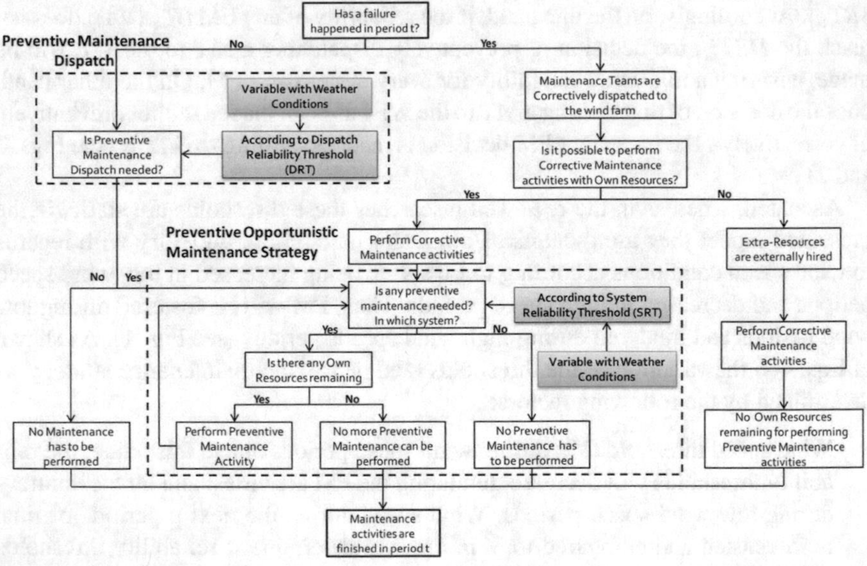

Fig. 2 Flowchart of the dynamic opportunistic maintenance model

$$SRT_{ikjt} = SRT_{ikj} + (2W_t - 1) \cdot SRT_{ikj} \cdot w_{ik} \cdot \left(\frac{RP}{GP_t + RP \cdot W_t} \right)^{(2W_t - 1)} \quad (4)$$

$$DRT_{ikt} = DRT_{ik} + (2W_t - 1) \cdot DRT_{ik} \cdot w_{ik} \cdot \left(\frac{RP}{GP_t + RP \cdot W_t} \right)^{(2W_t - 1)} \quad (5)$$

$$W_t = \begin{cases} 1 \; \sum_{l=t}^{t+p} \frac{v_l}{p} \le V \\ 0 \; \sum_{l=t}^{t+p} \frac{v_l}{p} > V \end{cases} \quad (6)$$

It is not the aim of the dynamic opportunistic maintenance to eliminate the PM activities, but to plan them when weather conditions are more advantageous. Since WTs should be stopped during maintenance, downtime periods (time-based availability) will be similar in both dynamic and static maintenance strategies. However, energy losses will be reduced through dynamic opportunistic maintenance, maximizing WF's production and energy-based availability and minimizing opportunity cost derived from the unavailability periods. In other words, the WTs will be available during the most profitable periods. Likewise, as maintenance activities will be prone to be planned within low wind speed periods, safety of maintenance teams will also be improved [10].

Furthermore, on some occasions WFs must be stopped due to different reasons, such as for substation maintenance. Since no power can be generated during these periods, maintenance managers usually take the opportunity of performing PM. In order to consider such situations within the model, and help the manager on the maintenance decision to be made during these periods as well, they should be considered as no-wind periods $\left(\sum_{l=t}^{t+P} \frac{v_l}{p} = 0 \right)$, since no power can be generated. Consequently, according to the established maintenance policy, the thresholds will be increased during these periods, fostering PM.

2.3 Mathematical Model

In this section the mathematical formulation of the dynamic opportunistic maintenance model is developed according to the maintenance process shown in Fig. 2 (intermediate binary variables are specified in Table 3). To this aim, the standard approach of discretising the time in order to formulate stochastic programming models ($T = \{0, 1, 2, \ldots, T\}$) has been considered. Without loss of generality, some assumptions have been made for its formulation:

1. Degradation processes of the systems are considered independent from each other and they are associated to the operation time (ageing systems).
2. As commonly done in ageing systems, Increasing Failure Rate (IFR) is considered.

3. Installed WTs are the same model, composed by similar systems. Therefore, reliability distribution of the FMs of the systems is irrespective of the WT that contains them.
4. Reliability of the FMs follows the Weibull distribution, with scale parameter α and shape parameter β.
5. Maintenance activities should be finished during the period of time in which they are started.
6. A maintenance dispatch is considered per period of time, where several MTs can be dispatched.
7. The wake effect affection to WTs' production has been considered to be minimised during the WF layout design optimisation [48] and thus neglected.
8. PM is assumed to be less resource-consuming than CM since
    (a) extra damages in other systems because of failures reduction;
    (b) stock management can be planned in advance;
    (c) resources can be allocated to maintenance tasks in a balanced way.
9. WF maintenance managers make decisions in discrete time and frequently [39].

The principal objective of the model is to minimize the LCC due to O&M while providing the maximum WF energy-based availability. Accordingly, the optimal reliability thresholds that define the dynamic opportunistic maintenance policy (SRT_{ikjt}, DRT_{ikt}), i.e. the decision variables of the model, should be found.

To this aim, the objective function (see Eq. 7) of the model bears the main O&M costs to be faced in a WF, related to the failure occurrence (z_{hikt}) and the PM decision (y_{hikjt}): (1) own and externally hired human resources needs, both in terms of own maintenance teams (NT, c^{team}) and extra time needed (ET, c^{et}); (2) dispatching of MTs to the WF (c^{disp}); (3) material and tools requirements $\left(c^{c}_{ik}, c^{pr}_{ikj}\right)$; and (4) production losses (c^{na}) and penalty costs (c^{p}) due to WTs unavailability during maintenance, directly proportional to maintenance tasks duration $\left(m^{c}_{ik}, m^{pr}_{ikj}\right)$ and the lost power during maintenance (GP_t). The cost of the imperfect maintenance has been defined as a function of the restoration factor $\left(q^{c}_{ik}, q^{pr}_{ikj}\right)$ (as in [11]), as well as the maintainability. Due to the long-term nature of the study the maintenance cost at each period t has to be updated to present value according to an interest rate (k_a):

$$\begin{aligned}\min LCC\left(DRT_{ikt}, SRT_{ikjt}\right) = \Bigg[& \sum_{t} ET_t \cdot c^{et} + \sum_{t} NT \cdot c^{team} + \sum_{t} (\gamma_t + \theta_t) \cdot c^{disp} \\ & + \sum_{h}\sum_{i}\sum_{k\neq 1}\sum_{t} z_{hikt} \cdot c^{c}_{ik} \cdot \left(q^{c}_{ik}\right)^2 \\ & + \sum_{h}\sum_{i}\sum_{k\neq 1}\sum_{t} z_{hikt} \cdot m^{c}_{ik} \cdot \left(q^{c}_{ik}\right)^2 \cdot GP_t \cdot \left(c^{na} + c^{p}\right) \\ & + \sum_{h}\sum_{i}\sum_{k}\sum_{j}\sum_{t} y_{hikjt} \cdot \left(q^{pr}_{ikj}\right)^2\end{aligned}$$

Table 3 Intermediate binary variables utilised in the model

$Z_{hikt} = \begin{cases} 1 \text{ if } CM\ k \text{ is performed in system } i \text{ of } WT\ h \text{ in period } t \\ 0 \text{ otherwise} \end{cases}$	$\theta_t = \begin{cases} 1 \text{ if a } MT\ k \text{ is correctively dispatched to } WF \text{ in period } t \\ 0 \text{ otherwise} \end{cases}$
$y_{hikjt} = \begin{cases} 1 \text{ if } PM\ k \text{ is performed in } FM\ k \text{ of system } i \text{ of } WT\ h \text{ in period } t \\ 0 \text{ otherwise} \end{cases}$	$\sigma_{hikjt} = \begin{cases} 1 & \text{if a } PM\ j \text{ should be performed in } FM\ k \text{ of system } i \text{ of } WT\ h \text{ in period } t \\ 0 \text{ otherwise} \end{cases}$
$\theta_t = \begin{cases} 1 \text{ if a } MT\ k \text{ is correctively dispatched to } WF \text{ in period } t \\ 0 \text{ otherwise} \end{cases}$	$\varphi_t = \begin{cases} 1 \text{ if there are available resources for performing } PM \text{ in period } t \\ 0 \text{ otherwise} \end{cases}$

$$\cdot\left(m_{ikj}^{pr}\cdot GP_t\cdot\left(c^{na}+c^{p}\right)+c_{ikj}^{pr}\right)\Big]\cdot(1+k_a)^{-t} \quad (7)$$

The objective of maximizing WFs' energy-based availability-calculated attending to the lost power during maintenance tasks duration and total available power $\left(H\sum_t GP_t\right)$—is ensured through the establishment of a minimum energy-based availability requirement (Eq. 8). The time-based availability—calculated attending to the tasks duration and total operating time (OT)—is also calculated through Eq. 9.

$$\frac{H\cdot\sum_t GP_t-\left[\sum_h\sum_i\sum_k\sum_t m_{ik}^{c}\cdot z_{hikt}\cdot\left(q_{ik}^{c}\right)^2\cdot GP_t+\sum_h\sum_i\sum_k\sum_j\sum_t m_{ikj}^{pr}\cdot y_{hikjt}\cdot\left(q_{ikj}^{pr}\right)^2\cdot GP_t\right]}{H\cdot\sum_t GP_t}\geq D^o \quad (8)$$

$$\frac{OT-\left[\sum_h\sum_i\sum_k\sum_t m_{ik}^{c}\cdot z_{hikt}\cdot\left(q_{ik}^{c}\right)^2+\sum_h\sum_i\sum_k\sum_j\sum_t m_{ikj}^{pr}\cdot y_{hikjt}\cdot\left(q_{ikj}^{pr}\right)^2\right]}{OT} \quad (9)$$

The generation of power is modelled regarding the average wind speed at each period. Power is only generated in wind speeds between cut in (v^i) and cut out (v^o) wind speeds, increasing non linearly until the wind speed in which the rated power (RP) is reached v^r). The mathematical relationship has been defined as in Karki and Patel [49]:

$$GP_t=\begin{cases}0 & 0\leq v_t<v^i\\ RP\cdot\left(a+b\cdot v_t+c\cdot v_t^2\right) & v^i\leq v_t<v^r\\ RP & v^r\leq v_t<v^o\\ 0, & v^o\leq v_t\end{cases}\quad \forall t\epsilon T \quad (10)$$

where the parameters in Eq. 10 are obtained as follows [49]

$$a=\frac{1}{(v^i-v^r)^2}\left[v^i(v^i+v^r)-4v^iv^r\left(\frac{v^i+v^r}{2v^r}\right)^3\right] \quad (11)$$

$$b=\frac{1}{(v^i-v^r)^2}\left[4(v^i+v^r)\left(\frac{v^i+v^r}{2v^r}\right)^3-(3v^i+v^r)\right] \quad (12)$$

$$c=\frac{1}{(v^i-v^r)^2}\left[2-4\left(\frac{v^i+v^r}{2v^r}\right)^3\right] \quad (13)$$

As stated, the decision of preventively dispatching a MT to the WF relies on DRT_{ikt}, which varies according to the wind speed forecasting (see Eqs. 5, 6) between [0, 1] (Eq. 14). When the reliability of any FM of a system (R_{ik}) does not reach its required threshold (DRT_{ikt}), MTs are preventively dispatched to the WF (Eq. 15). Likewise, if a failure happens in t ($z_{hikt}=1$), the dispatch of the MT is compulsory (Eq. 16). Furthermore, dispatches of MTs are limited to one per period (Eq. 17), being able to send several teams in the same dispatch.

$$0\leq DRT_{ikt}\leq 1\quad \forall i\epsilon I,\forall k\epsilon K,\forall t\epsilon T \quad (14)$$

$$\gamma_t = \begin{cases} 1 \; R_{ik}(VA_{hikt}) \leq DRT_{ikt} \\ 0 \; otherwise \end{cases} \quad \forall h \epsilon H, \forall i \epsilon I, \forall k \epsilon K, \forall t \epsilon T \tag{15}$$

$$\theta_t = \begin{cases} 1 \sum_{\text{h}} \sum_{\text{i}} \sum_{\text{k} \neq 1} z_{hikt} \geq 1 \\ 0 \; otherwise \end{cases} \quad \forall t \epsilon T \tag{16}$$

$$\theta_t + \gamma_t \leq 1 \quad \forall t \epsilon T \tag{17}$$

The decision of performing PM activities is based on SRT_{ikjt} [0, 1] (see Eqs. 4, 6, 18). When the reliability of a system does not reach the required reliability threshold (Eq. 19), a system is susceptible of being preventively maintained; however, more conditions have to be met in order to perform PM (see Fig. 2): (1) only the most comprehensive maintenance is performed, i.e. if both imperfect and perfect maintenance are needed, perfect maintenance will be performed (Eq. 20); (2) a MT must have been previously dispatched to the WF (Eq. 21); (3) own human resources have to be available for performing PM (Eqs. 25, 26); and (4) only a maintenance activity is performed at each time on the WT (Eq. 22).

$$0 \leq SRT_{ikjt} \leq 1 \quad \forall i \epsilon I, \forall j \epsilon J, \forall k \epsilon K, \forall t \epsilon T \tag{18}$$

$$\sigma_{hikjt} = \begin{cases} 1 \; R_{ik}(VA_{hikt}) \leq SRT_{ikjt} \\ 0 \; otherwise \end{cases} \quad \forall h \epsilon H, \forall i \epsilon I, \forall k \epsilon K, \forall j \epsilon J, \forall t \epsilon T \tag{19}$$

$$y_{hikjt} \leq \sigma_{hikjt} - \sigma_{hik(j+1)t} \quad \forall h \epsilon H, \forall i \epsilon I, \forall k \epsilon K, \forall j \epsilon J, \forall t \epsilon T \tag{20}$$

$$y_{hikjt} \leq \theta_t + \gamma_t \quad \forall h \epsilon H, \forall i \epsilon I, \forall k \epsilon K, \forall j \epsilon J, \forall t \epsilon T \tag{21}$$

$$\sum_{\text{i}} \sum_{\text{k}} \sum_{\text{j}} y_{hikjt} + \sum_{\text{i}} \sum_{\text{k}} z_{hikt} \leq 1 \quad \forall h \epsilon H, \forall t \epsilon T \tag{22}$$

In fact, with regard to human resource limitations, some capacity constraints have been set (Eqs. 23–26). As shown in Fig. 2, each CM must be performed, and if no resources are available, extra time (*ET*) is externally hired at a higher cost (Eq. 23). Equation 24 ensures that no more than the maximum available MTs ($MT^{\max}$) are used for CM at each time $\left(NT_t^c \in \{1, 2, \ldots, NT^{\max}\}\right)$. Since PM can only be performed with own resources, Eq. 25 analyzes if there are still own resources available after CM; and if so, PM is performed if needed (Eq. 26).

$$\sum_{\text{h}} \sum_{\text{i}} \sum_{\text{k} \neq 1} z_{hikt} \cdot m_{ik}^c \cdot \left(q_{ik}^c\right)^2 \leq C \cdot NT_t^c + ET_t \quad \forall t \epsilon T \tag{23}$$

$$NT_t^c \leq NT^{\max} \quad \forall t \epsilon T \tag{24}$$

$$\varphi_t = \begin{cases} 1 \sum_{h} \sum_{i} \sum_{k \neq 1} z_{hikt} \cdot m_{ik}^{c} \cdot \left(q_{ik}^{c}\right)^2 \leq C \cdot NT^{\max} \\ 0 \sum_{h} \sum_{i} \sum_{k \neq 1} z_{hikt} \cdot m_{ik}^{c} \cdot \left(q_{ik}^{c}\right)^2 > C \cdot NT^{\max} \end{cases} \quad \forall t \epsilon T \tag{25}$$

$$\sum_{h} \sum_{i} \sum_{k} \sum_{j} y_{hikjt} \cdot m_{ikj}^{pr} \cdot \left(q_{ikj}^{pr}\right)^2 \leq C \cdot \left(NT^{\max} - NT_t^c\right) \cdot \varphi_t \ \forall t \epsilon T \tag{26}$$

When PM or CM are performed ($y_{hikjt} = 1, z_{ikht} = 1$), the system's virtual age associated to FM k (VA_{hikt}) is reduced according to the restoration factor $\left(q_{ikj}^{pr}, q_{ik}^{c}\right)$. If no action is performed in a system during a period, the virtual age should be increased by a period (Eq. 27).

$$\sum_{h} \sum_{i} \sum_{k} \sum_{j} y_{hikjt} \cdot m_{ikj}^{pr} \cdot \left(q_{ikj}^{pr}\right)^2 \leq C \cdot \left(NT^{\max} - NT_t^c\right) \cdot \varphi_t \quad \forall t \epsilon T \tag{26}$$

$$VA_{hikt} = \left(VA_{hik(t-1)} + 1\right) \cdot \left(1 - z_{hikt} \cdot q_{ik}^{c} - y_{hikjt} \cdot q_{ikj}^{pr}\right)$$
$$\forall h \epsilon H, \forall i \epsilon I, \forall k \epsilon K, \forall j \epsilon J, \forall t \epsilon T \tag{27}$$

3 Simulation Process

As shown in the analytical model, expected maintenance cost and production of the WF depend on the followed opportunistic maintenance policy, which is determined by the dynamic reliability thresholds. Therefore, in order to find profitable maintenance strategies, it is necessary to establish the correct set of thresholds (SRT_{ikj}, DRT_{ik}) and their variation according to the forecasted wind conditions (w_{ik}, V, p).

Due to the different stochastic processes that have to be considered within this complex system model, such as failure occurrence, repair processes, weather conditions, etc., it is hard to handle it analytically [11, 33]. Therefore, although most part of the problem has been analytically derived, simulation techniques have been used to handle the many random scenarios that can appear for each set of reliability thresholds, as commonly done in other researches about the topic [11, 31, 33–35]. Particularly, an agent-based simulation has been developed due to its suitability to handle engineering problems with multi-agent systems [50]. The simulation process developed follows 6 different steps (see Fig. 3).

Step 1. In the initialization of the simulation all the parameters needed for the simulation process are specified: the parameters needed for the dynamic reliability thresholds modelling, costs related to maintenance, reliability and maintainability distributions, number of MTs, maximum iteration period, etc.

Step 2. The simulation clock and virtual age of the FMs of the systems are updated, identifying their new reliability according to their age. Wind speed is also forecasted and reliability thresholds are accordingly updated.

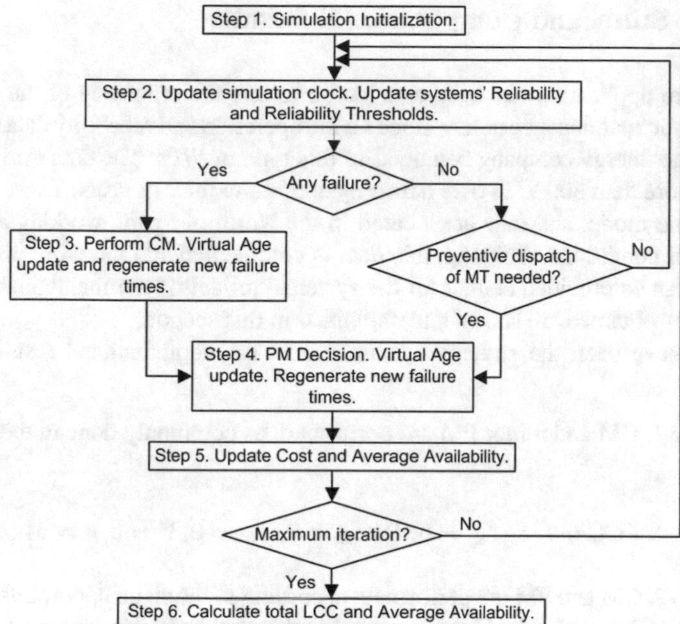

Fig. 3 Simulation process for LCC and energy-based availability evaluation

Step 3. If any failure has happened, needed CM is applied and the virtual age of the system is updated. After the CM, the new time to failure (*TTF*) is obtained through the Inverse Transform Technique [51], according to Eq. 28 (adopted from [33]), in which *R* is uniformly distributed between [0, 1). If no failure occurs, whether a MT has to be preventively dispatched or not is also decided in this step.

$$TTF_{hik} = \alpha_{ik}\left[\left(\frac{VA_{hik}}{\alpha_{ik}}\right)^{\beta_{ik}} - \ln(1-R)\right]^{\frac{1}{\beta_{ik}}} - VA_{hik} \quad (28)$$

Step 4. If a MT has been dispatched to the WF, PM decision is made according to the reliability thresholds. The virtual age and the new time to failure (TTF_{hik}) are updated for the maintained systems, according to Eq. 28.

Step 5. LCC and energy-based availability are updated. If actual period is equal to the maximum iteration period, step 6 is followed. Otherwise, steps 2, 3, 4 and 5 are repeated.

Step 6. The total expected LCC and the average energy-based availability are calculated for the established opportunistic maintenance policy, $LCC = f\,[SRT_{ikj}, DRT_{ik}, w_{ik}, V, p]$.

4 Case Study and Computational Results

An onshore application has been considered to test the efficiency of the dynamic opportunistic maintenance policy, since all the operation and reliability data provided by the wind energy company belonged to this type of WFs. The data available has been on more than 300 WTs over a time span of more than 12 years. These WTs are all the same-model and they are located in the North of Spain, working at similar operational conditions. Although this data is confidential and therefore no detailed numbers can be provided neither for the systems' reliability nor maintainability, the final results obtained are shown and explained in this section.

Three have been the strategies compared in the computational results shown below:

- *Strategy 1*. CM and minor PM are performed, as commonly done in the industry [11]

$$LCC = f\left[SRT_{ikj} = 0, DRT_{ik} = 0, w_{ik} = 0, V = 0, p = 0\right]$$

- *Strategy* 2. CM and PM are performed, according to the static opportunistic maintenance policy established by the reliability thresholds. In this strategy, thresholds will not vary according to the wind conditions.

$$LCC = f\left[SRT_{ikj}, DRT_{ik}, w_{ik} = 0, V = 0, p = 0\right]$$

- *Strategy* 3. CM and PM are performed, according to the presented dynamic opportunistic maintenance policy established by the reliability thresholds and their variation degree regarding the wind conditions.

$$LCC = f\left[SRT_{ikj}, DRT_{ik} w_{ik}, V, p\right]$$

Finally, a sensitivity analysis has been performed in order to discuss the most influential parameters within the model and to evaluate the different assumptions made. Despite the fact that the methodology has been applied to an Onshore WF, it could also be applied to an Offshore WF.

4.1 WF Profile

A recently installed virtual WF that consists of 40 WTs (H = 40) of a rated power of 1.67 MW is considered. For each WT the 4 most critical systems are considered (N = 4), regarding both their reliability and the consequences of their failures, according to the data available for the study. These 4 systems are: gearbox, blades, pitch system and yaw system. For each system three independent FMs are analyzed ($K = 3$). As stated in Subsection 2.1, the $k = 1$ FM does not have material requirements nor need of field-maintenance, since they are provoked by sensors' false alarms. The systems can also undergo two different PM levels ($J = 2$) associated to the FMs ($k = 2, 3$), with a restoration factor associated to the maintenance routine $\left(q_{ikl}^{pr} = 0.75 \text{ and } q_{ik2}^{pr} = 1\right)$ (see [11]).

The access cost to the WF is assumed to be 5000 €, own resources 800 €/day per maintenance team, 2 maintenance teams, extra resources 250 €/h per maintenance team, the total opportunity cost 105 €/MWh, the penalization cost 35 €/MWh, the interest rate 5% and the lead time to the WF one hour. Finally, the cost for the materials and the maintainability of PM has been set a 30% lower than for CM. Further information about material cost for the WT under study can be found on Martin-Tretton et al. [52].

Since wind conditions are a key factor within the methodology, real wind data has been utilised in order to feed the simulation and obtain as much realistic scenarios as possible. The wind turbines cut-in, cut-out and rated speeds are assumed to be 3, 25 and 13 m/s, respectively. Finally, daily wind average forecasting potential has been established to 5 periods ($p = 5$).

4.2 Sensitivity Analysis

In this section the different parameters that condition the dynamic reliability thresholds, and hence, the presented dynamic opportunistic maintenance model, are discussed through a sensitivity analysis, considering the following base scenario: $SRT_{ilj} = 0.0$, $SRT_{i21} = 0.8$, $SRT_{i22} = 0.4$, $SRT_{i31} = 0.8$, $SRT_{i32} = 0.4$, $DRT_{i2} = 0.2$, $\text{DRT}_{\text{i3}} = 0.2$, $w_{\text{ik}} = 0.5$, $V = 2.0$, $p = 5$.

Wind speed threshold (V). The opportunistic maintenance strategy, both in terms of LCC and loss of production, shows a better performance when the wind speed threshold is established at low values. If higher values are established, the reactivity of the reliability thresholds is higher and maintenance is prone to be over-sized, increasing LCC and production losses (Fig. 4).

Number of periods for which average wind speed can be forecasted (p). It is expected that if the average wind speed was forecasted for more periods of time, the performance of the maintenance strategy would show better results. However, the results show that the performance of the dynamic opportunistic maintenance model is quite regular from $p = 3$ on (Fig. 5).

Reactivity weight (w_{ik}). It helps determining the variation degree of the thresholds associated to each FM according to wind and it is expected that the optimal value will be different for each FM. For the defined base scenario, the best performance, both according to LCC and Production losses, is found at $w_{ik} = 0.5$ (Fig. 6).

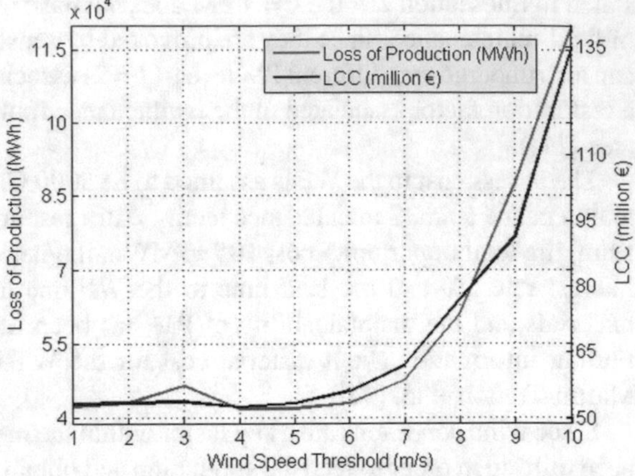

Fig. 4 Expected LCC and production losses with different wind speed thresholds

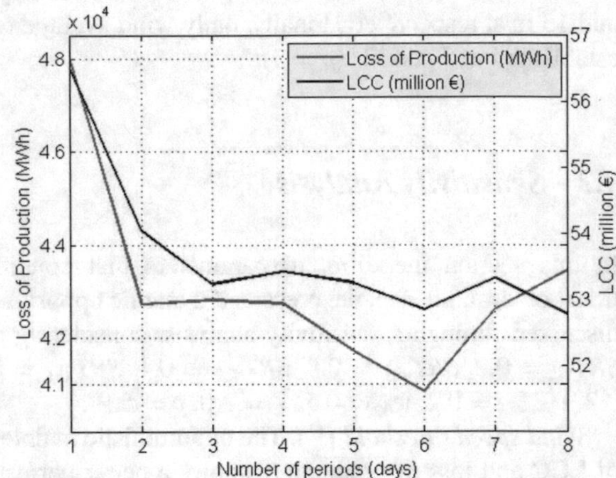

Fig. 5 Sensitivity analysis according to predicted periods

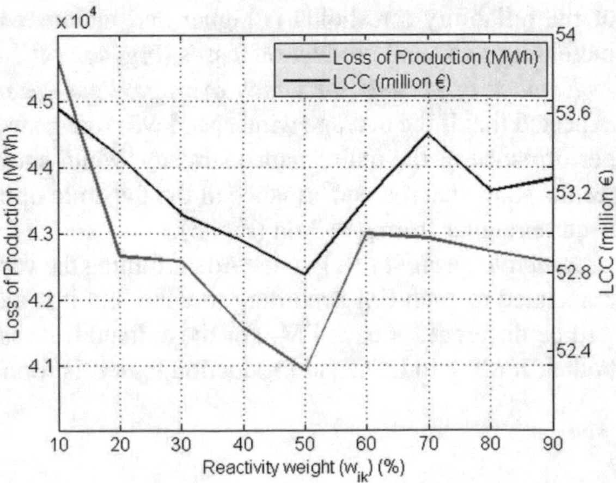

Fig. 6 Expected LCC and production losses with different reactivities to wind (w_{ik})

4.3 Optimisation Results and Discussion

The optimisation results presented in this section have been obtained through the OptQuest Engine, a commercial optimisation software developed by Fred Glover in OptTek Systems Inc. (Opttek Systems Inc. 2000) that has been proved to be robust and efficient on finding high-quality solutions [53]. The OptQuest Engine is based on scatter search and it also integrates successfully Tabu search, integer programming, and a procedure to configure and train neural networks for the optimisation of stochastic problems [54]. Especially, neural networks play an important role in order to avoid getting trapped in local minima, since they are able to remember good solutions and recombine them to guide the search towards the best solutions [55]. For further information about scatter search and the OptQuest Engine the reader is addressed to [53, 56, 57].

In order to compare the different strategies that are analyzed within the research, optimal values have been found for each one (see Table 4). The most remarkable results are:

LCC & Production Losses. Opportunistic maintenance strategies, both based on static and dynamic reliability thresholds, are proven to be economically effective compared to strategy 1. In fact, according to the obtained results, opportunistic maintenance policies can reduce LCC by a 25% (see Table 5 and Fig. 7). Likewise, optimisation results show that the use of the dynamic reliability thresholds (strategy 3) considerably outperforms the use of static reliability thresholds (strategy 2), minimizing the total production losses by almost a 27% and slightly improving the LCC as well. Furthermore, it is remarkable that the results in terms of production losses for strategy 3 are not achievable through strategy 2. In fact, if the static reliability thresholds of strategy 2 were optimised for minimizing production losses instead of being optimised for minimizing the LCC, production losses would still be a 24.7% lower for strategy 3 than 2, additionally increasing the LCC (see Table 5).

This result is mainly due to the fact that WTs must be stopped during PM. Thus, even if more PM implies a better reliability, it is not the key to reduce the wind energy production losses. Consequently, if production losses are to be minimized, PM should be planned during low wind energy periods. It is at this point where the presented dynamic opportunistic maintenance, which systematically takes advantage of the low wind energy periods for performing maintenance, maximizes the total production and outperforms strategy 2, where the PM is planned regardless of the operational context of the WTs.

Energy-based availability and Time-Based availability. If energy-based and time-based availability of strategies 2 and 3 are compared (Fig. 8), it is observed that the dynamic reliability thresholds improve the overall energy-based availability. Whereas following strategy 2 implies that maintenance activities have almost the same impact in both time-based and energy-based availability, through the dynamic reliability thresholds, the impact of maintenance activities on the energy-based availability is minimized, reaching an energy-based availability over 99.1%. This fact, along with the slight difference regarding LCC between strategies 2 and 3, reaffirms that whereas

Table 4 Optimised values for the Decision Variables

	Gearbox			Pitch			Yaw			Blades		
FM1	Dec. Var	Str. 2	Str. 3	Dec. Var	Str. 2	Str. 3	Dec. Var	Str. 2	Str. 3	Dec. Var	Str. 2	Str. 3
	SRT_{11j}	0.0	0.0	SRT_{21j}	0.0	0.0	SRT_{41j}	0.0	0.0	SRT_{41j}	0.0	0.0
	DRT_{11}	0.0	0.0	DRT_{21}	0.0	0.0	DRT_{31}	0.0	0.0	DRT_{31}	0.0	0.0
	w_{11}	0.0	0.0	W_{21}	0.0	0.0	W_{31}	0.0	0.0	W_{31}	0.0	0.0
	v	0.0	0.5	v	0.0	0.5	v	0.0	0.5	v	0.0	0.5
FM2	SRT_{121}	0.725	0.775	SRT_{221}	0.9	0.925	SRT_{321}	0.825	0.875	SRT_{421}	0.775	0.825
	SRT_{122}	0.175	0.675	SRT_{222}	0.175	0.225	SRT_{322}	0.0	0.125	SRT_{422}	0.675	0.6
	DRT_{12}	0.0	0.1	DRT_{22}	0.175	0.15	DRT_{32}	0.0	0.125	DRT_{42}	0.675	0.6
	w12	0.0	0.55	w22	0.0	0.2	w_{32}	0.0	1.0	w_{42}	0.0	0.6
	v	0.0	0.5	v	0.0	0.5	v	0.0	0.5	v	0.0	0.5
FM3	SRT_{131}	0.725	0.75	SRT_{232}	0.925	0.975	SRT_{331}	0.9	0.90	SRT_{431}	0.8	0.875
	SRT_{132}	0.3	0.325	SRT_{232}	0.575	0.525	SRT_{332}	0.6	0.55	SRT_{432}	0.45	0.575
	DRT_{13}	0.175	0.3	DRT_{23}	0.175	0.225	DRT_{33}	0.475	0.5	DRT_{43}	0.0	0.125
	w_{13}	0.0	0.35	w_{23}	0.0	0.15	w_{33}	0.0	0.7	w_{43}	0.0	0.45
	v	0.0	0.5	v	0.0	0.5	v	0.0	0.5	v	0.0	0.5

Table 5 Main optimisation results for each strategy

Strategy		LCC (€)	Production loss (MWh)
1		67,544,964	66,895
2	min LCC	51,856,606	50,239
	min Pr. loss	52,982,284	48,790
3		50,904,497	36,743

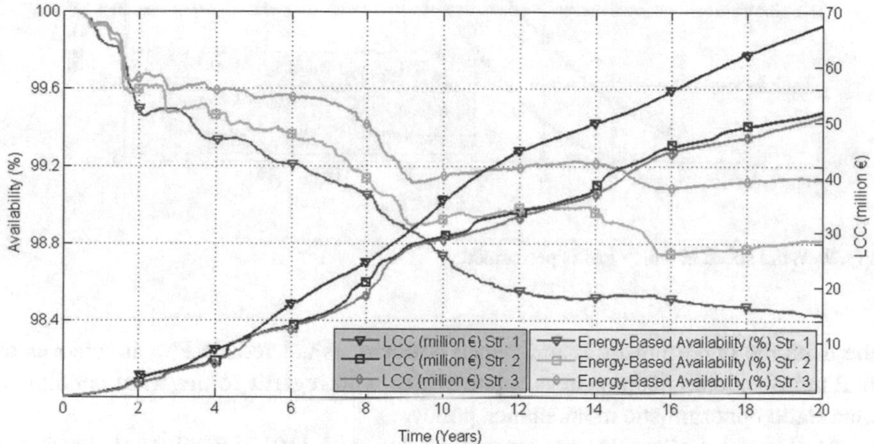

Fig. 7 Strategies' performance comparison according to energy-based availability and LCC

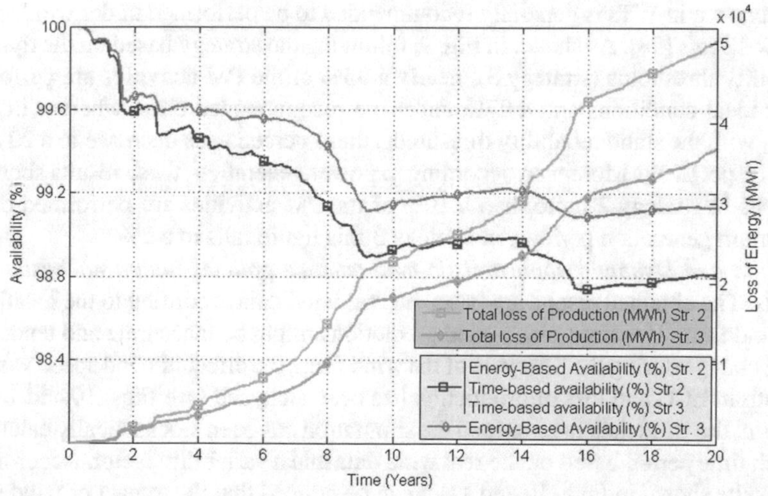

Fig. 8 Strategies' performance comparison according to energy-based and time-based availability

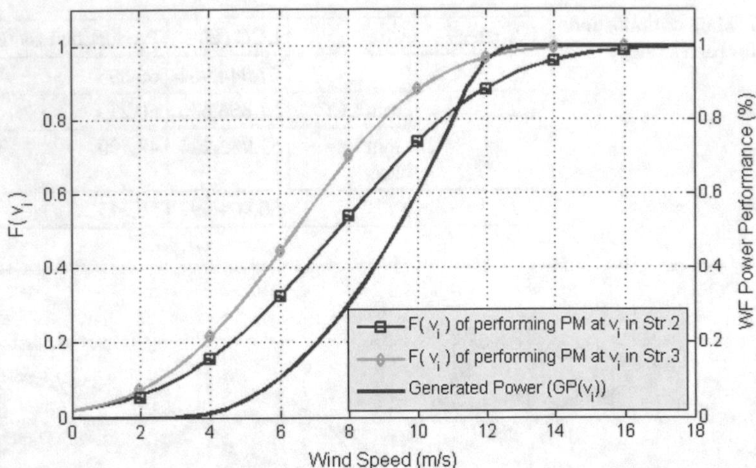

Fig. 9 Wind speed at which PM is performed

the dynamic opportunistic maintenance policy does not reduce PM, it achieves to find more suitable maintenance opportunities with regards to the wind conditions than static opportunistic maintenance policy.

Wind speed during PM. As stated in Carlos et al. [10] optimal wind speeds for performing maintenance activities are those bellow 5 m/s, not only in terms of production, but also in terms of workers' safety. In fact, regarding workers' safety, maintenance in WTs is generally recommended to be performed under wind speeds bellow 12 m/s [58]. As shown in Fig. 9, following the strategy based on the dynamic reliability thresholds (strategy 3), nearly a 35% of the PM activities are performed under ideal conditions and a 97% under the recommended ones, whereas in strategy 2, with the static reliability thresholds, these percentages decrease to a 22 and a 88%, respectively. Moreover, according to power generation, these results show that whereas in strategy 2 more than a 10% of the PM activities are performed during maximum generation periods, in strategy 3 this figure falls to a 2%.

Static and Dynamic opportunistic maintenance policies under stochastic wind speeds. The obtained results are based on real wind data according to the location of the wind farm. However, since wind predictions might be inaccurate and tendencies might change during the life cycle of the wind farm, the effect of wind speed variability both in LCC and loss of production has been analyzed (see Figs. 10 and 11). To this aim, the wind data used to feed the simulation has been stochastically calculated at each time period based on the real wind data and a variability factor. According to the results shown in Figs. 10 and 11, it can be noticed that the impact of wind speed variation in the static opportunistic maintenance is minor, since the reliability thresholds are optimised without considering the wind speed as an input. On the contrary, the wind speed variability directly affects the dynamic opportunistic maintenance, mostly in environments where the variability is higher than 60%. However, strategy

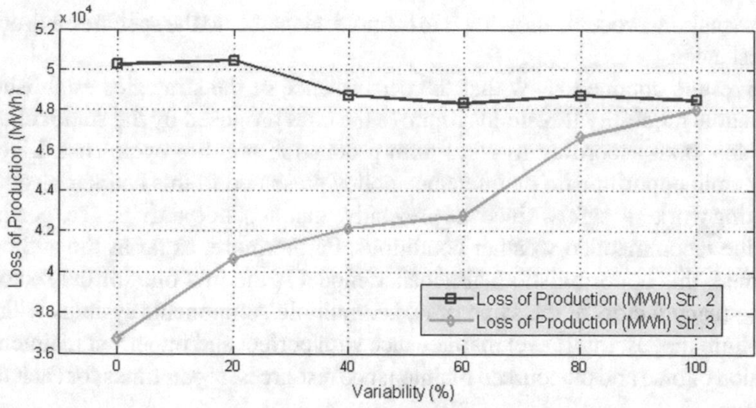

Fig. 10 Loss of production variability

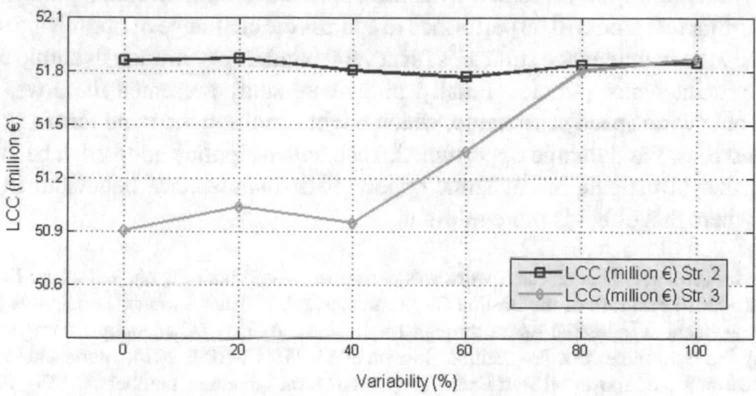

Fig. 11 LCC variability

3 still outperforms strategy 2, excluding the most variable environment, where the results of strategies 2 and 3 are similar.

5 Concluding Remarks

Different opportunistic maintenance policies have been proposed for the wind power industry during the past few years, with different focus, objectives and assumptions. However, all these studies meet at a point: decision making process is always based on static reliability or age thresholds. On the contrary, the reliability thresholds proposed in the presented opportunistic maintenance model are allowed to vary according to the weather conditions. Thus, a dynamic nature has been provided to the maintenance decision making process, allowing it to be more adaptable to the specific environment circumstances.

The results obtained show that the performance of the strategies established by the dynamic reliability thresholds improve the ones proposed by the static reliability thresholds, both according to wind farm production and life cycle cost. Likewise, the dynamic opportunistic maintenance policy presented in this chapter also allows improving workers' safety, since the preventive maintenance activities are performed under the recommended weather conditions. Furthermore, as far as the authors are concerned, this opportunistic maintenance model is the first one for the sector that bears so many factors at the same time, i.e. multiple components systems with multiple failure modes, multilevel maintenance with perfect and imperfect maintenance, utilisation of own and outsourced maintenance resources, repair times for each failure mode, etc.

Future efforts will concentrate on the integration of the condition based maintenance strategies in the presented dynamic opportunistic maintenance policy. Likewise, further research will be performed to address the challenge of optimising simultaneously the maintenance strategies for several wind farms through dynamic opportunistic maintenance policies. Finally, in the case study presented there was not a remarkable wind speed seasonality, which might condition the wind farm's production, and thus, the dynamic opportunistic maintenance policy adopted. The authors will further investigate the dynamic opportunistic maintenance behaviour in wind farms where this effect is more relevant.

Acknowledgements This research work was performed within both the context of the Emaitek-Plus 2015–2016 Program of the Basque Government and the SustainOwner ('Sustainable Design and Management of Industrial Assets through Total Value and Cost of Ownership'), a project sponsored by the EU Framework Programme Horizon 2020, MSCA-RISE-2014: Marie Skłodowska-Curie Research and Innovation Staff Exchange (RISE) (grant agreement number 645733—Sustain-Owner—H2020-MSCA-RISE-2014).

References

1. Ackermann, T. (2005). *Wind power in power systems*. Wiley-Blackwell. https://doi.org/10.1002/0470012684.
2. Anon. (2015). *Global status of wind power in 2014, Global wind energy council*. http://www.gwec.net/wp-content/uploads/2015/03/GWEC_Global_Wind_2014_Report_LR.
3. Tavner, P. J., Xiang, J., & Spinato, F. (2007). Reliability analysis for wind turbines. *Wind Energy, 10*(1), 1–18. https://doi.org/10.1002/we.204.
4. Shafiee, M. (2015). Maintenance logistics organization for offshore wind energy: Current progress and future perspectives. *Renewable Energy, 77,* 182–193. https://doi.org/10.1016/j.renene.2014.11.045.
5. Kaldellis, J., & Kapsali, M. (2013). Shifting towards offshore wind energy—Recent activity and future development. *Energy Policy, 53,* 136–148. https://doi.org/10.1016/j.enpol.2012.10.032.

6. Anon. (2015). The European offshore wind industry—Key trends and statistics 2014, *The European Wind Energy Association, EWEA*. http://www.ewea.org/fileadmin/files/library/publications/statistics/EWEA-European-Offshore-Statistics-2014.pdf.
7. Byon, E. (2013). Wind turbine operations and maintenance: A tractable approximation of dynamic decision making. *IIE Transactions, 45*(11), 1188–1201. https://doi.org/10.1080/0740817x.2012.726819.
8. Krokoszinski, H. (2003). Efficiency and effectiveness of wind farms-keys to cost optimized operation and maintenance. *Renewable Energy, 28*(14), 2165–2178. https://doi.org/10.1016/s0960-1481(03)00100-9.
9. Sen, Z. (1997). Statistical investigation of wind energy reliability and its application. *Renewable Energy, 10*(1), 71–79. https://doi.org/10.1016/0960-1481(96)00021-3.
10. Carlos, S., Sánchez, A., Martorell, S., & Marton, I. (2013). Onshore wind farms maintenance optimization using a stochastic model. *Mathematical and Computer Modelling, 57*(7–8), 1884–1890. https://doi.org/10.1016/j.mcm.2011.12.025.
11. Ding, F., & Tian, Z. (2012). Opportunistic maintenance for wind farms considering multi-level imperfect maintenance thresholds. *Renewable Energy, 45,* 175–182. https://doi.org/10.1016/j.renene.2012.02.030.
12. Márquez, F. P. G., Tobias, A. M., Pérez, J. M. P., & Papaelias, M. (2012). Condition monitoring of wind turbines: Techniques and methods. *Renewable Energy, 46,* 169–178. https://doi.org/10.1016/j.renene.2012.03.003.
13. Yang, W., Tavner, P. J., Crabtree, C. J., Feng, Y., & Qiu, Y. (2014). Wind turbine condition monitoring: Technical and commercial challenges. *Wind Energy, 17*(5), 673–693. https://doi.org/10.1002/we.1508.
14. Garcia, M. C., Sanz-Bobi, M. A., & del Pico, J. (2006). Simap: Intelligent system for predictive maintenance. *Computers in Industry, 57*(6), 552–568. https://doi.org/10.1016/j.compind.2006.02.011.
15. Horenbeek, A. V., Ostaeyen, J. V., Duflou, J. R., & Pintelon, L. (2013). Quantifying the added value of an imperfectly performing condition monitoring system-application to a wind turbine gearbox. *Reliability Engineering & System Safety, 111,* 45–57. https://doi.org/10.1016/j.ress.2012.10.010.
16. Anon. (2005). Managing the wind. *Refocus, 6,* 48–51. https://doi.org/10.1016/s1471-0846(05)70402-9.
17. Amirat, Y., Benbouzid, M., Al-Ahmar, E., Bensaker, B., & Turri, S. (2009). A brief status on condition monitoring and fault diagnosis in wind energy conversion systems. *Renewable and Sustainable Energy Reviews, 13*(9), 2629–2636. https://doi.org/10.1016/j.rser.2009.06.031.
18. Kusiak, A., Zhang, Z., & Verma, A. (2013). Prediction, operations, and condition monitoring in wind energy. *Energy, 60,* 1–12. https://doi.org/10.1016/j.energy.2013.07.051.
19. Wang, H., & Pham, H. (2006). Optimal preparedness maintenance of multi-unit systems with imperfect maintenance and economic dependence. In *Springer series in reliability engineering* (pp. 135–150), Springer. https://doi.org/10.1007/1-84628-325-6_7.
20. Ding, F., & Tian, Z. (2011). Opportunistic maintenance optimization for wind turbine systems considering imperfect maintenance actions. *International Journal of Reliability, Quality and Safety Engineering, 18*(05), 463–481. https://doi.org/10.1142/s0218539311004196.
21. Nicolai, R., & Dekker, R. (2007). A review of multi-component maintenance models. In *Proceedings of European Safety and Reliability Conference* (pp. 289–296). http://www.dimat.unina2.it/marrone/dwnld/Proceedings/ESREL/2007/Pdf/CH036.pdf.
22. Dekker, R., Wildeman, R. E., & van der Duyn Schouten, F. A. (1997). A review of multi-component maintenance models with economic dependence. *Mathematical Methods of Operations Research, 45*(3), 411–435. https://doi.org/10.1007/bf01194788.
23. Sasieni, M. W. (1956). A markov chain process in industrial replacement. *OR, 7*(4), 148. https://doi.org/10.2307/3007561.
24. Nakagawa, T., & Murthy, D. (1993). Optimal replacement policies for a two-unit system with failure interactions. *Revue française d'automatique, d'informatique et de recherche opérationnelle. Recherche opérationnelle, 27*(4), 427–438. http://archive.numdam.org/ARCHIVE/RO/RO_1993__27_4/RO_1993__27_4_427_0/RO_1993__27_4_427_0.pdf.

25. Wang, H. (2002). A survey of maintenance policies of deteriorating systems. *European Journal of Operational Research, 139*(3), 469–489. https://doi.org/10.1016/s0377-2217(01)00197-7.
26. Ritchken, P., & Wilson, J. G. (1990). (m, t) group maintenance policies. *Management Science, 36*(5), 632–639. https://doi.org/10.1287/mnsc.36.5.632.
27. Vergin, R. C., & Scriabin, M. (1977). Maintenance scheduling for multicomponent equipment. *IIE Transactions, 9*(3), 297–305. https://doi.org/10.1080/05695557708975158.
28. Crocker, J., & Kumar, U. (2000). Age-related maintenance versus reliability centred maintenance: A case study on aero-engines. *Reliability Engineering & System Safety, 67*(2), 113–118. https://doi.org/10.1016/s0951-8320(99)00052-6.
29. Dagpunar, J. S. (1996). A maintenance model with opportunities and interrupt replacement options. *The Journal of the Operational Research Society, 47*(11), 1406. https://doi.org/10.2307/3010206.
30. Zheng, X., & Fard, N. (1991). A maintenance policy for repairable systems based on opportunistic failure-rate tolerance. *IEEE Transactions on Reliability, 40*(2), 237–244. https://doi.org/10.1109/24.87134.
31. Tian, Z., Jin, T., Wu, B., & Ding, F. (2011). Condition based maintenance optimization for wind power generation systems under continuous monitoring. *Renewable Energy, 36*(5), 1502–1509. https://doi.org/10.1016/j.renene.2010.10.028.
32. Besnard, F., Patrikssont, M., Strombergt, A.-B., Wojciechowskit, A., & Bertling, L. (2009). An optimization framework for opportunistic maintenance of offshore wind power system. In *IEEE Bucharest PowerTech, IEEE Institute of Electrical & Electronics Engineers*. https://doi.org/10.1109/ptc.2009.5281868.
33. Abdollahzadeh, H., Atashgar, K., & Abbasi, M. (2016). Multi-objective opportunistic maintenance optimization of a wind farm considering limited number of maintenance groups. *Renewable Energy, 88,* 247–261. https://doi.org/10.1016/j.renene.2015.11.022.
34. Atashgar, K., & Abdollahzadeh, H. (2016). Reliability optimization of wind farms considering redundancy and opportunistic maintenance strategy. *Energy Conversion and Management, 112,* 445–458. https://doi.org/10.1016/j.enconman.2016.01.027.
35. Sarker, B. R., & Faiz, T. I. (2016). Minimizing maintenance cost for offshore wind turbines following multi-level opportunistic preventive strategy. *Renewable Energy, 85,* 104–113. https://doi.org/10.1016/j.renene.2015.06.030.
36. Zhu, W., Fouladirad, M., & Bérenguer, C. (2016). A multi-level maintenance policy for a multicomponent and multifailure mode system with two independent failure modes. *Reliability Engineering & System Safety, 153,* 50–63. https://doi.org/10.1016/j.ress.2016.03.020.
37. Iqbal, M., Azam, M., Naeem, M., Khwaja, A., & Anpalagan, A. (2014). Optimization classification, algorithms and tools for renewable energy: A review. *Renewable and Sustainable Energy Reviews, 39,* 640–654. https://doi.org/10.1016/j.rser.2014.07.120.
38. Pham, H., & Wang, H. (1996). Imperfect maintenance. *European Journal of Operational Research, 94*(3), 425–438. https://doi.org/10.1016/s0377-2217(96)00099-9.
39. Byon, E., Ntaimo, L., & Ding, Y. (2010). Optimal maintenance strategies for wind turbine systems under stochastic weather conditions. *IEEE Transactions on Reliability, 59*(2), 393–404. https://doi.org/10.1109/tr.2010.2046804.
40. Tavner, P., Greenwood, D. M., Whittle, M. W. G., Gindele, R., Faulstich, S., & Hahn, B. (2012). Study of weather and location effects on wind turbine failure rates. *Wind Energy, 16*(2), 175–187. https://doi.org/10.1002/we.538.
41. Wilson, G., & McMillan, D. (2014). Modeling the relationship between wind turbine failure modes and the environment. *Gas (CCGT), 22,* 6–4. https://pure.strath.ac.uk/portal/files/30166175/Modeling_the_relationship_between_wind_turbine_failure_modes_and_the_environment.pdf.
42. Kahrobaee, S., & Asgarpoor, S. (2013). A hybrid analytical-simulation approach for maintenance optimization of deteriorating equipment: Case study of wind turbines. *Electric Power Systems Research, 104,* 80–86. https://doi.org/10.1016/j.epsr.2013.06.012.
43. Andrawus, J., Watson, J., Kishk, M., & Adam, A. (2006). The selection of a suitable maintenance strategy for wind turbines. *Wind Engineering, 30*(6), 471–486. https://doi.org/10.1260/030952406779994141.

44. Byon, E., & Ding, Y. (2010). Season-dependent condition-based maintenance for a wind turbine using a partially observed Markov decision process. *IEEE Transactions on Power Systems, 25*(4), 1823–1834. https://doi.org/10.1109/tpwrs.2010.2043269.
45. Carroll, J., McDonald, A., & McMillan, D. (2015). Failure rate, repair time and unscheduled O&M cost analysis of offshore wind turbines. *Wind Energy*. https://doi.org/10.1002/we.1887.
46. Yañez, M., Joglar, F., & Modarres, M. (2002). Generalized renewal process for analysis of repairable systems with limited failure experience. *Reliability Engineering & System Safety, 77*(2), 167–180. https://doi.org/10.1016/s0951-8320(02)00044-3.
47. Karyotakis, A. (2011). *On the optimisation of operation and maintenance strategies for offshore wind farms*. Ph.D. thesis, University College London (UCL). http://discovery.ucl.ac.uk/id/eprint/1302066.
48. González, J. S., Rodriguez, A. G. G., Mora, J. C., Santos, J. R., & Payan, M. B. (2010). Optimization of wind farm turbines layout using an evolutive algorithm. *Renewable Energy, 35*(8), 1671–1681. https://doi.org/10.1016/j.renene.2010.01.010.
49. Karki, R., & Patel, J. (2008). Reliability assessment of a wind power delivery system. *Proceedings of the Institution of Mechanical Engineers, Part O: Journal of Risk and Reliability, 223*(1), 51–58. https://doi.org/10.1243/1748006xjrr218.
50. Niazi, M., & Hussain, A. (2011). Agent-based computing from multi-agent systems to agent-based models: A visual survey. *Scientometrics, 89*(2), 479–499. https://doi.org/10.1007/s11192-011-0468-9.
51. Banks, J. (2003). Discrete event simulation. In *Encyclopedia of information systems* (pp. 663–671). Elsevier BV. https://doi.org/10.1016/b0-12-227240-4/00045-9.
52. Martin-Tretton, M., Reha, M., Drunsic, M., & Keim, M. (2012). Data collection for current us wind energy projects: Component costs, financing, operations, and maintenance. *Contract, 303,* 275–3000.
53. Laguna, M., & Martí, R. (2003). The optquest callable library. In *Optimization software class libraries* (pp. 193–218). Springer. https://doi.org/10.1007/0-306-48126-x_7.
54. Glover, F., Kelly, J. P., & Laguna, M., The optquest approach to crystal ball simulation optimization. *Graduate school of Business, University of Colorado 16*. https://www.researchgate.net/profile/Fred_Glover/publication/267771945_THE_OPTQUEST_APPROACH_TO_CRYSTAL_BALL_SIMULATION_OPTIMIZATION/links/545d0a330cf295b5615e665f.pdf.
55. Yun, I., & Park, B. (2006). Application of stochastic optimization method for an urban corridor. In *Proceedings of the 2006 Winter Simulation Conference, Institute of Electrical & Electronics Engineers (IEEE)*. https://doi.org/10.1109/wsc.2006.322918.
56. Glover, F. (1998). A template for scatter search and path relinking. *Lecture Notes in Computer Science, 1363,* 13–54. https://doi.org/10.1007/bfb0026589.
57. Laguna, M., & Martí, R. (2005). Experimental testing of advanced scatter search designs for global optimization of multimodal functions. *Journal of Global Optimization, 33*(2), 235–255. https://doi.org/10.1007/s10898-004-1936-z.
58. Rodríguez, P. C., & Simón, N. J. (2014). Wind Turbines: Safety measures in maintenance activities. *Instituto Nacional de Seguridad e Higiene en el Trabajo.*

Part IV
Emerging Value-Based and Intelligent Asset Management Processes

Chapter 13
A Review of the Roles of Digital Twin in CPS-Based Production Systems

Elisa Negri, Luca Fumagalli and Marco Macchi

Abstract The Digital Twin (DT) is one of the main concepts associated to the Industry 4.0 wave. This term is more and more used in industry and research initiatives; however, the scientific literature does not provide a unique definition of this concept. The chapter aims at analyzing the definitions of the DT concept in scientific literature, retracing it from the initial conceptualization in the aerospace field, to the most recent interpretations in the manufacturing domain and more specifically in Industry 4.0 and smart manufacturing research. DT provides virtual representations of systems along their lifecycle. Optimizations and decisions making would then rely on the same data that are updated in real-time with the physical system, through synchronization enabled by sensors. The chapter also proposes the definition of DT for Industry 4.0 manufacturing, elaborated by the European H2020 project MAYA, as a contribution to the research discussion about DT concept.

Keywords Digital Twin · Cyber-Physical Systems · Industry 4.0 · Production systems

1 Introduction

In the recent years, the traditional manufacturing industry is challenged worldwide with the amazing growth and advancements in digital technologies that allow easy integration of interconnected intelligent components inside the shopfloor, that is at the basis of the so-called Industry 4.0 and that is made possible by the widespread adoption of information and communication technologies by manufacturing compa-

E. Negri (✉) · L. Fumagalli · M. Macchi
Department of Management, Economics and Industrial Engineering, Politecnico di Milano, Piazza Leonardo da Vinci, 32, 20133 Milan, Italy
e-mail: elisa.negri@polimi.it

L. Fumagalli
e-mail: luca1.fumagalli@polimi.it

M. Macchi
e-mail: marco.macchi@polimi.it

A. Crespo Márquez et al. (eds.), *Value Based and Intelligent Asset Management*,
https://doi.org/10.1007/978-3-030-20704-5_13

nies. Industry 4.0 has been recognized at international level as one of the strategical responses of the manufacturing companies to the economic crisis, to the tendency to delocalize production and to the increased market complexity [1]. The technological basis of Industry 4.0 roots back in the Internet of Things (IoT) [2], which proposed to embed electronics, software, sensors, and network connectivity into devices (i.e. "things"), in order to allow the collection and exchange of data through the internet [3]. As such, IoT can be exploited at industrial level: devices can be sensed and controlled remotely across network infrastructures, allowing a more direct integration between the physical world and virtual systems, and resulting in higher efficiency, accuracy and economic benefits. Although it is a recent trend, Industry 4.0 has been widely discussed and its key technologies have been identified [4], among which Cyber-Physical Systems (CPS) have been proposed as smart embedded and networked systems within production systems [1, 5]. They operate at virtual and physical levels interacting with and controlling physical devices, sensing and acting on the real world [6]. According to scientific literature, in order to fully exploit the potentials of CPS and IoT, proper data models should be employed, such as ontologies [7, 8], which are explicit, semantic and formal conceptualizations of concepts in a domain [9]. They are the core semantic technology providing intelligence embedded in the smart CPS [10] and could help the integration and sharing of big amounts of sensed data [11, 12]. Through the use of Big Data analytics, it is possible to access sensed data, through smart analytics tools, for a rapid decision making and improved productivity [13, 14].

With the use of these technologies, Industry 4.0 opens the way to real-time monitoring and synchronization of the real world activities to the virtual space thanks to the physical-virtual connection and the networking of CPS elements [1]. The Digital Twin (DT) is meant as the virtual and computerized counterpart of a physical system that can be used to simulate it for various purposes, exploiting a real-time synchronization of the sensed data coming from the field; such a synchronization is possible thanks to the enabling technologies of Industry 4.0 and, as such, the DT is deeply linked with it. The DT was first born in the aerospace field and only recently has been adopted also in manufacturing contexts: such a term is used in industrial environments and in governmental research initiatives; however, scientific literature that describes the contextualisation of the concept in the manufacturing domain is still at its infancy. A review of the contributions on this would be highly beneficial, in order to pave the way and clarify the conceptual foundations for future research works on the topic.

In this sense, the objective of the chapter is to contribute to offer a deeper understanding of the proposed definitions of DT in scientific literature and to help in the identification of the role of the DT for manufacturing in the Industry 4.0 era. The chapter is structured as follows: Sect. 2 deals with the research objectives and methodology; Sect. 3 presents the literature analysis of the Digital Twin concept and applications; Sect. 4 shapes the new role of the DT for Industry 4.0 manufacturing; finally, Sect. 5 proposes some concluding remarks.

2 Research Statement

According to the research motivations outlined in the Introduction, the chapter aims at contributing to the shaping of the DT definition in scientific literature. This is in the direction of clarifying and extrapolating a unique definition and relevance of DT for the manufacturing sector, starting from different instances of research works in various sectors and contexts. To this aim, the chapter answers the following Research Questions: "What is the definition of Digital Twin in scientific literature?" and "What is its role within Industry 4.0?".

The used methodology is a thorough literature review on the concept. In particular, the authors have searched on the Scopus Database all publications released in the five years between 2012 and 2016, that had the term "Digital Twin" in the title, abstract or in the keywords. The type of publications (Journal articles, conference papers and others) was not a filtering criteria, the only considered language was English. The reason for excluding publications before 2012 was that the main focus of the current study was the DT in relation to the Industry 4.0 that was first defined at Hannover Messe, Germany, in 2011 [15], therefore no relevant publications are to be found prior to the year 2012 with respect to these topics.

3 Literature Analysis

3.1 The Digital Twin Concept

The first definition of the DT was forged by the NASA as "an integrated multi-physics, multi-scale, probabilistic simulation of a vehicle or system that uses the best available physical models, sensor updates, fleet history, etc., to mirror the life of its flying twin. It is ultra-realistic and may consider one or more important and interdependent vehicle systems": this definition first appeared in the draft and after in the final release of the NASA Modeling, Simulation, Information Technology & Processing Roadmap in 2010 [16, 17]. From that moment on, aerospace researchers started referring to the said NASA roadmap as the seminal work to define the DT (as an example [18]). As it is evident, the main scope of the original definition of the DT was to mirror the life of air vehicles with a series of integrated sub-models that reflected different aspects and vehicle systems, by considering stochasticity, historical data and sensor data, including in this way interactions of the vehicle with the real world. Only in subsequent research works, published in the same year, other aspects emerged such as the life-cycle view [19], the check on mission requirements [19, 20] and the use of the DT for prognostics and diagnostics activities [21], that then remained as core characteristics of the concept in future works. In 2015 with the work by Rios et al. [22], the definition of DT comprised a generic "Product", opening the way to the use of such a concept in other fields rather than only air vehicles, even though their work was still inserted in research about aircraft structures. Initial works in other sectors appeared even before. In fact, alongside the research in the aerospace field, in 2013 the first works reporting research on DT in manufacturing sector appeared. In particular, Lee and colleagues considered it to be the virtual counterpart of production resources, and not only of the product, setting the basis for a debate about the role of the DT in advanced manufacturing environments, such as the envisioned Industry 4.0 with its core technologies, big data analytics and cloud platforms [23]. This debate continues still today, and this work is inserted in such a stream.

To have a complete view of the definitions of the DT appeared in literature, please refer to Table 1.

Research on the DT in manufacturing is an evolution of the already ongoing research stream about Virtual Factories (VF). These are defined as the digitalization of the plant integrated with the real system coming in help to the production during all the lifecycle of each asset [34]. In fact, information generated at design phase of a production system is not exploited during the operational phases, even though they could be highly valuable. This information can be used for easier performances' evaluation and management decisions during production operations. Already, works on the VF underlined the importance assumed by a proper semantic meta data model to support the necessary information structuring [35]. This was also in line with a rich manufacturing research stream dealing with the role and benefits of ontologies for production systems [36–38]. The semantic Virtual Factory Data Model (VFDM) has been developed [35], which establishes a coherent and extensible standard for the common representation of the factory entities, such as buildings, processes, products and resources, giving in this way a holistic view of the production environment [34]. The DT goes beyond the VF to include a real time synchronization with the physical system, thanks to which the user or the autonomous system can take the right decision about the actual and the future production, based on a wide range of available information. Also the DT must therefore be supported by a proper data model structuring information about the system operations, its history, its behaviour and its current state.

3.2 The Digital Twin History

As emerges from Table 1, despite the DT concept is recent, it has been used for different sectors and it has been linked to various aspects. A deeper investigation was deemed necessary. The authors chose to adopt a systematic search approach as described in the Sect. 2, "Research Statement". It emerged that the English language publications released from 2012 to 2016 listed on the Scopus Database are in total 26, and are presented in the Table 2.

From a temporal point of view, Table 2 shows that there has been an increasing interest in the DT: in fact, all references from 2012 and 2013 come from the same conference (respectively from the 53rd and 54th editions of the AIAA/ASME/ASCE/AHS/ASC Structures, Structural Dynamics and Materials Con-

Table 1 Definitions of Digital Twin in literature

No.	Ref.	Year	Definition of Digital Twin
1	[16–18]	2010 and 2012	An integrated multi-physics, multi-scale, probabilistic simulation of a vehicle or system that uses the best available physical models, sensor updates, fleet history, etc., to mirror the life of its flying twin. The Digital Twin is ultra-realistic and may consider one or more important and interdependent vehicle systems
2	[19]	2012	A cradle-to-grave model of an aircraft structure's ability to meet mission requirements, including submodels of the electronics, the flight controls, the propulsion system, and other subsystems
3	[20]	2012	Ultra-realistic, cradle-to-grave computer model of an aircraft structure that is used to assess the aircraft's ability to meet mission requirements
4	[23]	2013	Coupled model of the real machine that operates in the cloud platform and simulates the health condition with an integrated knowledge from both data driven analytical algorithms as well as other available physical knowledge
5	[21]	2013	Ultra-high fidelity physical models of the materials and structures that control the life of a vehicle
6	[24]	2013	Structural model which will include quantitative data of material level characteristics with high sensitivity
7	[25]	2015	Very realistic models of the process current state and its behavior in interaction with the environment in the real world
8	[22]	2015	Product digital counterpart of a physical product
9	[26]	2015	Ultra-realistic multi-physical computational models associated with each unique aircraft and combined with known flight histories
10	[27]	2015	High-fidelity structural model that incorporates fatigue damage and presents a fairly complete digital counterpart of the actual structural system of interest
11	[28]	2016	Virtual substitutes of real world objects consisting of virtual representations and communication capabilities making up smart objects acting as intelligent nodes inside the internet of things and services
12	[29]	2016	Digital representation of a real world object with focus on the object itself
13	[30]	2016	The simulation of the physical object itself to predict future states of the system

(continued)

Table 1 (continued)

No.	Ref.	Year	Definition of Digital Twin
14	[31]	2016	Virtual representation of a real product in the context of Cyber-Physical Systems
15	[32]	2016	An integrated multi-physics, multi-scale, probabilistic simulation of an as-built system, enabled by Digital Thread, that uses the best available models, sensor information, and input data to mirror and predict activities/performance over the life of its corresponding physical twin
16	[33]	2016	A unified system model that can coordinate architecture, mechanical, electrical, software, verification, and other discipline-specific models across the system lifecycle, federating models in multiple vendor tools and configuration-controlled repositories

ference), suggesting that initially the research communities working on it were few. In 2014 there is only one paper on Scopus, while in 2015 and in 2016 the number of publications—along with the number of application fields—are increasing, respectively 5 and 12.

Publications were mainly conference papers (20), with a low number of Journal articles (5) and 1 Book Chapter. This is in line with the expectation that scientific literature on DT is very recent.

Most recent publications go beyond the boundaries of the aerospace field to reach the smart manufacturing environments, in particular related to the Industry 4.0. New technologies, such as the CPS, open the way to new uses, beyond the diagnostics and prognostics purposes, and meanings of the DT for the production systems domain. From Table 2, different possible uses emerge that have been categorized as follows:

1. The initial intended use in literature is to support health analyses for an improved maintenance activity and planning; in particular, the DT has been proposed to:
   a. Monitor anomalies, fatigue, crack paths in the physical twin [18, 19, 21, 27, 31, 39, 40];
   b. Monitor geometric and plastic deformation on the material of the physical twin [41];
   c. Model reliability of the physical system [20].
2. A second use with a wider scope is to digitally mirroring the life of the physical entity:
   a. To study the long-term behaviour of the system and predict its performances by keeping into account the different synergistic effects of environmental conditions [24, 26, 40, 42];
   b. To provide information continuity along the different phases of the lifecycle [25, 43];
   c. For the Virtual Commissioning of the system [28, 44];

Table 2 Results of the systematic research on Scopus

No.	Ref.	Year	Type	Field	Use	Diagnostics Prognostics	Simulation	Simulation Software
1	[49]	2012	C	AS	Not available	Not available	Not available	Not available
2	[19]	2012	C	AS	As virtual health sensor, it forecasts maintenance needs	Prediction of cracking	FEM and Montecarlo simulations	DDSim (Damage and Durability Simulator)
3	[45]	2012	C	AS	Prediction confidence; decision-making in condition-based maintenance	Life prediction including failure	Numerical simulation (accumulation models for damage and life prediction)	Not available
4	[18]	2012	C	AS	Monitor manufacturing anomalies	DT for Health management and maintenance history	DT are simulations	Not available
5	[20]	2012	C	AS	Model reliability	DT integrated in Condition-based maintenance systems	FEM and CFD simulations	Rockstar Sim. Suite, Stick-to-Stress Real Time Dynamic Flight Simulator (S2S DFS)
6	[41]	2013	C	AS	Analyze plastic deformation behavior	Fatigue crack growth prediction	No	No
7	[21]	2013	C	AS	Digital mirror of life of physical twin	Predictions of early warnings of microcracks	Simulation integrated with on-board health management system	Not available
8	[24]	2013	C	AS	Investigate long-term behaviour of structure under multiple environmental conditions	Prognosis of structural composites and their synergistic response in a multi-physical environment	DT are simulations	Not available

(continued)

Table 2 (continued)

No.	Ref.	Year	Type	Field	Use	Diagnostics Prognostics	Simulation	Simulation Software
9	[39]	2014	J	AS	Monitor crack paths by filling information gaps	Prediction of crack path	Production-level simulation	Not available
10	[25]	2015	C	M	Information continuity along product lifecycle	Not available	Complex behavior of production	Not available
11	[22]	2015	C	AS	Simulating, predicting, optimizing the product lifecycle	Maintenance services, especially related to usage predictions	No	No
12	[26]	2015	C	AS	Monitor and predict performances	Damage and cracks detection and monitoring	No	No
13	[48]	2015	J	AS	Not available	Diagnostics and prognostics of aircrafts	FEM & Montecarlo Simulations	Not available
14	[27]	2015	J	AS	Fatigue-damage prediction		FEM simulation	Not available
15	[28]	2016	C	R	Virtual Commissioning	No	Simulations at system level during design	Matlab-Simulink
16	[29]	2016	C	I	IoT lifecycle management	No	No	No
17	[30]	2016	C	M	Optimize system behavior at design	No	DT are simulations	Not available
18	[31]	2016	C	M	Monitor the physical entity	No	Data exchange simulation	Not available
19	[32]	2016	C	AS	Engineering analyses and decision making	Prognosis of crack propagation	Manned flight simulators	Not available

(continued)

Table 2 (continued)

No.	Ref.	Year	Type	Field	Use	Diagnostics Prognostics	Simulation	Simulation Software
20	[42]	2016	BC	AS	Defining system behavior	No	Not Available	Not available
21	[46]	2016	J	AS	Support design	Not available	Aircraft mock-up	Dassault Systemès V6
22	[47]	2016	C	M	Layout optimization	No	HMI interactions	Not available
23	[43]	2016	J	M	Data Manag. in lifecycle	No	No	No
24	[33]	2016	C	AS	Systems engineering and mechanical design integr.	No	CAE-based simulations	Mathematica and Matlab/Simulink
25	[44]	2016	C	R	Virtual Commissioning	No	To implement and optimize the algorithm for control of robots	VEROSIM (Virtual Environment and Robotic Simulation)
26	[40]	2016	C	AS	Detecting failures; definition of performance	Product failures detection	No	No

Type: C = Conference, J = Journal, BC = Book Chapter; Field: AS = Aeronautics and Space, R = Robotics, M = Manufacturing, I = Informatics

    d. To manage the lifecycle of the Internet of Things devices [29].

3. DT have been proposed to support decision making through engineering and statistical analyses [32, 45]:

    a. Optimization of system behaviour during design phase [30, 33, 43, 46, 47];
    b. Optimization of product lifecycle, knowing the past and present states, it is possible to predict and optimize the future performances [22, 45].

Many of the mentioned uses were targeted at improving maintenance activities, such as condition-based maintenance, diagnostics and prognostics activities. This was not only related to the use number one (health analyses), but also the other uses [24, 26, 32, 40, 42, 45]. From this the clear vocation of the DT in its first meaning emerges as an instrument to support better prediction of failures during the system lifecycle (in particular air vehicles in the aerospace field) based on field data coming from sensors [16–18, 27, 31, 32].

Another aspect that appeared since the very first interpretations of the DT concept was the intimate connection to simulation, that is seen in two different ways:

i. For most of the authors, the DT is a model that represents the system that different types of simulations can be based upon [19–21, 25, 27, 28, 31, 33, 39, 44–48],
ii. Others consider the DT as the simulation of the system itself [16–18, 24, 30, 32].

Although the simulation seems to be a key aspect related to the DT concept, there are authors who do not mention it [22, 26, 29, 40–43, 49]. Irrespective of the connection between the two concepts, the authors mention different purposes and characteristics of the simulations and the DT. In the aerospace, the most mentioned simulations replicate the continuous time history of flights (with historical data, maintenance history information), generating enormous databases of simulations to understand what the aircraft has experienced and to forecast future maintenance needs and interventions, with the use of Finite Element Methods (FEM), Computational Fluid Dynamics (CFD), Montecarlo and Computer-Aided Engineering (CAE) applications-based simulations [19–21, 27, 33, 46, 48]. Some authors highlight the fact that these simulations should be connected with on-board devices and sensors to have a continuous synchronization with the field condition [21]. In the robotics field, the simulations are mainly performed for the Virtual Commissioning to optimize the control algorithms for robots during development phase [28, 44]. In manufacturing, the main objective of simulations are to represent the complex behaviour of the system, also considering the possible consequences of external factors, human interactions and design constraints [25, 30, 47].

Also, different simulation tools have been proposed, as it is shown in Table 2.

3.3 *Digital Twin in the Industrial Engineering*

This section is devoted to a more detailed analysis of a subset of the papers presented in Table 2, namely those applying the DT concept to the industrial engineering: thus comprising all the papers in the manufacturing and robotics sectors. Also the informatics papers were considered in this analysis because the application domain was the industrial IoT. The schematic results of this analysis are presented in Table 3, where the papers are confronted with the main aspects emerged from the previous literature on DT.

1. The important connection between the DT concept and the Industry 4.0, mentioned in Sect. 1, is confirmed also by the considered papers: in fact, only two papers do not mention it [44, 47], all the others name the Industry 4.0 [25, 28, 31] or one of the connected concepts (IoT [29], Smart CPS [30], Smart Product [43]).
2. The Big Data topic is not strongly recognized by the authors as a key aspect of the DT modeling. In fact, only Rosen recognizes that the DT model requires a huge digital data storage [25] and Schroeder carries on this argumentation stating that Big Data management and analytics become an issue in a DT context [31]. Abramovici mentions Big Data as an analysis method to elaborate data for DT-based optimizations [43].

Table 3 Analysis of the papers about Digital Twin in industrial engineering

No.	Ref.	Field	Industry 4.0	Big data	Lifecycle	CPS	Semantic data model
11	[25]	M	Industry 4.0, IoT	Yes	Production system lifecycle	Yes	Meta-information and semantics
16	[28]	R	Industry 4.0	No	Complex technical systems lifecycle	No	No
17	[29]	I	Industrial IoT	No	IoT lifecycle	No	No
18	[30]	M	Smart CPS	No	Production system lifecycle	Yes	No
19	[31]	M	Industry 4.0, IoT	Yes	Product lifecycle	Yes	AutomationML model for Data Exchange
23	[47]	M	No	No	No	No	Database with CAD models
24	[43]	M	Smart Products	Yes	Product lifecycle	Yes	Semantic data management
26	[44]	R	No	No	No	No	No

3. The lifecycle perspective is also not shared by all authors. Some authors see the DT employed in the lifecycle of the production system [25, 30] or of the product [31, 43]. Others do not adhere to the product-production system dichotomy, by considering technical system lifecycle as focus: the system is both a product (during design phase) and a production system equipment (during operations) [28, 29]. The idea is that design, service and recycling/disposal phases would benefit from information generated during operations [29], stored in a unique data source [43], thus reducing effort in decision making, and in collecting data to perform simulations [28].
4. Despite the fact that the first conceptualizations of the DT did not deal with advanced manufacturing, it can be stated that the connection with the CPS is getting stronger. In fact, the papers in the manufacturing field mention the use of the DT to simulate a CPS system or product (Smart products "are cyber-physical products/systems (CPS) which use and integrate internet-based services to perform a required functionality" [43]) [25, 30, 31].
5. Some authors stress the importance of having a proper data model to support the named information collection and continuity during all lifecycle phases. Generally, semantic meta-data models are proposed also along product lifecycle, where semantic data management covers product lifecycle data both from the virtual and real lifecycles and their related information flow [25, 43]. Schroeder sees the DT as composed of different models and data, which are aggregated in an AutomationML model for data exchange [31]. Arisoy mentions that the DT runs on a database with Computer-Aided Design (CAD) models to decide grasping point locations [47].

4 Shaping the New Role of Digital Twins for Industry 4.0

Within the European H2020 project MAYA, the research on DT of CPS-based factories has received a new impulse. MAYA proposes multi-disciplinary integrated simulation and forecasting tools, empowered by digital continuity and continuous real-world synchronization, towards reduced time to production and optimization. MAYA project aims at developing a plant DT supporting activities in all factory lifecycle phases: from the design, through the optimization of the operational life, to the dismissal phase. The central concept of the MAYA innovations is the combination of the virtual and physical dimensions with the simulation domain. The key to empower the DT representation and simulation of the actual factory lifecycle is a semantic meta-data model, describing exhaustively the CPS features. This is done through the Centralized Support Infrastructure, a platform that supports:

- The semantic meta data model, that structures information to ensure the digital continuity of data generated at all lifecycle phases of the production system;
- The simulation framework, that connects different simulation methodologies and tools for a multi-disciplinary replica of the physical system;

- The communication layer, that ensures the seamless connection of physical CPS to the digital world, to achieve real-time synchronization and update with the huge amount of field data.

In this sense, MAYA covers all aspects related to the DT that appeared in a stronger or feebler connection in the analysed literature, in a unique framework: lifecycle perspective, information continuity through semantic meta data model, Big Data and real time synchronization with the field. Thanks to the research works performed within MAYA project, it is possible to identify the main characteristics that the DT for Industry 4.0 manufacturing systems. The DT consists of a virtual representation of a production system that is able to run on different simulation disciplines that is characterized by the synchronization between the virtual and real system, thanks to sensed data and connected smart devices, mathematical models and real time data elaboration. This is in line with the role also suggested in the aerospace field. The topical role within Industry 4.0 manufacturing systems is to exploit these features to forecast and optimize the behaviour of the production system at each life cycle phase in real time. This is fully enabled by the Industry 4.0 technologies and it is in line with the view of [50].

The MAYA approach to the DT will be demonstrated with the use of two industrial use cases: the first is a big company in the automotive sector that will exploit the DT for a quicker Virtual Commissioning of its assembly lines; the second is an SME producing cutting, bending and shearing lines for metal sheets, where the DT is useful to improve line performance evaluation and optimization.

5 Conclusions

The chapter proposes a literature analysis about the concept of DT, in order to clarify its definition and its history, started from the aerospace field and then translated also in manufacturing applications. Although it is a highly relevant concept at industrial and research initiatives level, on the DT many elaborations and works have been proposed that did not necessarily reflect the same definition of DT. These publications are recent and mostly polarized on conference proceedings, suggesting the fact that the scientific literature is still at its infancy. For this reason, the chapter contribution comes at the right moment, by offering a systematic review of what has been written on the topic and by creating a first basis for future research works on the topic. This chapter in fact is aimed at clearing out the definitions given in the scientific literature and at shaping the role of the DT for Industry 4.0 manufacturing contexts, in order to motivate and analyze why and how a concept, originally born in the aerospace sector, could be beneficial to the manufacturing domain, in the presence of the technologies associated to Industry 4.0. In particular, it emerged that the relevance of DT for manufacturing industry lies in their definition as virtual counterparts of physical devices. These are digital representations based on semantic data models that allow running simulations in different disciplines, that support not only a prognostic assessment at design stage (static perspective), but also a continuous update of the virtual representation of the object by a real time synchronization with sensed data. This allows the representation to reflect the current status of the system and to perform real-time optimizations, decision making and predictive maintenance according to the sensed conditions.

The presence of already available or soon upcoming commercial software tools to support the DT creation demonstrates its importance for industry, to name a few: Predix (GE digital) (www.predix.com) and Simcenter 3D by Siemens (www.plm.automation.siemens.com/it_it/products/simcenter/3d/).

Research on Digital Twins is still at the beginning, there is a need for future research works on relevant industrial applications to investigate and demonstrate the wide range of applications and benefits where the DT could express their potential. An interesting application where the role of DT could be fundamental is the demonstration of new production control methods, see e.g. the so called Synchro-push production control policy, which has been recently formulated by [51] and that needs real-time synchronized simulations of the production system operations to be fully demonstrated, such as the DT.

Acknowledgements This project has received funding from the European Union's Horizon 2020 research and innovation programme under grant agreement No 678556.

References

1. Lee, J., Bagheri, B., & Kao, H. (2015). A Cyber-Physical Systems architecture for Industry 4.0-based manufacturing systems. *Manufacturing Letters, 3*, 18–23. https://doi.org/10.1016/j.mfglet.2014.12.001.
2. Ashton, K. (2009). That 'Internet of Things' thing. *RFiD Journal, 22,* 97–114.
3. Sarma, S., Brock, D. L., & Ashton, K. (2000). The networked physical world.
4. Brettel, M., Friederichsen, N., Keller, M., & Rosenberg, M. (2014). How virtualization, decentralization and network building change the manufacturing landscape: An Industry 4.0 perspective. *International Journal Mechanical, Aerospace, Industrial, Mechatronics Engineering, 8,* 37–44.
5. Jazdi, N. (2014). Cyber Physical Systems in the context of Industry 4.0. In *2014 IEEE International Conference on Automation, Quality and Testing, Robotics* (pp. 1–4). https://doi.org/10.1109/aqtr.2014.6857843.
6. Baheti, R., & Gill, H. (2011) Cyber-Physical Systems. In T. Samad & A. Annaswamy (Eds.), *Impact Control Technology, IEEE Control Systems Society* (pp. 161–166).
7. Garetti, M., Fumagalli, L., & Negri, E. (2015). Role of ontologies for CPS implementation in manufacturing. *MPER—Management and Production Engineering Review, 6,* 26–32. https://doi.org/10.1515/mper-2015-0033.
8. Negri, E., Fumagalli, L., Garetti, M., & Tanca, L. (2016). Requirements and languages for the semantic representation of manufacturing systems. *Computers in Industry, 81,* 55–66. https://doi.org/10.1016/j.compind.2015.10.009.
9. Gruber, T. (1995). Toward principles for the design of ontologies used for knowledge sharing. *International Journal of Human-Computer Studies, 43*, 907–928. http://www.sciencedirect.com/science/article/pii/S1071581985710816. Accessed September 29, 2014.

10. Legat, C., Seitz, C., Lamparter, S., & Feldmann, S. (2014). Semantics to the shop floor: Towards ontology modularization and reuse in the automation domain. In *Proceedings of 19th IFAC World Congress 2014* (pp. 3444–3449). http://www.researchgate.net/publication/261361140_Semantics_to_the_Shop_Floor_Towards_Ontology_Modularization_and_Reuse_in_the_Automation_Domain.
11. Borgo, S. (2014). An ontological approach for reliable data integration in the industrial domain. *Computers in Industry, 65,* 1242–1252. https://doi.org/10.1016/j.compind.2013.12.010.
12. Heymans, S., Ma, L., Anicic, D., Ma, Z., Steinmetz, N., Pan, Y., et al. (2008). Ontology reasoning with large data repository. In: *Ontology Management* (pp. 89–128). Springer US.
13. Davis, J., Edgar, T., Porter, J., Bernaden, J., & Sarli, M. (2012). Smart manufacturing, manufacturing intelligence and demand-dynamic performance. *Computers & Chemical Engineering, 47,* 145–156. https://doi.org/10.1016/j.compchemeng.2012.06.037.
14. Lee, J., Kao, H., & Yang, S. (2014). Service innovation and smart analytics for Industry 4.0 and big data environment. *Procedia CIRP*, 3–8. https://doi.org/10.1016/j.procir.2014.02.001.
15. Jasperneite, J. (2012). Was hinter Begriffen wie Industrie 4.0 steckt. *Internet Und Automation, 12*, 12.
16. Shafto, M., Conroy, M., R. Doyle, E. Glaessgen, C. Kemp, LeMoigne, J., et al. (2010). DRAFT Modeling, Simulation, Information Technology & Processing Roadmap. *Technology Area 11.*
17. Shafto, M., Conroy, M., Doyle, R., Glaessgen, E., Kemp, C., LeMoigne, J., et al. (2012). Modeling, simulation, information technology & processing roadmap. *Technology Area 11.*
18. Glaessgen, E. H., & Stargel, D. S. (2012). The digital twin paradigm for future NASA and U. S. Air Force vehicles. In *53rd AIAA/ASME/ASCE/AHS/ASC Structures, Structural Dynamics and Materials Conference* (p. 1818). https://doi.org/10.2514/6.2012-1818.
19. Tuegel, E. J. (2012). The airframe digital twin : Some challenges to realization. In *53rd AIAA/ASME/ASCE/AHS/ASC Structures, Structural Dynamics and Materials Conference* (p. 1812). https://doi.org/10.2514/6.2012-1812.
20. Gockel, B. T., Tudor, A. W., Brandyberry, M .D., Penmetsa, R. C., & Tuegel, E. J. (2012) Challenges with structural life forecasting using realistic mission profiles. In *53rd AIAA/ASME/ASCE/AHS/ASC Structures, Structural Dynamics and Materials Conference* (p. 1813). https://doi.org/10.2514/6.2012-1813.
21. Reifsnider, K., & Majumdar, P. (2013). Multiphysics stimulated simulation digital twin methods for fleet management. In *AIAA/ASME/ASCE/AHS/ASC Structures, Structural Dynamics and Materials Conference* (p. 1578). https://doi.org/10.2514/6.2013-1578.
22. Ríos, J., Hernandez, J. C., Oliva, M., & Mas, F. (2015). Product avatar as digital counterpart of a physical individual product : Literature review and implications in an aircraft. In I*SPE CE* (pp. 657–666).
23. Lee, J., Lapira, E., Bagheri, B., & Kao, H. (2013). Recent advances and trends in predictive manufacturing systems in big data environment. *Manufacturing Letters, 1*, 38–41. https://doi.org/10.1016/j.mfglet.2013.09.005.
24. Majumdar, P., FasalHaider, M., & Reifsnider, K. (2013). Multi-physics response of structural composites and framework for modeling using material geometry. In *54th AIAA/ASME/ASCE/AHS/ASC Structures, Structural Dynamics and Materials Conference* (p. 1577). https://doi.org/10.2514/6.2013-1577.
25. Rosen, R., Von Wichert, G., Lo, G., & Bettenhausen, K. D. (2015) About the importance of autonomy and digital twins for the future of manufacturing. In *IFAC-PapersOnLine* (pp. 567–572). Elsevier Ltd. https://doi.org/10.1016/j.ifacol.2015.06.141.
26. Bielefeldt, B., Hochhalter, J., & Hartl, D. (2015). Computationally efficient analysis of SMA sensory particles embedded in complex aerostructures using a substructure approach. In *ASME Proceedings of Mechanics and Behavior of Active Materials* (pp. V001T02A007–10 pp).
27. Bazilevs, Y., Deng, X., Korobenko, A., Lanza di Scalea, F., Todd, M. D., & Taylor, S. G. (2015). Isogeometric fatigue damage prediction in large-scale composite structures driven by dynamic sensor data. *Journal of Applied Mechanics, 82,* 1–12. https://doi.org/10.1115/1.4030795.
28. Schluse, M., & Rossmann, J. (2016). From simulation to experimentable Digital Twins. In *2016 IEEE International Symposium on Systems Engineering* (pp. 1–6).

29. Canedo, A. (2016). Industrial IoT lifecycle via digital twins. In *Proceedings of the Eleventh IEEE/ACM/IFIP International Conference on Hardware/Software Codesign and System Synthesis* (p. 29).
30. Gabor, T., Belzner, L., Kiermeier, M., Beck, M. T., & Neitz, A. (2016). A simulation-based architecture for smart Cyber-Physical Systems. In: *2016 IEEE International Conference on Autonomic Computing (ICAC)* (pp. 374–379). https://doi.org/10.1109/icac.2016.29.
31. Schroeder, G. N., Steinmetz, C., Pereira, C. E., & Espindola, D. B. (2016). Digital twin data modeling with AutomationML and a communication methodology for data exchange. In *IFAC-PapersOnLine* (pp. 12–17). Elsevier B.V. https://doi.org/10.1016/j.ifacol.2016.11.115.
32. Kraft, E. M. (2016). The US air force digital thread/digital twin—Life cycle integration and use of computational and experimental knowledge II. The evolution of integrated computational/experimental fluid dynamics. In: *54th AIAA Aerospace Sciences Meeting* (pp. 1–22). https://doi.org/10.2514/6.2016-0897.
33. Bajaj, M., Zwemer, D., & Cole, B. (2016). Architecture to geometry—Integrating system models with. In *AIAA SPACE Forum* (pp. 1–19). https://doi.org/10.2514/6.2016-5470.
34. Sacco, M., Pedrazzoli, P., & Terkaj, W. (2010). VFF : Virtual Factory Framework. In *2010 IEEE International Technology Management Conference (ICE)* (pp. 1–8).
35. Terkaj, W., & Urgo, M. (2014). Ontology-based modeling of production systems for design and performance evaluation. In *2014 12th IEEE International Conference on Industrial Informatics (INDIN)* (pp. 748–753).
36. Garetti, M., & Fumagalli, L. (2012). Role of ontologies in open automation of manufacturing systems. In *Proceedings of XVII Summer School of Industrial Mechanical Plants*—12/9/2012–14/9/2012, Venice, Italy.
37. Negri, E., Fumagalli, L., Macchi, M., & Garetti, M. (2015). Ontology for service-based control of production systems. In *APMS 2015, Part II, IFIP AICT 460* (pp. 484–492). http://www.springerlink.com/index/10.1007/978-1-4020-9783-6.
38. Fumagalli, L., Pala, S., Garetti, M., & Negri, E. (2014). Ontology-based modeling of manufacturing and logistics systems for a new MES architecture. In *APMS 2014, IFIP Advances in Information and Communication Technology 438 (PART I)* (pp. 192–200).
39. Cerrone, A., Hochhalter, J., Heber, G., & Ingraffea, A. (2014). On the effects of modeling as-manufactured geometry: Toward digital twin. *International Journal of Aerospace Engineering, 2014,* 1–10.
40. Fourgeau, E., Gomez, E., Adli, C. Fernandes, H., & Hagege, M. (2016). System engineering workbench for multi-views systems methodology with 3DEXPERIENCE platform. The aircraft RADAR use case. *Complex System Design & Management Asia, 426*, 269–270. https://doi.org/10.1007/978.
41. Yang, J., Zhang, W., & Liu, Y. (2013). Subcycle fatigue crack growth mechanism investigation for aluminum alloys and steel (Special Session on the Digital Twin). In *54th AIAA/ASME/ASCE/AHS/ASC Structures, Structural Dynamics and Materials Conference* (p. 1499). https://doi.org/10.2514/6.2013-1499.
42. Grieves, M., & Vickers, J. (2016). Digital twin : Mitigating unpredictable, undesirable emergent behavior in complex systems. In *Transdisciplinary Perspectives on Complex Systems* (pp. 85–113). https://doi.org/10.1007/978-3-319-38756-7.
43. Abramovici, M., Gobel, J. C., & Dang, H. B. (2016). Semantic data management for the development and continuous reconfiguration of smart products and systems. *CIRP Annals—Manufacturing Technology, 65,* 185–188. https://doi.org/10.1016/j.cirp.2016.04.051.
44. Grinshpun, G., Cichon, T., Dipika, D., & Roßmann, J. (2016). From virtual testbeds to real lightweight robots: Development and deployment of control algorithms for soft robots, with particular reference to industrial peg-in-hole insertion tasks. In *Proceedings of ISR 2016: 47st International Symposium on Robotics* (pp. 208–214).
45. Smarslok, B. P., Culler, A. J., & Mahadevan, S. (2012). Error quantification and confidence assessment of aerothermal model predictions for hypersonic aircraft. In *53rd AIAA/ASME/ASCE/AHS/ASC Structures, Structural Dynamics and Materials Conference* (p. 1817). https://doi.org/10.2514/6.2012-1817.

46. Ríos, J., Morate, F. M., Oliva, M., & Hernández, J. C. (2016). Framework to support the aircraft digital counterpart concept with an industrial design view. *International Journal of Agile Systems and Management, 9,* 212–231. https://doi.org/10.1504/IJASM.2016.079934.
47. Arisoy, E. B., Ren, G., Ulu, E., Ulu, N. G., & Musuvathy, S. (2017). A data-driven approach to predict hand positions for two-hand grasps of industrial objects. In *ASME 2016 International Design Engineering Technical Conferences and Computers and Information in Engineering Conference* (pp. V01AR02A067–11 pp). https://doi.org/10.1115/detc2016.
48. Wang, H.-K., Haynes, R., Huang, H.-Z., Dong, L., & Atluri, S. N. (2015). The use of high-performance fatigue mechanics and the extended Kalman/ particle filters, for diagnostics and prognostics of aircraft structures. *CMES: Computer Modeling in Engineering & Sciences, 105,* 1–24.
49. Holzwarth, P., Tuegel, R., & Kobryn, E. (2012). Airframe digital twin: An overview. In *Prognosis Health Management Solutions Conference MFPT 2012* (p. 20).
50. Garetti, M., Rosa, P., & Terzi, S. (2012). Life cycle simulation for the design of product–service systems. *Computers in Industry, 63,* 361–369. https://doi.org/10.1016/j.compind.2012.02.007.
51. Garetti, M., Macchi, M., Pozzetti, A., Fumagalli, L., & Negri, E. (2016). Synchro-push: A new production control paradigm. In *Summer School Francesco Turco 2016* (pp. 150–155).

Chapter 14
A Social Network of Collaborating Industrial Assets

Hao Li, Adrià Salvador Palau and Ajith Kumar Parlikad

Abstract The IoT (Internet of Things) concept is being widely regarded as the fundamental tool of the next industrial revolution—Industry 4.0. As the value of data generated in social networks has been increasingly recognised, Social Media and the Internet of Things have been integrated in areas such as product-design, traffic routing, etc. However, the potential of this integration in improving system-level performance in industrial environments has rarely been explored. This chapter discusses the feasibility of improving system-level performance in industrial systems by integrating Social Networks into the IoT concept. We propose the concept of a Social Internet of Industrial Assets (SIoIA) which enables the collaboration between assets by sharing status data. We also identify the building blocks of SIoIA and characteristics of one of its important components—Social Assets. A sketch of the general architecture needed to enable a Social Network of Collaborating Industrial Assets is proposed and two illustrative application examples are given.

Keywords Internet of Things · Social networks · Asset management · Diagnostics and prognostics · Condition-based maintenance · Collaborating decision-making

1 Introduction

The large amounts of data generated by well-instrumented assets, together with the rapid development of Information Communication Technologies, has led to growing applications of the IoT concept in industry. During the past years, quite a few IoT applications have been seen in various aspects of the current industrial practices

H. Li · A. S. Palau · A. K. Parlikad (✉)
Institute for Manufacturing, 17 Charles Babbage Rd., Cambridge CB3 0FS, UK
e-mail: aknp2@cam.ac.uk

H. Li
e-mail: hl433@cam.ac.uk

A. S. Palau
e-mail: as2636@cam.ac.uk

A. Crespo Márquez et al. (eds.), *Value Based and Intelligent Asset Management*,
https://doi.org/10.1007/978-3-030-20704-5_14

including environmental monitoring, inventory and production management, food supply chains (FSC), transportation, security, and surveillance [1].

As an important aspect of industrial management, effective asset management is key to reducing the total cost of asset ownership while improving machine availability, guaranteeing security, and increasing productivity. In recent years, the IoT has been increasingly regarded as an effective framework to improve asset management policies, allowing asset managers to have a much broader knowledge of their asset fleet [2, 3]. As a result of this, the notion of SIoT (Social Internet of Things), which results from integrating Social Media into the IoT, has been implemented in application areas such as product lifecycle management [4], traffic routing [5], and workplace help and support [6].

Although quite a few circumstances exist where enhancing social behaviour of industrial assets are likely to be beneficial, the potential of this integration in improving system-level performance in asset fleets has rarely been explored. For instance, in a Social Network, a fleet of assets with similar characteristics could share their diagnostics and prognostics knowledge gained by learning from their own condition data. This could help assets to improve their prognostics accuracy, and also to identify latent problems which would be difficult to notice with only the information available to an asset itself.

Following the SIoT concept, this work attempts to explore the possibility of improving asset management performance by developing a Social Network of Collaborating Industrial Assets for knowledge and data sharing between machines. The paradigm presented in this chapter was first published in the Journal of Risk and Reliability.

Section 2 reviews recent developments in SIoT, use of distributed decision-making systems in different aspects of maintenance optimisation, architectures and frameworks proposed for IoT and SIoT, as well as evolution of what we call Smart Objects. In Sect. 3, our vision of Social Assets are presented and the fundamental properties needed to transform Smart Assets into Social Assets are discussed. Section 4 outlines the building blocks for a Social Internet of Industrial Assets (SIoIA) and presented a general architecture for SIoIA. Subsequently, two illustrative examples of SIoIA applications are presented in Sect. 5. Section 6 provides a guideline on future work. Conclusion of the chapter is given in Sect. 7.

2 Literature Review

The term IoT was first coined in 1999 by The Auto-ID Labs, within the context of supply chain management enabled by RFID (radio-frequency identifications) technology [7]. However, its current definition has been extended to include a dynamic global network infrastructure with self-configuring capabilities, where physical and virtual things have identities, physical attributes, and virtual personalities, use intelligent interfaces, and are seamlessly integrated into the information network [8]. Today, vast amounts of data are generated and shared across the IoT [9]. Examples range from self-driving cars [10] and continuously monitored gas turbines to the immensely popular smart phones, GPS enabled wristbands and other wearables.

Smart phones differ from other IoT-powered devises in that they are designed, operated and marketed as a prominently social tool. Profiting from this social dimension, the data generated by such phones has been widely used in the consumer world to benchmark and optimise product quality and customer experience. For instance, companies like Garmin, Nike and Microsoft have provided platforms for consumers to share and compare exercise data collected via smart phones and other smart gadgets [11].

Besides smart phones, the notion of incorporating social elements to the IoT has been around for approximately a decade, leading to the development of SIoT. One of the early ideas associated with SIoT is "Blogject", a neologism meaning "things that blog". An example was a flock of pigeons that were equipped with telematics for wireless communication, a GPS device for track tracing, and sensors to record the content of air pollutants [12]. The potential of combining social and technical networks has also been tested on service provision to both human users and technical systems. A use-case of a socio-technical network—The Cognitive Office, was reported, where Twitter was used to enable an online social network for objects in a smart office to post events from selected sensors and listen for Tweets from other devices [6]. The exploitation of SIoT can also be found in traffic routing problems, such as opportunistic communication enabled by social networks in dynamic traffic networks [5]. Extending beyond objects socialising with each other, the integration of humans into SIoT has also been discussed, adding the human-to-thing element to achieve the complete vision of SIoT [13].

The growing volumes of available data generated by modern industrial assets equipped with sensors, and the concurrent development of advanced data analytic tools and artificial intelligence has naturally led to discussions of the SIoT in the context of industrial assets. An incentive application of the SIoT focuses on trying to improve industrial system performance by making use of distributed decision-making with the help of the IoT, giving a social dimension to industrial assets. For example, the attempts of using distributed decision-making in production scheduling, maintenance scheduling, and inventory management can be found in [14–16], respectively.

In some cases, the social element in these approaches is not the human operator, like in the case of smart phones, but the asset itself. In other cases both humans and machines act socially. In this chapter, we focus in the cases where most of the social collaboration is performed by industrial assets, and where human agents are limited to setting the system constraints. In order to set up such systems, one must first address the understanding of distributed artificial intelligence techniques.

The introduction of distributed artificial intelligence, mainly based on agent-based systems and holonic manufacturing systems paradigms, is usually aimed to satisfy production requirements such as customisation, agility, flexibility, and robustness. A common distributed artificial intelligence approach is to use Multi-Agent Systems (MAS), which can be defined as "distributed systems of independent actors, called agents, that cooperate or compete to achieve a certain objective" [17]. Broadly, agent-based systems have been designed for one of the following purposes: maintenance resource integration, machinery fault diagnostics and prognostics, and maintenance scheduling. The rest of the section surveys recent advance of agent-based systems in the two latter areas which are of more relevance to the purpose of this chapter.

A Multi-Agent System-based reference model for fault management system (FMS) has been developed by [18]. The FMS can be integrated with the supervision applications to support the decision-making on the controlled processes including component monitoring, failure detection, failure prediction, maintenance scheduling, and maintenance plan execution. A prototype of the reference model has been implemented on Java Development Framework (JADE) to a pool pumping system as a case study. Cerrada's paper had followed from a line or research starting in the early 2000s: arguing that distributed fault detection and handling is a more suited paradigm for fault management systems, Ouelhadj [19] described a multi-agent architecture for distributed and real-time monitoring. The major functions of the monitoring system are performed via the information exchange and co-ordination based on Contract Net Protocol (CNP) between a set of Resource Monitoring Agents (RMA) each responsible for a manufacturing resource. Focusing on data interpretation and condition monitoring applications [20], introduced a hierarchical decentralised multi-agent architecture named COMMAS. Unlike the work of [19] where one agent is assigned to one production resource to perform a wide range of information finding tasks, the agents in COMMAS each represent one aspect of application so as to distribute the responsibilities of information processing. Three hierarchical categories of agents responsible for data fusion, cross sensor corroboration, and reasoning and decision-support functions, respectively, have been proposed. In a later work [21], the proposed COMMAS architecture was implemented for the design of a multi-agent transformer condition monitoring system using K-means clustering, rule induction, and a back-propagation neural network. Another work of his [22] developed an anomaly detection system employing an extended COMMAS architecture. An infotronics-based prognostic prognostics tool called The Watchdog Agent was proposed for product performance degradation assessment and prediction [23]. The Watchdog Agent is capable of diagnosing the current state and prognosticating the future state of its objective component based on the readings from multiple sensors.

The multi-agent paradigm, commonly used in reactive and dynamic production scheduling, can also be adopted for maintenance scheduling problems. Coudert et al. [24] proposed a production and maintenance integrated scheduling system called the RAMSES-II (Reactive Multi-agent System for Scheduling) based on fuzzy logic. In this framework, every manufacturing resource is associated with a machine agent in charge of bidding for production tasks, a maintenance agent that creates maintenance task orders, and a negotiation agent that reconciles the conflicts between the machine and the maintenance agents to get the maximal aggregated degree of satisfaction of two parties. Blackboards are employed to provide a virtual Gantt chart view of the ongoing negotiation process at different conceptual levels. Targeting at the same problem in a flow shop, Khelifati and Benbouzid-Sitayeb [25] proposed a similar agent-based approach comprised of machine agents and maintenance agents to simultaneously schedule production and periodic maintenance. A bus maintenance scheduling method based on multi-agent systems was proposed by Zhou et al. [15] to make up for the downsides of centralised scheduling while dealing with unforeseen events. The model is formulated using three layers and four types of agents to heuristically schedule incoming maintenance tasks cooperatively using CNP. An agent-based system for dynamic scheduling of maintenance tasks in the petroleum industry using reinforcement learning was developed by Aissani et al. [26], and put into experimentation to an Algerian petroleum refinery. The system consists of 'resource agents' for the pumps, 'parts agents' for the tanks containing oil, and an 'observer agent' that has a global view of the system. Their work differs from the aforementioned researches in that there is no centralised decision-making mechanism since each resource and part agent decides its next action depending on its own knowledge base. Feng et al. [27] employed MAS with a two-layer structure for online CBM decision-making among a mission-oriented aircraft fleet considering the constraint of limited maintenance resources. The coordination takes place at two levels both following a heuristic rule-based negotiation mechanism: the local scheduling decision is made via negotiation between Aircraft agents and Maintenance Agents representing maintenance teams while the global scheduling is done by the Management and Coordination Agent. Also dealing with aircraft fleets [28], collaborated with Boeing in order to enable its assets to become "Self-Serving Assets", with the goal "for assets to autonomously plan their own service and maintenance while collaborating with service and maintenance providers and other assets".

As the underlying technologies of the IoT concept have taken shape, research efforts towards the integration of these various technologies have started to produce architectures and frameworks for the IoT and SIoT. Sánchez López et al. [29] designed an IoT architecture within a Smart Object framework. The proposed architecture components include Smart Objects, network protocols, interfaces, and events and repository databases. A prototype of the architecture for the real-time monitoring of goods in supply chains was implemented using Wireless Sensor Networks and Web Services to show its feasibility and flexibility. Zhang et al. [30] extended the techniques of the IoT to the manufacturing field and developed a four-layer architecture mainly for real-time information capturing and dynamic monitoring and controlling for the manufacturing execution stage. Similarly aimed at industrial environment, Ungurean et al. [31] presented an IoT architecture composed of a data server module and a client application module based on OPC.NET specifications. A five-level Cyber-Physical Systems structure has been proposed for Industry 4.0 manufacturing systems [3]. The corresponding algorithms and technologies at each system layer have been suggested for the desired functionalities of the overall system. Guo et al. [32] has proposed a reference architecture for opportunistic IoT which exploits the potential benefits of human social behaviour in the IoT. Focusing on a specific instance of SIoT, an architecture based on vehicular ad hoc networks has been proposed that identifies social structures and related interactions of vehicles in the machine-to-machine social networks [33]. An SIoT architecture following the three-layer model made of the sensing, network, and application layers has been presented by Atzori et al. [34], where the Social components belong to the application layer.

The existing IoT and SIoT architectures and frameworks all have the same constituent element—what is called 'Smart Object' or 'Intelligent Object'. The very first architectures of the IoT are based on the success of RFID technology. While this approach is ideal in tracking physical objects within a confined space, it is insufficient in complex situations as the objects themselves have no analytic or decision-making capabilities. This has led to enhanced requirements for objects to be smart. For instance, in the work of Kosmatos et al. [35], RFID-tagged objects are integrated with its online abstraction positioned with application logic. Arguing that the characteristics of software agents are very similar to those of smart objects, Fortino et al. [36] propose a multi-layered agent-based architecture for the development of Smart Objects. Kortuem et al. [37] categorised Smart Objects into three levels according to their degree of awareness, representation, and interaction.

As shown in this literature review, "Smart Objects" have been around for some time, and its connection to Asset Management has been proposed before by, for example, Brintrup et al. [28] associated to the idea of Self-Serving Assets. However, this connection has only recently been made, and specifics for how to enable inter-asset collaboration are missing in literature. In this chapter, we address the problem of inter-asset collaboration in asset fleets, giving specific examples and detailing the properties of the building blocks of such a Social Network of Collaborating Industrial Assets.

3 Social Assets

Smart Objects can be described as autonomous physical/digital objects augmented with sensing, processing, and network capabilities [37]. What it takes for objects to be 'smart' still applies to the basic components for a Social Network of Collaborating Industrial Assets in the manufacturing domain (i.e., machines or assets). However, as modern production process is often complex and requires co-efforts from a fleet of various assets, instead of just a single asset, interactions and mutual understanding between assets inevitably play a vital role. For instance, in a quarrying process where rocks are crushed and extracted from the mountains and transported by loaders to a conveyor belt, a loader without social capabilities will continue loading rocks to a conveyor belt close to failure, this may lead to severe disruptions if there is a lack of maintenance resource when the failure happens. Thus, apart from being 'smart', these components also need to be 'social'.

In order to characterise Social Assets, it is useful to first review the properties assets are supposed to possess to be 'smart'. This corresponds to the five characteristics of Smart Products/Objects proposed in the new generation of object-centric industrial paradigm [29]:

- Identity—A unique identity is needed for the Smart Asset to be tracked, recorded, and referred to in the system.

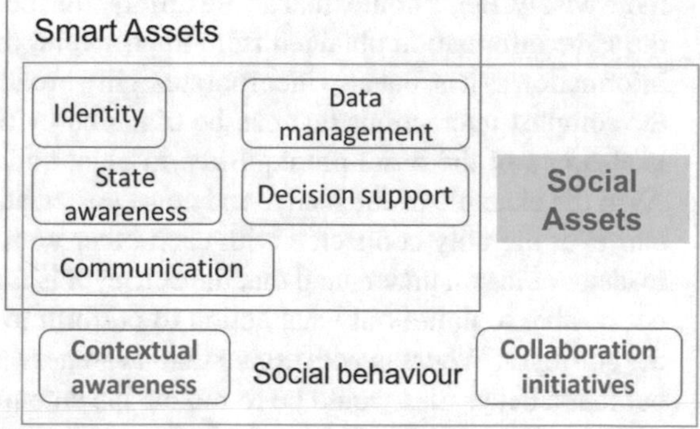

Fig. 1 Fundamentals for Smart Assets and Social Assets

- State awareness—Smart Assets should be able to sense its own operating status and measure performance indexes, such as vibration, temperature, and concentration of a particular chemical element, etc.
- Communication—They should be capable of exchanging information with other Smart Assets, OEMs, and centralised servers.
- Data management—It is necessary to classify and store measurements made by sensor transducers associated with the assets, while removing some history data, if appropriate.
- Decision support—Smart Assets should have a decision support mechanism to generate apt actions in response to different situations that have arisen. Moreover, they should be able to exhibit opportunistic, goal-directed behaviour and act proactively.

One of the earlier works to systematically discuss the characteristics of Social Objects and policies needed to establish a navigable SIoT is done by Atzori et al. [34]. In a later work, Atzori et al. [38] discussed the main features that a SIoT will possess once the aforementioned basic relationships have been established. However, the fundamental properties needed for objects to socialise and collaborate, in particular in industrial environment, are not sufficiently explored.

In our vision, in order for machines to cooperate, as a means to optimising system-level performance, two more fundamental properties are needed—contextual awareness and collaboration initiatives. This transforms Smart Assets into Social Assets, as shown in Fig. 1.

- **Contextual awareness**—A piece of asset with contextual awareness not only investigates its own status, but also perceives its environment, which may be the physical world, a user, a collection of other assets, or the Internet [36]. Assets should have an understanding of the consequence of any action they take that might cause changes to the surroundings, and also the impacts that the behaviour of other participants in the Social Network would have on themselves. Contextual perception implies understanding of a problem domain much broader than a single asset could than the boundaries of the asset itself, thus enabling the asset to act more wisely. Being contextual aware differs from being only connected in that with the same information obtained from the network, the asset not only perceives the information as it is, but also incorporates it in a broader context to better understand the complex interactions that can be triggered by this piece of information. This is also one of the basic prerequisites to plant collaboration initiatives in assets. As in the example of the loader and conveyor mentioned earlier, if the loader and conveyor are only connected with each other with no contextual awareness, the loader would still understand that the conveyor is failing. However, the loader will not be able to figure out what action to perform to lower the degradation rate of the conveyor. Whereas with contextual awareness, the loader would easily figure out that a better idea would be to reduce the amount of rocks to be loaded on the conveyor belt, which will keep the quarry operating until maintenance resource is in place.

- **Collaboration initiatives**—Collaboration initiatives here refers to the fact that an asset views itself as closely related to other assets and is constantly seeking the opportunity to collaborate with other machines to achieve a system-level goal. Enabling assets to have collaboration initiatives is considered necessary in manufacturing systems for two reasons. First, the local goals of a single asset are very likely to be contradictory to the goal of a plant manager or another asset (e.g., two machines push production tasks to each other so as to avoid deterioration and wear and tear while the plant manager wants maximum productivity). If assets are open for cooperation, these conflicts can be resolved through exchange of knowledge, negotiation between machines, and collective decision making. In the example given, the machine with less severe degradation for now could offer to take on more load in exchange for less workload when later its degradation level has risen. Second, just as humans help and give advice to each other, through cooperation initiatives, assets will be inclined to recommend assets of similar status, based on their own experience, the best possible action path to take for the asset itself, as well as for the overall improvement of system-level performance.

4 The Building Blocks for Social Network of Industrial Assets

While Social Assets act as the main building block, other components should also be added during the actual implementation of SIoIA. The building components of SIoIA and their function blocks are shown in Fig. 2. A detailed description is given as follows:

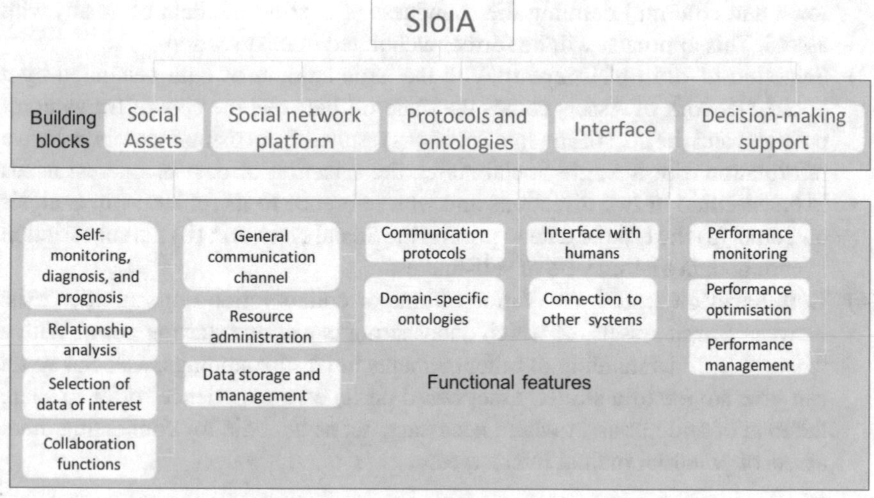

Fig. 2 SIoIA building blocks and their functional features

4.1 *Social Assets*

A software agent is assigned to each asset, augmenting it with intelligent and social behaviours. These assets should be actively engaged throughout the production process. Specifically, Social Assets should be able to carry out:

(1) **Self-monitoring, diagnostics, and prognostics**: Social Assets should be equipped with algorithms to effectively process status data obtained by sensors and events data. These algorithms should be able to store the analysis outcome in an easily accessible way. There are two potential benefits of doing so: (a) providing a model for near real-time machine status evaluation, fault diagnostics, and remaining useful life prediction; (b) allowing fast improvement of the model when new information channel is available. In particular, for the case of a network of Social Assets, new information channels can be built instantly since all machines are actively engaged in communication with others.

(2) **Relationship analysis**: Just like human beings interacting with each other in different ways on the basis of relationships, Social Assets also act differently when faced with other Social Assets based on dynamic metrics quantifying their closeness and similarity. The ability to perform relationship analysis is necessary in certain situations if improved asset performance is to be achieved. For instance, machines of the same type and configuration operating under similar conditions can be assumed to have similar degradation behaviour and thus they can choose to share status data with each other to build more accurate diagnostics and prognostics models of their own. One approach of defining similarity can be via a set of inter-asset similarity metrics with the element in

row i and column j defining the soundness of sharing the data of asset j with asset i. This approach will be further elaborated in next section.

(3) **Selection of data of interest**: With the large amount of data generated by a Social Network of Assets comes the trade-off between the cost of heavy computation, and the cost of myopic decisions resulted from using incomprehensive information. For a single Social Asset, the selection of data of interest needs to be executed in two directions and with reference to its relationship analysis outcome: (a) the type of data to post on the Social Network; (b) the subscription to certain data that may be of self-interest.

(4) **Collaboration functions**: Various forms of collaboration opportunities exist between Social Assets, of which data sharing is just one starting point. With a reasonable understanding of other elements in its surroundings, a Social Asset can give advice to a similar asset based on its own experience, offer to share the load of another asset where necessary, act as a media for connecting other assets of potential mutual interest, etc.

4.2 *Social Network Platform*

A social platform is where the actual social behaviours of assets occur. It should be equipped with necessary hardware and software to function as a channel for:

(1) **General communication**: It should enable (a) machines to post production-related data, (b) machines to subscribe to their own information of interest, and (c) M2M (machine-to-machine) communication.

(2) **Resource administration**: Acting like a yellow page to manage assets registration data and provide a directory service.

(3) **Data storage and management**: Efficiently storing and deleting data shared on the platform.

Research work dedicated to the development of SIoT and IoT architectures that can be extended to meet these demands can be found in [29, 31, 34].

4.3 *Protocols and Ontologies*

Protocols are needed to guarantee a unified data transport format while ontologies provide a mutually understandable communication context. A detailed review of existing protocols in support of the IoT is given in [39]. Examples of ontologies that can be extended or modified to suit this purpose can be found in [40].

4.4 Interface

An interface is needed for data sharing with other systems as well as human operators:

(1) **Interaction with humans**: Although ideally human interventions with the system would be minimal, an interface is needed for human-computer interaction such as receiving parameter inputs, getting commands, sending alerts, and suggesting actions;
(2) **Connection to other systems**: Channels for information exchange with other systems regarding service, finance, and logistics are needed for decision making at the organisation level.

Similar requirements for interfaces have also been mentioned in [41].

4.5 Decision-Making Support

A series of decision-making support functions as listed below should be available. However, it still remains to be explored whether the function should reside in individual assets, at a supervisory level, in a hybrid combining these two, or through other manners.

(1) **Performance monitoring**: Production data sent to the social platform is used by the performance monitor to assess individual asset performance and system performance. The performance monitor can serve to, for instance, identify the bottleneck of the manufacturing plant. At a higher level, by connecting with other relevant databases of the organisation, potential risks to the organisation's objectives can be identified;
(2) **Performance optimisation**: The purpose of the performance optimiser is to generate optimal action plans based on knowledge provided by the performance monitor. The consequent system management plans are then passed on to the performance manager for further consideration;
(3) **Performance management**: The major function of performance management is to select the best system management plan proposed by the performance optimiser and post it on the Social Network Platform to pass to relevant assets to implement.

The rest of the section presents a sketch of the general architecture needed to enable a Social Network of Collaborating Assets, as shown in Fig. 3. The sketch gives the relationships between the building blocks shown in Sect. 4 and how they are operating. The Social Assets here are composed of: (1) physical assets that directly interact with the environment, perform production tasks, and generate condition data [2]; social agents that act as representatives of physical assets in the cyber world and possess the characteristics described in Sect. 3. The interface with humans and connection to other systems can be realised through placing agents representing humans or

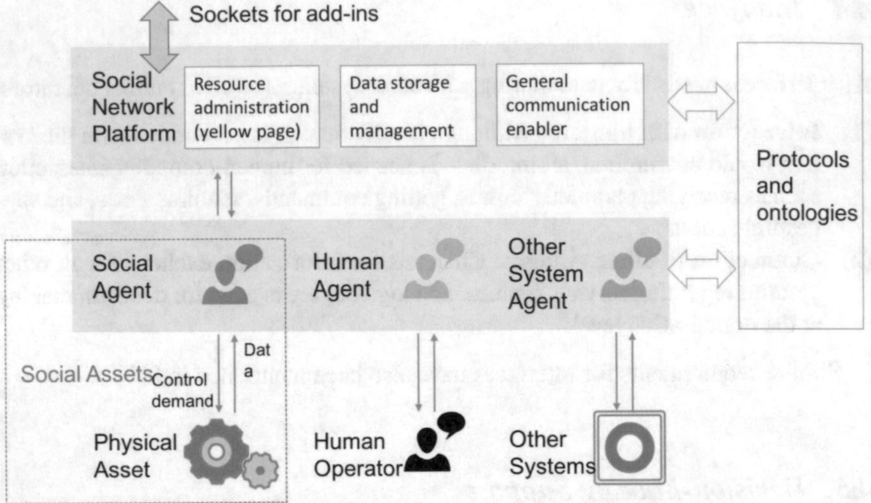

Fig. 3 Sketch of the architecture for a Social Network of Collaborating Assets

other systems in the Social Network Platform. One way of enabling decision-making support, as add-ins, is illustrated here as an example. Here, the Social Network Platforms provide sockets to allow installation of add-ins with different decision-support functions such as plotting higher-dimensional data analytics.

5 Application Examples

5.1 Collaborative Industrial Assets for Workload Sharing

Directly following the two enhancing characteristics of Social Assets—contextual awareness and cooperation initiative, here we present an example of applications of collaborating industrial assets aiming at improving system-level performance by sharing the workload assigned to each other. The two main motivations for this application are:

(1) Parallel machines capable of performing the same type of task are commonly seen in manufacturing plants, enabling the shifting of workload from one machine to the other while still fulfilling the production demand;
(2) The degradation rate of an asset as it operates is closely related by the type and amount of tasks assigned to it [42, 43], making it possible to actively control the residual life of an asset by dynamically changing its workload.

An instance of the proposed architecture in the previous section is presented in Fig. 4 that serves the purpose of workload sharing.

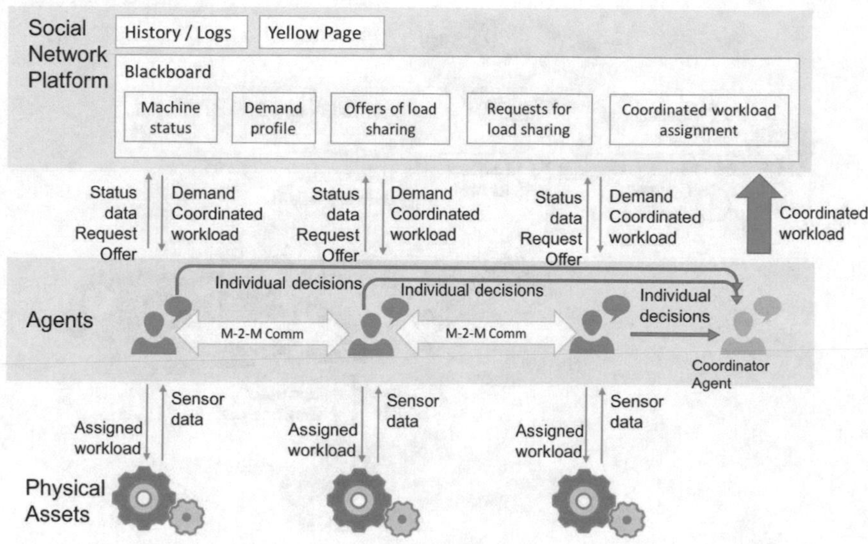

Fig. 4 Instance of the Social Network for Collaborating Assets architecture for workload sharing

The functions of the two types of Agents, the Coordinator Agent and Machine Agents, and the Social Network Platform are described as follows:

Social Network Platform

The Social Network Platform in this case has dual roles: a blackboard and a yellow page. As a blackboard, it displays the up-to-date status of production resources participating in the network, any request for help posted by close-to-failure machines, and production-related information such as the demand profile in the near future. As a yellow page, it maintains a list of agents registered in the Social Network and keeps logs of their activities if necessary.

Coordinator Agent

For computational efficiency, the complex optimisation is performed by machines themselves whereas the Coordinator Agent mainly resolves conflicts of Machine Agents decisions following simple rules or through simple computation.

Machine Agents

Each physical machine has an abstracted representative, a Machine Agent in the virtual world. By checking its own status and the demand profile, the Machine Agents make decisions on the amount of workload to take on, and on the optimal preventive maintenance threshold for itself. The workload decided by Machine Agents individually together with other necessary information needed for coordination is sent to the Coordinator Agent for finalisation. Two other important behaviours of Machine Agents are: seeking help from other machines and offering help to other machines.

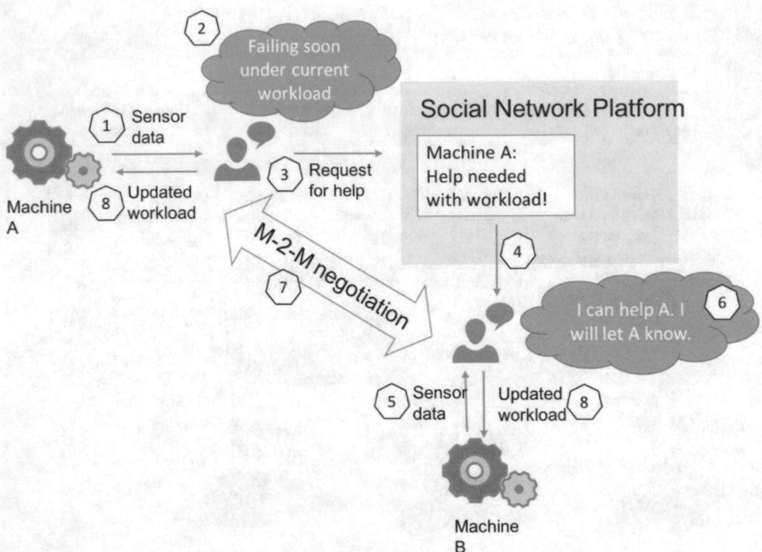

Fig. 5 Data and action flow of a machine seeking help for workload sharing

For illustration purposes, the flow of data exchange and actions in the scenario where a machine seeks help for workload sharing is presented in Fig. 5.

5.2 Distributed Collaborative Prognostics for Industrial Assets

As discussed previously in this chapter, a Social Network of Collaborating Industrial Assets enables assets to take independent decisions in order to optimise performance and monitoring. A common application enabled by continuous monitoring is failure prognostics, which consists of accurately predicting the time to failure of industrial assets [44]. The analysis of estimated probabilities for the asset's time to failure allows asset managers to implement optimised maintenance policies which leverage the risk of failure with the costs of preventive and corrective maintenance, and the associated loss of operational time [45]. Currently, the optimisation of maintenance policies for large asset fleets is usually done within a centralised architecture, with exceptions such as the Watchdog Agent [23], which implements a hybrid approach using human agents and software agents. A social network of collaborating industrial assets enables the implementation of a new kind of architecture designed for improving prognostics in large asset fleets: distributed collaborative learning, which is aimed at reducing human input as much as possible. A detailed insight into the mathematical underpinnings of this concept is also provided in the next chapter.

5.3 *Machine Agents*

In this architecture, assets are able to increase their prediction accuracy by using the knowledge gathered by other similar assets in the asset fleet. Each industrial asset has an agent installed that learns separately from the other assets in the network, and collaborates with them sharing relevant data (see Fig. 6). In practice, each asset must be able to continuously compute the following parameters and vectors:

(1) A list of inter-asset similarity metrics within the asset fleet. A similarity metric is a positive real number that defines the estimated difference between two assets, and that is assumed to be directly related to the statistical soundness of sharing their data points. This is represented by d_{ij} where i is the asset that will receive the data and j is the asset that will send it. The N assets with smaller metric values will be connected to this asset, the set of assets and inter-asset connections is defined as an Asset Network [46].
(2) Failure diagnostics performed by evaluation of the sensor data obtained from the Smart Asset.

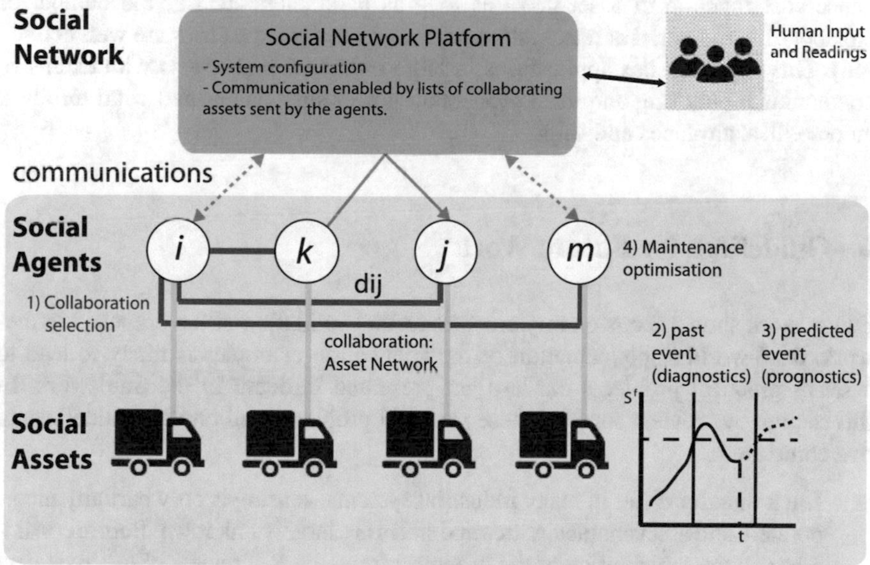

Fig. 6 A schematic of the architecture used for distributed collaborative learning and the functions that each agent must be able to perform. In this case, only Social and Human agents and Social assets are needed, as the coordination is performed in a distributed way. Each Social Asset hosts a Social Agent labelled by a letter marked in italics, and each agent calculates its asset similarity metric by means of the parameter d_{ij} (in purple, shown for the asset (i). Each agent must compute four tasks that are described in Sect. 5.2 and illustrated in the figure. Communications are performed through the social network platform. Here, discontinuous lines indicate communication between the agents and the platform, while the continuous line indicate inter-asset communication enabled by the platform

(3) An estimate of its own time to failure with an associated error, both performed using a combination of the asset's own historical data and data from other collaborating assets.
(4) An optimisation of its own maintenance policy based on the estimated time to failure considering the constraints given by the Social Network Platform.

5.4 *Social Network Platform*

Inter-asset communications and general parameter settings such as the number of collaborating assets, N, and constraints in the maintenance policy are set by the Social Network Platform (see Fig. 6). This means that each list of N collaborating assets will be sent and stored in the Social Network Platform in real time. This corresponds, as commented in Sect. 5.1 to a yellow page function.

This architecture combines the benefits of distributed decision making with the importance of proper sampling in predictive statistics. Preliminary studies have shown that distributed collaborative learning has a positive impact in the maintenance cost function of asset fleets as long as parameters such as the number of collaborating industrial assets, system noise and inter-asset metrics are well defined [46]. This approach has applications in large industrial asset fleets with extensive continuous monitoring and local processing power such as gas and wind turbines, automobiles, airplanes and ships.

6 Guidelines for Future Work

Future work should focus on implementation and validation of the proposed framework. Real-world implementation of the application examples is likely to lead to a set of practical problems that are not presented in detail in the framework. In this section, we review some of these practical problems and provide guidelines to overcome them.

(1) **Lack of asset data**: in many industrial systems, sensing is only partially incorporated, and asset characteristics and states are largely unknown. Before considering an implementation of the proposed framework, a review of the asset fleet must be undertaken. Within the fleet, the asset typologies incorporating a condition monitoring system, are readily adaptable to the proposed architecture by either assigning to them a cloud-based agent or endowing them with processing power. Management must then consider which additional asset typologies in the fleet are candidate for the implementation of a condition maintenance system. This typically corresponds to complex assets with high maintenance costs.
(2) **Real-time diagnostics and prognostics**: the transition to real-time health management is not without problems. Aposteriori diagnostic techniques are often

difficult to adapt dynamically, as the dynamic nature of asset condition is usually not considered. A good enabler of cluster-based dynamic diagnostics is the method presented by Dr. Edzel R. Lapira in his Ph.D. thesis [47]. Regarding prognostics, the main challenge is including censored data in the algorithms and deciding on the frequency of re-training and prediction. An example of how to overcome these issues can be found in [48].

(3) **Appropriate cost metrics**: determining repairing or replacement cost in real-time health management systems is not a trivial problem. If the particular industry is used to rigid maintenance scheduling, the proposed approach may seem operationally impossible. It is important to weight the particularities of the industry and asset type in question to decide on whether a real-time approach makes sense. If so, maintenance and replacement costs should be assessed in conversation with suppliers and clients, as all actors involved in the maintenance life-chain may have to change in order to adapt to the new framework.

(4) **Integrated maintenance and production planning**: at the asset level, the trade-off between keeping the asset in a satisfactory condition as well as maximizing its profitability must be considered. Resolving this issue requires a systematic approach to exhaust, classify, and quantify the cost and benefit caused by each operation or maintenance action.

(5) **Workshop-level decision making**: at the workshop-level, since a fleet of assets of different types and configurations are involved in the production process, it would be worth exploring coordination and negotiation strategies to resolve conflicts of interest between assets to improve system-level performance.

7 Concluding Remarks

In this chapter we have presented what we believe that will be the future paradigm of Asset Management: a Social Network of Collaborating Industrial Assets. In such a network, autonomous assets will take independent decisions and collaborate with each other in a distributed way, coordinated by a central platform. Human input has been reduced as much as possible: setting global constraints and target functions. Therefore, the role of the asset manager will not be any more to choose and optimise performance and maintenance policies but instead to set system constrains and monitor statistics obtained by the Social Network Platform from the asset fleet. In our proposed system, the Social Network Platform will perform the tasks of allowing collaboration, communicating constraints and objective functions to the assets and summarizing fleet information. All the other tasks, such as system optimisation, decision making, collaboration and self-monitoring will be performed by agents installed in the assets. The proposed system is then a bottom to top approach, where assets will have certain independence in choosing how to satisfy the constraints and requirements of the Asset Manager. This work forms part of the first steps towards empowering industrial assets with highly autonomous agents in order to reduce costs and increase efficiency.

Acknowledgements The authors wish to thank co-workers in DIAL in the Institute for Manufacturing of University of Cambridge.

Funding Statement
This research was supported by Sustain-Owner (Sustainable Design and Management of Industrial Assets through Total Value and Cost of Ownership), a project sponsored by the EU Framework Programme Horizon 2020, MSCA-RISE-2014: Marie Skodowska-Curie Research and Innovation Staff Exchange (RISE) (grant agreement number 645733 Sustain-owner H2020-MSCA-RISE-2014). This research was also supported by Cambridge Commonwealth, European and International Trust and China Scholarship Council.

Declaration of Conflicting Interests
The authors declare no potential conflicts of interest with respect to research, authorship and/or publication of this article.

References

1. Li, Y., Hou, M., Liu, H., & Liu, Y. (2012). Towards a theoretical framework of strategic decision, supporting capability and information sharing under the context of Internet of Things. *Information Technology and Management, 13*(4), 205–216.
2. Zhang, Y., Wang, W., Liu, S., & Xie, G. (2014). Real-time shop-floor production performance analysis method for the internet of manufacturing things. *Advances in Mechanical Engineering*.
3. Lee, J., Bagheri, B., & Kao, H. A. (2015). A Cyber-Physical Systems architecture for Industry 4.0-based manufacturing systems. *Manufacturing Letters* [Internet], *3*, 18–23. Society of Manufacturing Engineers (SME). Available from http://dx.doi.org/10.1016/j.mfglet.2014.12.001.
4. Cai, H., Da Xu, L., Member, S., Xu, B., Xie, C., Qin, S., et al. (2014). IoT-based configurable information service platform for product lifecycle management. *IEEE Transactions on Industrial Informatics, 10*(2), 1558–1567.
5. Schurgot, M. R, Comaniciu, C., Jaffr, K. (2012, July 1–8). *Beyond traditional DTN routing : Social networks for opportunistic communication*.
6. Kranz, M., Roalter, L., Michahelles, F. (2010). Things that Twitter: Social networks and the Internet of Things. In *What can Internet Things do Citiz CIoT Work Eighth International Conference Pervasive Computing Pervasive 2010*.
7. Gubbi, J., Buyya, R., Marusic, S., Palaniswami, M. (2013) Internet of Things (IoT): A vision, architectural elements, and future directions. *Future Generation Computing Systems* [Internet], *29*(7), 1645–60. Available from http://dx.doi.org/10.1016/j.future.2013.01.010.
8. Van Kranenburg, R. (2006). *The Internet of Things* [Internet]. Vol. 4, Communications Engineer, 20 p. Available from http://www.csa.com/partners/viewrecord.php?requester=gs&collection=TRD&recid=320585AN.
9. Zaslavsky, A., Perera, C., Georgakopoulos, D. (2013). Sensing as a service and big data. In *Proceedings of the International Conference on Advances in Cloud Computing (ACC)*. Bangalore, India, July, 2012.
10. Gerla, M., Lee, E.-K., Pau, G., Lee, U. (2014). internet of vehicles : From intelligent grid to autonomous cars and vehicular clouds. In *2014 IEEE World Forum on Internet of Things (WF-IoT) Internet* (pp. 241–246).
11. De Saulles, M. (2016). *The Internet of Things and business*. Routledge.
12. Bleecker, J. A. (2006). Manifesto for networked objects —Cohabiting with Pi- geons, Arphids and Aibos in the Internet of Things short title : Why things matter what' s a Blogject ? What about Spimes ? In *Proceedings of the 13th International Conference on Human–Computer Interaction with Mobile Devices and Services, MobileHCI* [Internet] (pp. 1–17). Available from http://research.techkwondo.com/files/WhyThingsMatter.pdf.

13. Ortiz, A. M., Hussein, D., Park, S., Han, S. N., Member, S., Crespi, N., et al. (2014). The cluster between Internet of Things and social networks: Review and research challenges. *IEEE Internet Things Journal, 1*(3), 206–215.
14. Lim, M. K., Zhang, Z., Goh, W. T. (2009). An iterative agent bidding mechanism for responsive manufacturing. *Engineering Applications of Artificial Intelligence* [Internet], *22*(7), 1068–1079. Available from http://dx.doi.org/10.1016/j.engappai.2008.12.003.
15. Zhou, R., Fox, B., Lee, H. P., & Nee, A. Y. C. (2004). Bus maintenance scheduling using multi-agent systems. *Engineering Applications of Artificial Intelligence, 17*(6), 623–630.
16. Jiang, C., Sheng, Z. (2009). Case-based reinforcement learning for dynamic inventory control in a multi-agent supply-chain system. *Expert Systems with Applications* [Internet], *36*(3), 6520–6526. Available from http://dx.doi.org/10.1016/j.eswa.2008.07.036.
17. Tuyls, K., & Weiss, G. (2012). Multiagent learning: Basics, challenges, and prospects. *AI Magazine, 33*(3), 41–52.
18. Cerrada, M., Cardillo, J., Aguilar, J., & Faneite, R. (2007). Agents-based design for fault management systems in industrial processes. *Computers in Industry, 58*(4), 313–328.
19. Ouelhadj, D. (2000, April). Multi-agent architecture for distributed monitoring in flexible manufacturing systems (FMS). *Robot Autom ...* [Internet]. Available from http://ieeexplore.ieee.org/xpls/abs_all.jsp?arnumber=846389.
20. Mangina, E. E., McArthur, S. D. J., McDonald, J. R. (2000). Autonomous agent for distributed problem solving in condition monitoring. In *Intelligent Problem Solving Methodology Approaches 13th International Conference on Industrial Engineering of Applied Artificial Intelligent Expert System* (pp. 683–93).
21. McArthur, S. D. J., Strachan, S. M., & Jahn, G. (2004). The design of a multi-agent transformer condition monitoring system. *IEEE Transactions on Power Systems, 19*(4), 1845–1852.
22. McArthur, S. D. J., Booth, C. D., McDonald, J. R., & McFadyen, I. T. (2005). An agent-based anomaly detection architecture for condition monitoring. *IEEE Transactions on Power Systems, 20*(4), 1675–1682.
23. Djurdjanovic, D., Lee, J., & Ni, J. (2003). Watchdog agent—An infotronics-based prognostics approach for product performance degradation assessment and prediction. *Advanced Engineering Informatics, 17*(3–4), 109–125.
24. Coudert, T., Grabot, B., Archimède, B. (2002). Production/maintenance cooperative scheduling using multi-agents and fuzzy logic. *International Journal of Production Research* [Internet], *40*(18), 4611. Available from http://search.ebscohost.com/login.aspx?direct=true&db=bth&AN=9303866&site=ehost-live.
25. Khelifati, S. L., Benbouzid-Sitayeb, F. (2011). A multi-agent scheduling approach for the joint scheduling of jobs and maintenance operations in the flow shop sequencing problem. *Lecture Notes in Computer Science* (including Subser Lecture Notes in Artificial Intelligence Lecture Notes Bioinformatics), 6923 LNAI (PART 2), 60–69.
26. Aissani, N., Beldjilali, B., Trentesaux, D. (2009). Dynamic scheduling of maintenance tasks in the petroleum industry: A reinforcement approach. *Engineering Applications of Artificial Intelligence* [Internet], *22*(7), 1089–103. Available from http://dx.doi.org/10.1016/j.engappai.2009.01.014.
27. Feng, Q., Li, S., Sun, B. (2012). An intelligent fleet condition-based maintenance decision making method based on multi-agent. *International Journal of Prognostics and Health Management* [Internet], 1–11. Available from https://www.phmsociety.org/sites/phmsociety.org/files/phm_submission/2012/ijphm_12_012.pdf.
28. Brintrup, A., McFarlane, D., Ranasinghe, D., Sánchez López, T., & Owens, K. (2011). Will intelligent assets take off? Toward self-serving aircraft. *IEEE Intelligent System, 26*(3), 66–75.
29. Sánchez López, T., Ranasinghe, D. C., Harrison, M., & McFarlane, D. (2012). Adding sense to the Internet of Things: An architecture framework for Smart Object systems. *Personal and Ubiquitous Computing, 16*(3), 291–308.
30. Zhang, Y., Zhang, G., Wang, J., Sun, S., Si, S., & Yang, T. (2014). Real-time information capturing and integration framework of the internet of manufacturing things. *International Journal of Computer Integrated Manufacturing* [Internet], *3052*(December), 1–12. Available from http://www.tandfonline.com/doi/abs/10.1080/0951192X.2014.900874#.VI9aA4qsWZ4.

31. Ungurean, I., Gaitan, N. C., Gaitan, V. G. (2014). An IoT architecture for things from industrial environment. *International Conference* on *Communications.*
32. Guo, B., Yu, Z., Zhou, X., Zhang, D. (2012). Opportunistic IoT: Exploring the social side of the Internet of Things. In *Proceedings of 2012 IEEE 16th International Conference on Computer Support* Cooperative *Work Design CSCWD* (pp. 925–929).
33. Alam, K. M., Saini, M., & Saddik, A. E. L. (2015). Toward social internet of vehicles: Concept, architecture, and applications. *IEEE Access, 3,* 343–357.
34. Atzori, L., Iera, A., Morabito, G., Nitti, M. (2012). The Social Internet of Things (SIoT)—When social networks meet the Internet of Things: Concept, architecture and network characterization. *Computer Networks* [Internet], *56*(16), 3594–608. Available from http://dx.doi.org/10.1016/j.comnet.2012.07.010.
35. Kosmatos, E. A., Tselikas, N. D., Boucouvalas, A. C. (2011, April). Integrating RFIDs and smart objects into a unified Internet of Things architecture. *Advances in Internet Things* [Internet], 5–12. Available from http://www.scirp.org/journal/PaperDownload.aspx?paperID=4696.
36. Fortino, G., Guerrieri, A., Russo, W., Bucci, V. P., Cs, R. (2012). Agent-oriented smart objects development. In *Computer Supported Cooperative Work in Design (CSCWD), 2012 IEEE 16th International Conference on Wuhan* (pp. 907–912).
37. Kortuem, G., Kawsar, F., Sundramoorthy, V., Fitton, D. (2010). Smart objects as building blocks for the Internet of Things. *IEEE Internet Computer* [Internet], *14*, 44–51. Available from http://oro.open.ac.uk/31631/.
38. Atzori, L., Iera, A., & Morabito, G. (2014). From, "smart objects" to "social objects": The next evolutionary step of the Internet of Things. *IEEE Communications Magazine, 52*(1), 97–105.
39. Al-fuqaha, A., Guizani, M., Mohammadi, M., Mohammed, A., & Ayyash, M. (2015). Internet of Things: A survey on enabling technologies, protocols, and applications. *IEEE Communications Surveys and Tutorials, 17*(4), 2347–2376.
40. Geerts, G. L., Leary, D. E. O. (2014). A supply chain of things : The EAGLET ontology for highly visible supply chains. *Decision Support System* [Internet], *63*, 3–22. Available from http://dx.doi.org/10.1016/j.dss.2013.09.007.
41. Iung, B., Levrat, E., Marquez, A. C., & Erbe, H. (2009). Conceptual framework for e-Maintenance: Illustration by e-Maintenance technologies and platforms. *Annual Review in Control, 33*(2), 220–229.
42. Celen, M., & Djurdjanovic, D. (2015). Integrated maintenance decision-making and product sequencing in flexible manufacturing systems. *Accept Publication ASME Journal Manufacturing Science and Engineering, 2016*(137), 1–15.
43. Hao, L., Liu, K., Gebraeel, N., Shi, J. (2015, October). Controlling the residual life distribution of parallel unit systems through workload adjustment. *IEEE Transactions on Automation Science* and *Engineering* [Internet], 1–11. Available from http://ieeexplore.ieee.org/lpdocs/epic03/wrapper.htm?arnumber=7302091.
44. Goyal, D., Pabla, B. S. (2015). CIRP Journal of Manufacturing Science and Technology Condition based maintenance of machine tools—A review. *CIRP Journal of Manufacturing Science and Technology* [Internet]. CIRP, *10*, 24–35. Available from http://dx.doi.org/10.1016/j.cirpj.2015.05.004.
45. Dabney, T., Hernandez, L., Scandura, P. A., Vodicka, R. (2008). Enterprise health management framework—A holistic approach for technology planning, R & D collaboration and transition. In *Proceedings—2008 International Conference on Prognostics and Health Management.*
46. Salvador Palau, A., Liang, Z., Parlikad, A. K, Lütgehetmann, D. (2017). Collaborative prognostics in social asset networks. *Future Generation Computing Systems.*
47. Lapira, E. R. (2012). *Fault detection in a network of similar machines using clustering approach.* University of Cincinnati.
48. Martinsson, E. (2016). *WTTE-RNN : Weibull time to event recurrent neural network a model for sequential prediction of time-to-event in the case*. Chalmers University of Technology, University of Gothenburgh.

Chapter 15
Collaborative Prognostics in Social Asset Networks

Adrià Salvador Palau, Zhenglin Liang, Daniel Lütgehetmann and Ajith Kumar Parlikad

Abstract With the spread of Internet of Things (IoT) technologies, assets have acquired communication, processing and sensing capabilities. In response, the field of Asset Management has moved from fleet-wide failure models to individualised asset prognostics. Individualised models are seldom truly distributed, and often fail to capitalise the processing power of the asset fleet. This leads to hardly scalable machine learning centralised models that often must find a compromise between accuracy and computational power. In order to overcome this, we present a novel theoretical approach to collaborative prognostics within the Social Internet of Things. We introduce the concept of Social Asset Networks, defined as networks of cooperating assets with sensing, communicating and computing capabilities. In the proposed approach, the information obtained from the medium by means of sensors is synthesised into a Health Indicator, which determines the state of the asset. The Health Indicator of each asset evolves according to an equation determined by a triplet of parameters. Assets are given the form of the equation but they are not aware of their parametric values. To obtain these values, assets use the equation in order to perform a non-linear least squares fit of their Health Indicator data. Using these estimated parameters, they are interconnected to a subset of collaborating assets by means of a similarity metric. We show how by simply interchanging their estimates, networked assets are able to precisely determine their Health Indicator dynamics and reduce maintenance costs. This is done in real time, with no centralised library, and without the need for extensive historical data. We compare Social Asset Networks with the typical self-learning and fleet-wide approaches, and show that Social Asset Networks

A. Salvador Palau (✉) · Z. Liang · A. K. Parlikad
Department of Engineering, Institute for Manufacturing, University of Cambridge, Cambridge CB3 0FS, UK
e-mail: as2636@cam.ac.uk

Z. Liang
e-mail: zhenglin_liang@sina.com

A. K. Parlikad
e-mail: aknp2@cam.ac.uk

D. Lütgehetmann
Institut für Mathematik, AG Topologie, Arnimallee 7, Raum 212, 14195 Berlin, Germany
e-mail: daniel.lutgehetmann@fu-berlin.de

A. Crespo Márquez et al. (eds.), *Value Based and Intelligent Asset Management*,
https://doi.org/10.1007/978-3-030-20704-5_15

have a faster convergence and lower cost. This study serves as a conceptual proof for the potential of collaborative prognostics for solving maintenance problems, and can be used to justify the implementation of such a system in a real industrial fleet.

Keywords Maintenance · Prognostics · Networks · SIoT · Asset management

1 Introduction

Asset health and performance management can be summarised as the task of managing assets in order to reduce the risk of failures and performance reduction while minimising cost [1]. In recent decades, this field of study has received attention due to the increased focus on services undertaken by manufacturing firms [2], and the rapid spread of Internet of Things technologies in industrial systems [3–7]. Such technologies allow assets to record, store and process data in situ, and in some cases also allow them to undertake independent actions [8]. The connection between asset management and the Social Internet of Things can be drawn on previous work by our group, where we presented the fundamental building blocks for the SIOIT (Social Internet of Industrial Things). We named one of these blocks "Social Assets", defined as assets with an assigned software agent augmenting them with intelligent and social behaviours [7]. In this chapter we present a novel approach for collaborative prognostics by applying the concept of Social Networks of Collaborating Industrial Assets presented in the previous chapter to the study of the dynamics of a network of such assets coupled to an asset health management policy.

A prolific field of health management is Prognostics, which is defined as "Predictive Diagnostics which includes determining the remaining useful life or time span of useful operation for a component" [9]. Effective Prognostics depend on a combination of precise evaluation of the current asset state together with reliable prediction models [10]. Prognostics have become increasingly feasible over the last decade, in which deployment of networked sensors has become the norm in many maintenance-intensive industries [11, 12]. This has enabled the development of Condition Based Maintenance, where the sensor data is gathered and studied in order to obtain a continuous assessment of the state of the asset [10]. Such state is often characterised using a synthetic variable known as Health Indicator, which is formed by a combination of inputs of the many sensors installed in the asset [13, 14].

The definition of a Health Indicator, as well as the choice of the model used for prognostics, is tied to multiple layers of error and uncertainty. For example, a typical method for sensor selection, multiple regression, relies heavily on assumptions such as linear independence of the variables, and always carries associated uncertainties. Prognostics models often rely on a limited sample which is taken to be representative of the population, and from which a general model is inferred. Regardless of these limitations, for assets with a well-understood deterioration process, it is possible to find a parametric model that can be used to predict the future state of the Health Indicator [15]. In this chapter, we study assets of this kind, where the time series of the

Health Indicator is defined by an inverse exponential model. This is a popular model, as it presents an initially slow decay followed by a period of faster deterioration [16].

Systems composed of several assets are referred to as multi-unit or multi-component systems. Multi-component systems are those treated as a multiplicity of assets forming a larger asset (for example, airplanes). Alternately, multi-unit systems usually refer to fleets of several assets of the same kind, sharing a subset of common failure modes. In this chapter we study the latter. Usually, multi-unit systems are reduced to systems where all units are assumed to be fundamentally equal [17, 18]. However, this assumption is known in many cases to not be true and different maintenance policies may apply better to each individual asset. A novel approach to deal with multi unit systems is the Social Internet of Things [7], which provides a paradigm for objects to connect with each other and exploit asset individualities to improve global performance [19].

In the Social Internet of Things, the fleet of connected objects form a network. Research exists using the technologies enabling the Social Internet of Things to conceptualise a network of assets in order to perform diagnostics. For instance, Edzel Lapira developed clustering techniques in order to obtain individualised fault-detection in a network of similar machines [20]. Lapira's approach, despite being adaptable to nonstationary regimes, does not consider a truly distributed architecture as most computations are done centrally. Other research exists where the concept of cluster-informed diagnostics is exploited in real-life experiments [21].

Despite these advances in the field of diagnostics, asset networks for prognostics have remained largely unexplored. Most existing multi-unit health prognostics methods are centralised and dependent on a rich set of historical data. Typically, the time series measurements of the Health Indicator are registered and saved in a database. Then, the trajectory of each asset is compared with previously recorded trajectories through a similarity metric, and the closest matches are used to obtain reliable prognostics [16, 22]. In such an approach it is common to use the k nearest trajectories and weight them in order to increase precision [23]. Other approaches consist of using machine learning to classify and predict failures, providing an estimate of the useful time to failure [24].

Multi-unit systems have also been studied in the field of Distributed Decision Making, which deals with problems involving multiple decisions aimed to optimise an objective function [25]. However, Distributed Decision Making is only scarcely incorporated into maintenance, usually coupled to human decisions [26], and artificial immune systems [27]. A largely unexplored field of Distributed Decision Making is collaborative health management, aimed to study how assets could collaborate with each other in order to improve maintenance and operational policies. This chapter aims to help bridging this gap, providing the first theoretical treatment of the mathematical implications of distributed collaborative prognostics. The research presented in this chapter was first published in the Future Generation Computer Systems journal.

Multi-unit centralised systems depending on large amounts of data are likely to encounter computational limitations, and can only be implemented once a comprehensive data set is available. So far, existing studies have largely overlooked this issue by utilising increasingly powerful centralised computers, and providing prognostic results from already established failure data libraries. Therefore, no theoretical computation of the cost drawbacks and benefits of de-centralising predictive maintenance decision exists in literature. In this chapter we explore how through collaborative prognostics, the Social Internet of Things can provide a solution for these known drawbacks. Thus providing theoretical grounding to implement the proposed approach in a real industrial system.

In Sect. 2 we describe the theoretical foundation of the problem. We begin by describing the system in Sect. 2.1, and continue by discussing the collaborative prognosis algorithm in Sect. 2.2. A detailed description of the maintenance policy follows in Sect. 2.3, followed with a discussion of the importance of uncertainty and error in the system (Sect. 2.4). In Sect. 3 we describe a simulation of the proposed approach for the case of a fleet of assets corresponding to three different families of assets, the results of the simulation are presented in Sect. 3.1. We show how the cost per time depends on the number of collaborating assets, the weighting of the data, the noise of the system and the maintenance policy in a non-trivial way. We also propose a small modification of the approach for large time steps (Sect. 3.2). In Sect. 4 we discuss the findings resulting from the simulation and the limitations of our approach. The conclusion follows in Sect. 5. The chapter concludes with a discussion of future work (Sect. 6), suggesting the computation of the case of multiple families of assets separated by different parametric distances.

2 Theoretical Foundation

We propose a distributed collaborative method which considers asset similarities in order to improve the reliability of prognostics. In order to do so, we introduce the concept of Social Asset Networks: a Social Asset Network is a set of assets with local computational and sensing capabilities which can communicate with each other through an IoT platform. In the system studied in this chapter, each asset has a digital replica, a software agent. From now on, when we refer to an asset performing an action, we actually mean the software agent associated with that given asset.

2.1 Description of the System

The system consists of a set of N assets connected to a network and able to communicate with each other. In practice, assets communicate through a Social Network Platform, which can be a single or several coordinated servers. Apart of communication, the Social Network Platform takes the role of system configuration, and allows for assets to join and leave the system at any time (see Fig. 1).

Inter-asset communications can be performed using any of the several existing IoT protocols [3, 28]. In the proposed system most of the data is treated locally by the

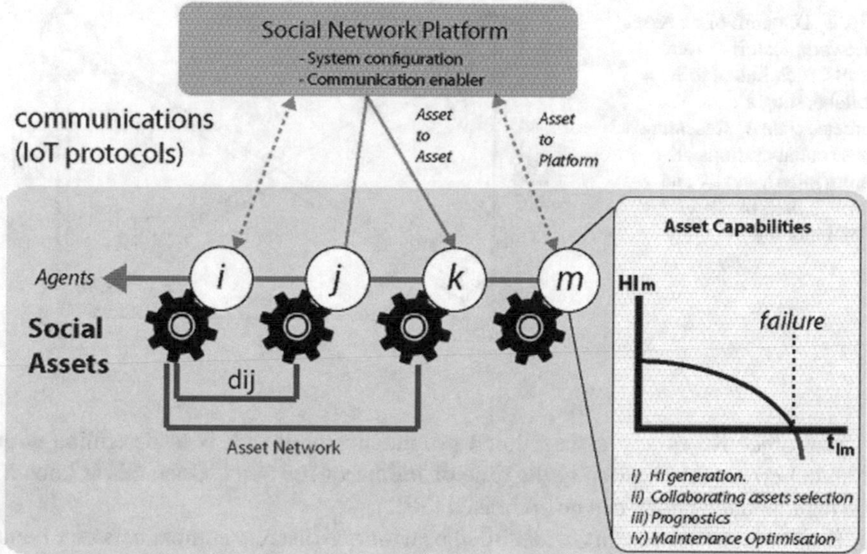

Fig. 1 Diagram of the proposed system, featuring the Social Assets, its embedded agents, the Social Network Platform, the different kinds of connections between the Assets and the platform, and the Assets' capabilities

assets and the data sent through communication is only a synthesised extraction of the sensor data. Therefore, communication costs are deemed negligible with respect of the cost optimisation presented in this chapter. In practice, this assumption must be carefully evaluated by studying the connection policy of each particular Asset Network. In a distributed system, processing costs only weakly depend on the utilisation of the distributed processing power of the assets through their power consumption. For the system proposed in this section, all assets are assumed to incur in a constant processing cost, regardless of the number of neighbours they are linked to. Therefore, such costs are not weighted in the calculations either.

In the proposed system, assets are able to measure the real value of their Health Indicator, which corresponds to a synthetic indicator formed by a compilation of sensor data as in [16]. When this Health Indicator reaches 0, the asset fails. The assets are assumed to know the form of the equation governing their Health Indicator, but not the concrete values of its parameters. The aim of the assets is to determine these parameters by collaborating with each other, in order to obtain precise prognosis and implement maintenance strategies based on such prognosis (see Fig. 1). Several methods can be used to obtain the Health Indicator value from the multi-sensor time-series data, some of the most common being logistic regression [29], and multi-linear regression. The Health Indicator model used in this chapter is characteristic of datasets conducive to multi-linear regression, as the logistic curve has asymptotes at 0 and 1 (we prefer a Health Indicator that crosses the abscissa axis in order to define a clear failure boundary).

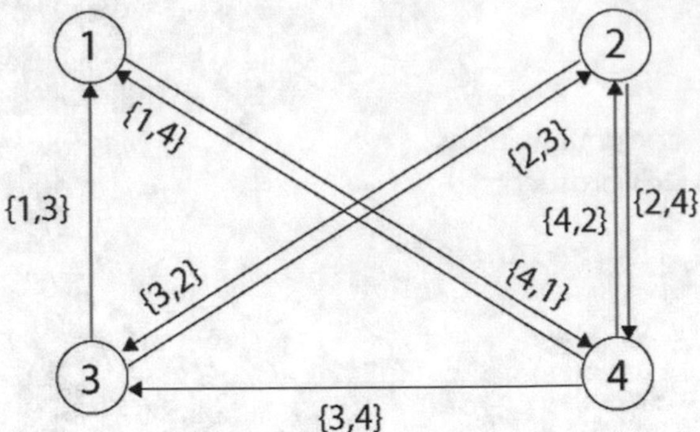

Fig. 2 Diagram of an Asset Network, featuring four assets, each linked to $N_i = 2$ collaborating assets. The directed links representing these collaborations are represented visually and through their binary identifiers i, j

The objective of any asset-tailored prognostics approach is to determine with minimal error and uncertainty the time of failure of the asset. Once this is known, the maintenance policy can be optimised [30].

Each asset in the network is identified by a unique discrete number between 1 and N. Although all the assets are connected to the network, at any given time each asset creates a list of N_i collaborating assets. There collaborations are represented using directed links (see Fig. 2). Such links are identified by binary sets of asset identifiers $\{i, j\}$, *i* is used to identify the asset on the receiving end of the link, and *j* to the asset on the sending end of the link.

The Health Indicator of each asset *i*, HI_i is chosen to follow the following equation (as in [16]):

$$\text{HI}_i(t_{li}) = a_i\left(1 - e^{-b_i(t_{fi} - t_{li})} + \varepsilon_{0,\sigma}\right). \quad (1)$$

where t_{li} is the local time of the asset: the time since the last repair or since the asset was installed. (a_i, b_i, t_{fi}) are assumed to be the parameters that govern the health of a particular asset. a_i is an amplitude parameter, determining the expected value of the Health Indicator at $t_{li} = 0$. t_{fi} is the expected time of failure. b_i is a curvature parameter: in practice the smaller b_i is the quicker the asset is deteriorating. This has no influence on t_{fi} but it affects the probability of failure at smaller values of t_{li}, which for smaller b_i increases significantly. $\varepsilon_{0,\sigma}$ is a Gaussian noise term with mean 0 and standard deviation σ which is included to represent realistic measurement limitations.

Assets can be in two different states: functioning and failure. An asset has failed if at any time its Health Indicator is smaller or equal to 0. For any other value of the Health indicator the asset is functioning. Therefore, at any time, the real state of the asset is assumed to be solely determined by the value of the Health Indicator.

At all times, the value of the Health Indicator is known by the manager and by the asset (supposedly through sensors installed within the asset). However, the values of (a_i, b_i, t_{fi}) are unknown to the assets. Instead, the assets compute at each time step a triplet of estimated values $\left(\bar{a}_i^e, \bar{b}_i^e, \bar{t}_{fi}^e\right)$ using the collaborative prognosis algorithm described in Sect. 2.2. These values can be used to predict the future state of the asset and implement maintenance strategies, therefore determining the prognostics model.

2.2 *Collaborative Prognosis Algorithm*

The agent associated with each asset is programmed to guess the real values from its own Health Indicator data and from the information conveyed by its asset neighbours. The agents execute the following steps:

1. During the first n time steps since the first installation of the asset, each asset i is connected to N_i collaborating assets as follows: first, a triplet of initial estimated parameters $\left(\bar{a}_i^e, \bar{b}_i^e, \bar{t}_{fi}^e\right)$ is randomly initialised. Second, the asset's collaborating assets are selected as the j elements giving the N_i smallest distances d_{ij}, defined as in [31]:

$$d_{ij} = \frac{1}{\sqrt{3}}\sqrt{\frac{(\bar{a}_i^e - \bar{a}_j^e)^2}{\bar{a}\text{max}} + \frac{(\bar{b}_i^e - \bar{b}_j^e)^2}{\bar{b}\text{max}} + \frac{(\bar{t}_{fi}^e - \bar{t}_{fj}^e)^2}{\bar{t}_f\text{max}}}. \tag{2}$$

   Where $\left(\bar{a}\text{max}, \bar{b}\text{max}, \bar{t}_f\text{max}\right)$ are the maximum values expected for their corresponding parameters [taken from the maxima of the randomly generated values of $(\bar{a}_i^e, \bar{b}_i^e, \bar{t}_{fi}^e)$]. Therefore, each asset has a vector of collaborating assets determined by random discrete numbers $\vec{N}_i = \vec{k}_i$ where $\vec{k}_i$ is a discrete vector composed by the indexes j indicating the identity of each near neighbour. Note how during this first n time steps, there is no learning. This is done in order to gather enough initial data to fit Eq. (1) in later steps. n has to be bigger or equal than the degrees of freedom of Eq. (1). In our algorithm, assets calculate their similarity to all other assets in the fleet individually. However, platform-based cluster methods can also be used, and have been shown to be scalable to large asset fleets [20, 21]. As long as the model used to calculate inter-asset similarities remains the same and operates in real time, the results obtained in this work are applicable to both cases.
2. At time n or higher, assets guess the real values from the available Health indicator data points by fitting Eq. (1) without the noise term using a trust-region-reflective nonlinear least squares fitting algorithm, for example MATLAB's lsqnonlin function [32]. This returns a triplet of estimated values $\left(a_i^e, b_i^e, t_{fi}^e\right)$.
3. Assets receive the triplets $\left(\bar{a}_i^e, \bar{b}_i^e, \bar{t}_{fi}^e\right)$ from the N_i nearest neighbours given by the vector $\vec{N}_i$, and re-evaluate their estimates using a weighted average from themselves and their neighbours:

$$\bar{a}_i^e = a_i^e w_{ii} + \sum_{\vec{N}_i} w_{ij}\bar{a}_j^e, \quad \bar{b}_i^e = b_i^e w_{ii} + \sum_{\vec{N}_i} w_{ij}\bar{b}_j^e$$
$$\bar{t}_{fi}^e = t_{fi}^e w_{ii} + \sum_{\vec{N}_i} w_{ij}\bar{t}_{fj}^e \tag{3}$$

where the weights are chosen to depend exponentially with the inverse of the distance:

$$w_{ij} = \frac{e^{-d_{ij}\gamma_j}}{\sum_{\vec{N}_i} e^{-d_{ij}\gamma_j}}. \tag{4}$$

This is set in order to enable a certain control of the influence of collaborating assets in the estimates of the asset itself. The parameter γ_j determines how strict the weighting is with respect to the estimated distance between assets. A system where all neighbours would be weighted equally corresponds to $\gamma_j = 0$. As γ_j increases, the information from collaborating assets is weighted with decreasing weights.

4. The assets send their guessed triplets to all the other assets in the system and calculate pairwise euclidean distances using Eq. (2).
5. Assets are linked to their N_i nearest neighbours, corresponding to the N_i smallest values of d_{ij} given by Eq. (2). These are the collaborating assets. Each asset stores the identifiers of his collaborating assets in a vector $\vec{N}_i$ formed by the indexes j of all near neighbours.
6. At each time step the assets are maintained according to the time-based maintenance policy described in Sect. 2.3.
7. Steps 2–6 are repeated iteratively at each time step.

The Social Network Platform featured in Fig. 1 takes the role of system configuration, setting system parameters such as γ_i, N_i, and allowing agents to join and leave the network. The platform also keeps track of the costs incurred upon implementation of the maintenance policy described in next section.

2.3 Maintenance Policy

We implement a Predictive Maintenance policy. Assets are preventively repaired when the asset's local time (or time since last repair), t_{li}, is bigger or equal than the estimated time of failure multiplied by a factor η_i, set by the Social Network Platform.

$$\frac{t_{li}}{\bar{t}_{fi}^e} \geq \eta_i, \quad 0 \leq \eta_i \leq 1. \tag{5}$$

Alternately, assets are correctively repaired immediately if $\mathrm{HI}_i \leq 0$. The cost of a preventive repair, C_p is much lower than the cost of a corrective repair, C_c. When an asset is repaired its age t_{li}, is set back to 0, and its Health Indicator is reset.

The cost per unit of time is optimised at every time step by assuming that the values of $\bar{t}^e_{fi}$ correspond to the real values of t_{fi}. Under such conditions, the problem reduces to a time based replacement policy problem. Under stability of the parameters $\left(\bar{a}^e_i, \bar{b}^e_i, \bar{t}^e_{fi}\right)$, optimising the system for one cycle would correspond to the long-term optimal solution [30]. However, in our system the optimisation is continuously updated as $\bar{t}^e_{fi}$ changes over time as the assets revise their estimates by learning from new data. For each asset with a triplet $\left(\bar{a}^e_i, \bar{b}^e_i, \bar{t}^e_{fi}\right)$, the total expected cost per unit time is given by:

$$C\left(\eta_i \bar{t}^e_{fi}\right) = \frac{C_p + C_c H\left(\eta_i \bar{t}^e_{fi}\right)}{\eta_i \bar{t}^e_{fi}}. \tag{6}$$

where $H\left(\eta_i \bar{t}^e_{fi}\right)$ is the expected number of failures in the interval $(0, \eta_i \bar{t}^e_{fi})$. Differentiating Eq. (6) with respect to $\eta_i t^e_{fi}$ and equating it to 0, we obtain:

$$\eta_i \bar{t}^e_{fi} h\left(\eta_i \bar{t}^e_{fi}\right) - H\left(\eta_i \bar{t}^e_{fi}\right) = \frac{C_p}{C_c}. \tag{7}$$

Assuming that the time step is arbitrarily small, the expected number of failures per cycle corresponds to the probability of failures during one cycle. First we calculate the probability of failure at time t.

$$P(t) = \int_{-\infty}^{0} N\left(\bar{\mathrm{HI}}_i(t), \sigma, x\right) dx = \frac{1}{2}\left(1 + \mathrm{erf}\left(\frac{-\bar{\mathrm{HI}}_i(t)}{\sigma\sqrt{2}}\right)\right). \tag{8}$$

Here $\bar{HI}_i$ is the Health Indicator without the random term, i.e.:

$$\bar{HI}_i(t_{li}) = \bar{a}^e_i\left(1 - e^{-\bar{b}^e_i\left(\bar{t}^e_{fi} - t_{li}\right)}\right) \tag{9}$$

The expected number of failures during one cycle corresponds to integrating $P(t)$ over time and normalising to the time period.

$$\begin{aligned} H\left(\eta_i \bar{t}^e_{fi}\right) &= \frac{1}{\eta_i \bar{t}^e_{fi}} \int_{0}^{\eta_i \bar{t}^e_{fi}} P(t) dt \\ &= \frac{1}{2\eta_i \bar{t}^e_{fi}}\left(\eta_i \bar{t}^e_{fi} + \int_{0}^{\eta_i \bar{t}^e_{fi}} \mathrm{erf}\left(\frac{-\bar{\mathrm{HI}}_i(t)}{\sigma\sqrt{2}}\right) dt\right) \end{aligned}$$

$$= \frac{1}{2\eta_i \bar{t}^e_{fi}} \left(\eta_i \bar{t}^e_{fi} + \int_0^{\eta_i \bar{t}^e_{fi}} \operatorname{erf} \left(\frac{-\bar{a}^e_i \left(1 - e^{-\bar{b}^e_i \left(\bar{t}^e_{fi} - t \right)} \right)}{\sigma \sqrt{2}} \right) dt \right). \tag{10}$$

Using the fundamental theorem of calculus, one can obtain $h\left(\eta_i \bar{t}^e_{fi}\right)$:

$$h\left(\eta_i \bar{t}^e_{fi}\right) = P\left(\eta_i \bar{t}_{fi}\right) = \frac{1}{2} \left(1 + \operatorname{erf} \left(\frac{-\mathrm{HI}_i \left(\eta_i \bar{t}^e_{fi} \right)}{\sigma \sqrt{2}} \right) \right). \tag{11}$$

Equation (7) then follows:

$$\eta_i \bar{t}^e_{fi} h\left(\eta_i \bar{t}^e_{fi}\right) - \frac{1}{2\eta_i \bar{t}^e_{fi}} \left(\eta_i \bar{t}^e_{fi} + \int_0^{\eta_i \bar{t}^e_{fi}} \operatorname{erf} \left(\frac{-\bar{a}_i \left(1 - e^{-\bar{b}_i \left(\bar{t}^e_{fi} - t \right)} \right)}{\sigma \sqrt{2}} \right) dt \right) = \frac{C_p}{C_c}. \tag{12}$$

Which can be solved for η_i numerically.

2.4 On the Trade-off Between Self-learning and Collaborative Learning

The problem of how to use information from other assets in order to predict the behaviour of a particular asset is not trivial. The fundamentals of this question can be traced back to the basic laws of measurement uncertainty. In effect, multiple assets can be viewed as multiple similar experiments with different degrees of variation. According to these laws, each asset will estimate its triplet of parameters $\left(\bar{a}^e_i, \bar{b}^e_i, \bar{t}^e_{fi}\right)$ with a level of uncertainty determined by its systematic and random components [33]:

$$\left(\bar{a}^e_i, \bar{b}^e_i, \bar{t}^e_{fi}\right) = \left(\bar{a}^e_i, \bar{b}^e_i, \bar{t}^e_{fi}\right) \pm \sqrt{U^2_{si} + U^2_{ri}}. \tag{13}$$

Each estimation will also carry some Error, given by the difference between the estimated value and the true value of the parameter triplet.

$$\left(Err^a_i, Err^b_i, Err^{t_f}_i \right) = \left(|\bar{a}^e_i - a_i|, |\bar{b}^e_i - b_i|, |\bar{t}^e_{fi} - t_{fi}| \right). \tag{14}$$

Typically, accumulating equivalent measurements governed by Gaussian uncertainty distributions, reduces the statistical uncertainty by a factor of $1/\sqrt{N_S}$ where N_S is the number of samples. Therefore, we expect the statistical uncertainty associated to the estimates to reduce by a factor of $1/\sqrt{N_i}$. Uncertainty reduction notwithstanding, if the near neighbours of an asset i correspond to very different asset types, weighting their data in the estimators of such asset will not always be beneficial, leading to an increase of Err_i larger to the reduction of the associated uncertainty. This phenomena can be observed in collaborative learning for high values of d_{ij}, and small values of γ_i. In other words, more data does not always lead to smaller errors (better estimates). An example of this is seen in Fig. 8 where high σ values imply suboptimal choice of the collaborating assets. This drives the optimal number of collaborating assets to be $N_i = 2$, instead of $N_i = 9$ as in the case of $\sigma = 0$.

3 Simulation Model

In order to simulate the behaviour of the system, we use the agent simulation software NetLogo [34]. NetLogo is an agent-base programming language used to simulate complex systems and emergent phenomena. In our simulations we connected NetLogo with MATLAB using MatNet, an open source extension [35]. MATLAB is a commercial numerical computation software that we use to perform the non linear fit of the health indicator described in Sect. 2.2.

In order to study the system we adopt the following assumptions:

- All assets have a common weight parameter $\gamma_i = \gamma$.
- The number of time steps run as step. 1 in Sect. 2.2 is set at $n = 10$.
- We simulate only three families of assets chosen represented by the triplets $F{:}(a, b, t_f) \rightarrow \{(0.25, 0.025, 25), (0.5, 0.05, 50), (0.75, 0.075, 75)\}$.
- Nine assets are simulated for each family, totalling 27 assets. This means that for $N_i > 9$, there will always be an asset forcibly connected to neighbours belonging to a different family.
- The cost of a preventive repair is set at $C_p = 1$ and the cost of corrective repair at $C_c = 100$.
- Due to computational constraints, the maintenance policy of individual assets is not optimised in real time. Instead, all assets share the same $\eta_i = \eta$, which is varied in steps.

System convergence is defined by the average difference between the estimated time of failure t^e_{fi} and the real time of failure of the family the asset belongs to:

$$\left\langle \left| \bar{t}^e_{fi} - t_{fi} \right| \right\rangle < \kappa. \tag{15}$$

where κ is chosen to be the smallest time step of the system, $\kappa = 1$.

The simulation focuses in exploring the dependence of maintenance cost with the number of collaborating assets N_i and the noise in the system σ. Additionally, we aim to explore the dynamics of the system coupled with the replacement policy (determined largely by η) and the weighting of the collaboration data (given by γ).

3.1 Results

The system parameters η, N_i, σ, γ are varied in steps. We examine the cost per unit time K, with respect to the different variables of the system:

$$K = \frac{\sum_T C_T}{T}. \tag{16}$$

where T is the total time of the simulation (the number of steps since initialisation). We see that the system follows a non-trivial behaviour with respect to all the studied variables, and results in different local minima for K. For a given η, a higher value of K implies a higher prediction error in the assets of the system, as the estimated time to failure t^e_{fi} is more likely to be higher than the true time to failure, leading to unwanted corrective repairs. K can then be seen in this chapter as a proxy for model accuracy. For example, for the case of $\sigma = 0.01$, $\gamma = 0.1$ and $\eta = 0.8$, increasing the number of collaborating assets results in smaller costs up until $N_i = 3$. For $N_i = 4$, the tendency reverses. In this case, assets that are too different from each other are allowed to exchange information, leading to a sub-optimal solution (see Figs. 3 and 4).

Figure 5 shows the dependence of K on η and N_i. The results suggest that $N_i = 7$, $\eta = 0.3$ is the best solution, given $\sigma = 0.1$. $\gamma = 0.1$. The dependence with respect to η follows the expected behaviour, with the solution being suboptimal for too

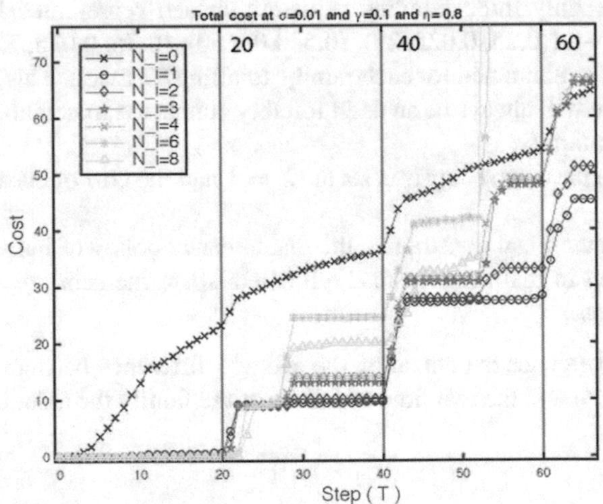

Fig. 3 Total cost of the system for different values of N_i averaged over 50 trajectories. The cost is plotted for $\sigma = 0.01$, $\gamma = 0.1$ and $\eta = 0.8$. Note how the cost function shows clear increases at $T = 25\,\eta = 20$, $T = 50\,\eta = 40$ and $T = 75\,\eta = 50$, points that correspond to the preventive maintenance times upon convergence. Note how the increase at $T = 25\eta$ is smaller than the increase at $T = 50\eta$, this is due to the fact that in the latter case the preventive maintenance time of the second cycle of assets with $t_{fi} = 25$ coincides with the maintenance time of the assets with $t_{fi} = 50$

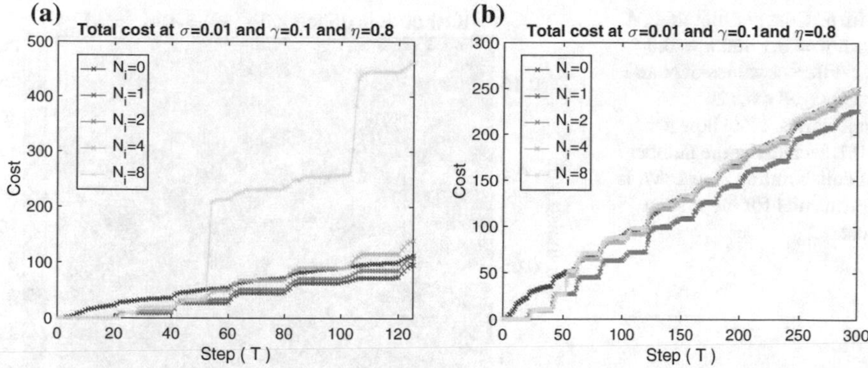

Fig. 4 Total cost of the system using collaborative learning after convergence (**a**), compared to restricting to self learning upon convergence (**b**) for $\varepsilon = 1$. The cost is plotted for different values of N_i averaged over 50 trajectories for $\sigma = 0.01$, $\gamma = 0.1$ and $\eta = 0.8$. Note how in (**a**) the cost functions for $N_i > 0$ exceed the values for the self-learning approach ($N_i = 0$) at long values of T. While in (**b**), even at long timesteps of the simulation, the cost of collaborative learning is always equal or smaller than the cost of self learning. With the case of $N_i = 2$ providing minimal costs. The minimum cost solution corresponds to the one reaching fastest convergence, as it implies less corrective repairs in the early phase of the simulation

Fig. 5 Cost per unit time, K with $\sigma = 0.1$ and $\gamma = 0.1$ for different values of N and η averaged over 20 trajectories. A global minimum is observed at $N = 7$, $\eta = 0.3$

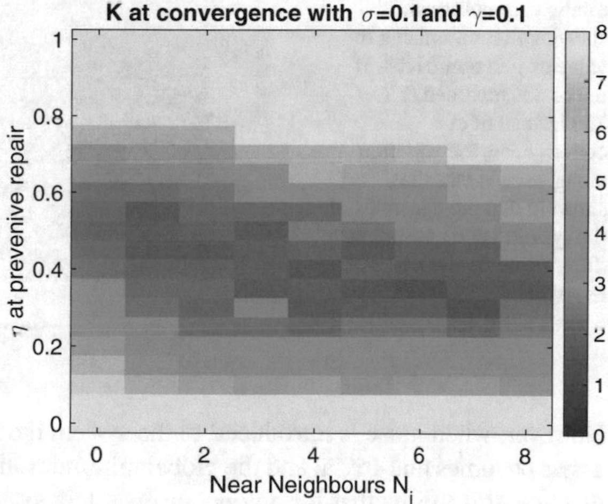

large or too small η. For large η the preventive repairs are done too close to the point of failure, leading to unwanted corrective repairs. For small η, the frequency of preventive repairs is too high and the assets in the system do not accumulate sufficient health indicator data to learn from themselves.

Figure 6 shows the dependence of K on N_i and σ for $\eta = 0.5$ and $\gamma = 0.1$. As expected, higher σ imply higher costs, and the optimal number of collaborating assets N_i will depend on the noise present in the system. It is interesting to see that for no noise, the dependency of K with N_i is monotonic (higher N_i means a better solution).

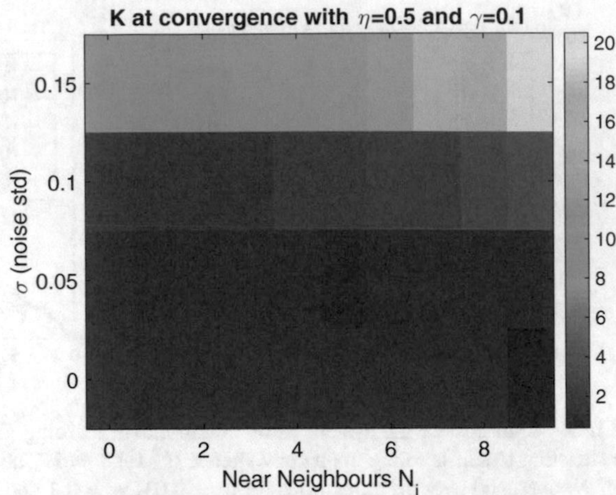

Fig. 6 Cost per unit time, K with $\gamma = 0.1$ and $\eta = 0.5$ for different values of N and σ averaged over 20 trajectories. Note how for >0.1, increasing the number of collaborating assets, N_i, is detrimental for the system cost

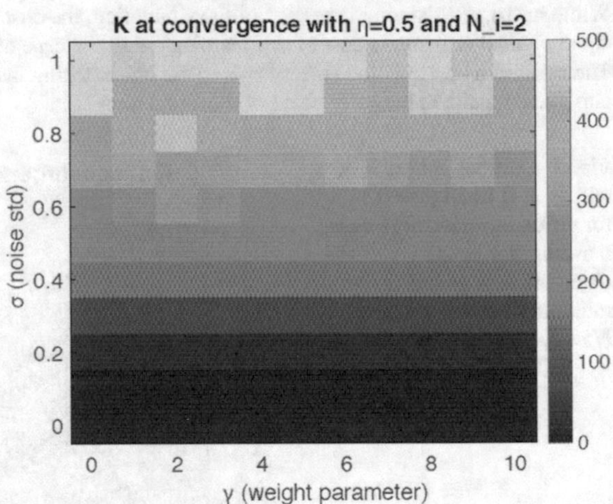

Fig. 7 Cost per unit time, K with $N = 2$ and $\eta = 0.5$ for different values of γ and σ averaged over 20 trajectories. Note how several local minima are observed, showing that optimising the value of γ is non-trivial. If the cost is recorded at T = 150 instead of at convergence, the variation along *gamma* vanishes, showing that the dynamics of the system upon convergence differ from the dynamics at large times

However, when noise is introduced to the system the ideal number of collaborating assets becomes non-trivial and the global minimum shifts to smaller values of N_i. In practice, this means that for a noisy or uncertain system collaborative learning has to be constrained in order to avoid an increase in prediction errors (see Sect. 2.4).

Figure 7 shows the dependence of K on γ and σ for $\eta = 0.5$ and $N_i = 2$. As expected, for the case of absence of noise ($\sigma = 0$), the weight parameter is better set at 0 as all information coming from the rest of the assets will be relevant. However, as the noise in the system increases, higher values of γ are more optimal as to constrain the information coming from different assets becomes a sensible idea.

3.2 System Behavior After Convergence

We have shown that collaborative prognostics ($N_i > 0$) reduces the maintenance cost upon convergence [satisfaction of Eq. (15)] for certain levels of system noise σ. However, when the system is run for a long period of time, for small values of γ, collaborative prognostics is costlier than self-learning ($N_i = 0$) (see Fig. 5). This is due to the fact that for the proposed model, self-learning able to estimate the real values of $\left(a_i, b_i, t_{fi}\right)$ with asymptotically small statistical uncertainty, given that the simulation is let run a long enough time. Adversely, collaborative learning always risks weighting information from neighbours from another asset family (see Sect. 2.4). This caveat is avoided by adding a condition that increases the value of γ upon convergence of the triplet $\left(\bar{a}_i^e, \bar{b}_i^e, \bar{t}_{fi}^e\right)$.

In order to do so we add the following condition: if $\left|\bar{t}_{fi}^e(t_{li}) - \bar{t}_{fi}^e(t_{li}+1)\right| < \varepsilon$, during five consecutive time steps, then $\gamma_i \to \infty$ (equivalent to $N_i = 0$). ε is chosen to be small (for example, $\varepsilon = \kappa$). In practice this means that once all the information has been extracted from the collaborating assets, the asset is allowed to learn only from itself. Once this condition is incorporated collaborative learning is less costly than self-learning at large values of T, even if very small values of γ are chosen (see Fig. 5).

3.3 Optimisation upon Convergence

In this chapter, the system has been simulated at global $\eta = \eta_i$ values, without real-time optimisation of the maintenance policy. In order to show that the network dynamics do not influence the validity of Eq. (12), we simulate the system for a family of assets and plot its cost per second depending on η and σ. In Fig. 8 we show how the theoretical optimised values of η correspond to simulation.

4 Discussion

In this chapter, we have introduced the idea of Social Asset Networks. The main difference between the approach we propose and alternative maintenance policies, is that ours does not require to work with a large library of previously recorded Health Indicator trajectories. Other popular multi-unit maintenance treatments rely on extensive datasets of the previous behaviour of the assets, which are then treated in a centralised computer and used in order to predict future behaviour. This implies a long period where sub-optimal maintenance policies are applied, increasing the total cost.

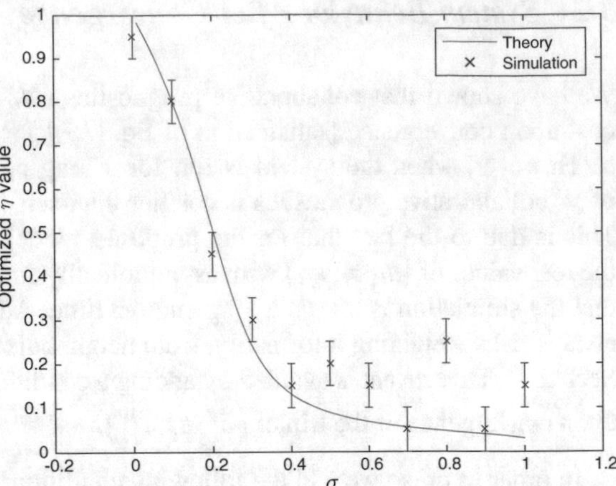

Fig. 8 Optimal (minimum cost) η value obtained from simulating a family of 9 assets interchanging information upon convergence, compared to the theoretical η value predicted by Eq. (12). Note how higher values of σ imply higher variance in the system as the estimated values of $\left(\bar{a}_i^e, \bar{b}_i^e, \bar{t}_{fi}^e\right)$ fluctuate due to the noise

As does any first attempt, our work has several limitations: the Health Indicator and noise distribution functions studied in this chapter are not general. For example, Weibull distributions are often found to describe better real time failures. Additionally, dependencies with the expected asset computational limitations are not considered except in the design of the system. Thus, the presented results only apply to a subset of well-understood industrial asset types, and care must be taken when generalising them. However, the concept of Social Asset Networks can be generalised to other diagnostics and prognostics models by adapting the failure and maintenance rules and the definition of the distance metric.

The results shown in Figs. 1, 2, 3, 4, 5, 6, 7, 8 clearly indicate that Social Asset Networks are multi-variable optimisation problems, with global and local minima following non-trivial dynamics. However, it is still possible to observe some general behaviours:

- *On the effect of noise*: as expected, an increase of the value of σ causes an increase of the total cost. At low values of σ, the optimal solution belongs to $N_i > 0$. However, at a certain value of σ, the optimal solution corresponds to assets only learning about themselves. This is explained because the number of collaborating assets not actually belonging to the same family induces an error Err_i that overcomes the reduction on uncertainty given by sharing the estimates. The system then never fulfils Eq. (15). By applying the fix proposed in Sect. 3.2, one can see that collaboration is more resilient to large values of σ as long as the definition of convergence considers the increase in noise.
- *On the effect of γ*: γ has a crucial effect on the long-term values of K for the simulation. For small values of γ (close to 0.1), the collaborative approach is sensitive to higher noise levels due to incorporation of unsuitable data. Increasing the value of γ either globally or in real time upon convergence of each individual asset eliminates this effect.

- *On the effect of the number of collaborating assets,* N_i: in general, increasing the number of collaborating assets helps convergence and reduces cost, as long as σ is kept sufficiently low. When σ reaches a given level, a value of N_i which differs from the maximum value of N_i can be found to correspond to the minimum in the maintenance cost function K with no additional local minima. If σ is further increased, the configuration of minimum cost corresponds to self-learning ($N_i = 0$). Care must be taken when N_i starts being on the order of the number of assets belonging to each asset family. If N_i overcomes that value, the cost starts increasing due to the increment of connections with assets belonging to different families. Fleet-wide learning (corresponding to $N_i = 8$), is significantly costlier than collaborative learning. This observation has practical relevance as in many industrial systems data abundance is often prioritised over specificity, leading to prediction errors.
- *On the effect of* η: η affects the cost as expected in a time-based maintenance system: there is a single value of η corresponding to the global minimum in the cost function. In practice, η_i can be optimised in real time for each asset in the network. This optimisation will result in a less costly system than choosing an optimal general value of η.
  The typical optimisation equations for single asset cases cannot be used to obtain the optimal η for an heterogeneous asset network. Instead, under convergence, each asset maintenance strategy must be optimised separately, obtaining individual values of η_i.

Apart from discussing the parametric dependencies, it is important to discuss the choice of the distance metric d_{ij}. Note how in our distance, all parameters receive the same weight when, in fact, the maintenance model is only determined by the time of failure $\bar{t}^e_{fi}$ (see Sect. 2.3). In principle, one could define the distance differently, weighting solely or mostly the time of failure and not the parameters $\bar{a}^e_i$ and $\bar{b}^e_i$. However, defining the metric solely based on $\bar{t}^e_{fi}$ would not work in a system where the whole data series is exchanged between assets (as in most realistic machine learning algorithms). In such a system, if each asset were to perform a weighted fit of the shared data, the differences in $\bar{a}^e_i, \bar{b}^e_i$ would propagate through the fit creating a big uncertainty in the estimated t_{fi}, resulting in a costlier maintenance system. We choose Eq. (2) because it can be generalised to a system where the assets exchange larger amounts of data with each other, apart from the triplet $\left(a^e_i, b^e_i, t^e_{fi}\right)$.

Previously published k near neighbour treatments focus on calculating the distance between measured Health Indicator trajectories [16], some assigning more weight to more recent measurements [36]. Focusing solely in recorded trajectories is reasonable, in the sense that similar failure modes are likely to be shared along the multi-unit system. Thus, the trajectory recorded for a certain asset may serve as a good predictor of the future behaviour of another particular asset. However, this does not always properly account for the differences between assets. Similar to our chapter, measuring asset similarity just by the similarity between observations disregards the effect of the operating environment. This is acknowledged by [36], who define an ideal reference system to be “a system physically identical with the operating system, which worked in the same operating conditions”.

The balance between asset similarity and observed trajectory similarity is an important one, that could be compared to the difference between changing the experimental conditions and changing the experiment itself. Controlling for asset similarity will correspond to observing different trajectories of the same process influenced by the individual conditions of each asset. When observing sufficiently similar assets, trajectory differences will correspond to several different processes or failure modes that may be shared between the assets. In practice, failure mode similarities shall be used to classify failure modes and asset similarities to weight different failure data in a predictive trend. This issue has been studied before in the field of cooperative agent information, in which Puuronen and Terziyan proposed similarity evaluation in three layers: problems, agents and solutions [37]. In a typical multi unit maintenance case, the problems would be the failure modes, the agents would correspond to the assets and the solutions would be the prognostic models generated by each asset for each failure mode.

The issue of uncertainty in Social Asset Networks is complex. The approach presented in this chapter stands mostly on international standards for measurement uncertainty. Concretely we focus on the Guide to the Expression of Uncertainty in Measurement (GUM) rules of measurement uncertainty and propagation [33]. Other approaches, such as belief functions and empirical Bayes methods can also be used to investigate the effects of data sharing between dissimilar assets. For the intend of our work -to pinpoint the opportunities and problems offered by this approach—the GUM rules suffice.

5 Conclusion

We present a novel approach to optimise preventive maintenance policies in multi unit systems connected to an IoT platform: Social Asset Networks. The approach consists on weighting the similarities between different assets in order to connect assets to collaborating assets, and consecutively update asset predictions according to the neighbours’ data. Social Asset Networks are real time, distributed smart maintenance systems that do not depend on large datasets of historic data. Using simulations within NetLogo, a multi-agent software, we prove the validity of such approach in a simple but generalisable case based on known Health Indicator models with a noise term. We conclude:

- Collaborative learning outperforms self-learning, as long as collaborative learning is reduced once assets have reached convergence and as long as σ is kept low. In other words, collaborative learning helps reaching convergence to the real values faster, but once the system has converged self-learning can outperform collaborative learning for arbitrarily large time scales.

- Maintenance cost decreases with increased number of neighbours until a tipping point when it increases again.
- Upon convergence, network dynamics do not greatly influence the optimisation of the maintenance policy. A maintenance policy can be defined for each individual asset in a multi asset system.

Additionally, we show non-trivial dependencies between the cost function and the different variables of the system. Social Asset Networks enable an asset manager to optimise the maintenance of large fleets of assets equipped with sensing, communication and processing technology without the need of centralised servers.

6 Future Work

We have only simulated an ad hoc case with assets belonging to three different families. However, in reality, depending on the system the asset types may be several and their relative distances may be much closer or further away than the proposed case. Future work should focus on the change of the validity of the approach when the relative distances between families decrease and the system is much more sensitive to measurement noise.

The effect on maintenance cost of the presented approach has been computed assuming negligible communication costs. That is, the results only show the effect of Social Asset Networks in improving prognostics accuracy. In order to completely assess the industrial applicability of the proposed approach, communication costs must be weighted in the calculations too. For this, the connection policy must be rigorously and specifically outlined. This remains as one of the fundamental pieces of future work to be undertaken.

In this chapter, we have considered that differences between assets fully determine differences between their Health Indicator models. Future work also should consider differences between failure modes and the different trajectories followed by these modes determined by the environment and the asset state.

In practice, distance metrics are unlikely to stem solely from parametric components. Indicators such as the production model of the asset, customer information, etcetera, are likely to be nominal attributes. In order to take into account such attributes, one could use metrics that combine parametric and nominal components, such as the Heterogeneous Value Difference Metric or the Interpolated Value Difference Metric [31].

The applicability of the presented approach should be demonstrated in an industrial test case. Ideal cases would be industries where the assets are expected to have a well-defined range of similarity. An example is the car industry, where car models vary across years and across variants of the same model, and where the deterioration of the cars depends greatly of environmental variables such as weather and driving culture. Another example is the gas turbine industry, where subsequent turbine models show very different maintenance histories depending on the workload, history of repairs and placement.

Acknowledgements We acknowledge Bang Ming Yong and the VR team at the Institute for Manufacturing for providing the computer used to perform the simulations. This research was supported by SustainOwner (Sustainable Design and Management of Industrial Assets through Total Value and Cost of Ownership), a project sponsored by the EU Framework Programme Horizon 2020, MSCA-RISE-2014: Marie Skodowska-Curie Research and Innovation Staff Exchange (Rise) (grant agreement number 645733 "Sustain-owner" H2020-MSCA-RISE-2014).

References

1. Dabney, T., Hernandez, L., Scandura, P. A., & Vodicka, R. (2008). Enterprise health management framework—A holistic approach for technology planning, R&D collaboration and transition. In *International Conference on Prognostics and Health Management, PHM*.
2. Medina-Oliva, G., Weber, P., & Iung, B. (2015). Industrial system knowledge formalization to aid decision making in maintenance strategies assessment. *Engineering Applications of Artificial Intelligence, 37,* 343–360.
3. Xu, L. D., He, W., & Li, S. (2014). Internet of things in industries: A survey. *IEEE Transactions on Industrial Informatics, 10*(4), 2233–2243.
4. Gubbi, J., Buyya, R., Marusic, S., & Palaniswami, M. (2013). Internet of Things (IoT): A vision, architectural elements, and future directions. *Future Generation Computer Systems, 29*(7), 1645–1660.
5. Tuwanut, P., & Kraijak, S. (2015). A survey on IoT architectures, protocols, applications, security, privacy, real-world implementation and future trends. In *11th International Conference on Wireless Communications, Networking and Mobile Computing (WiCOM 2015)* (Vol. 6, p. 6).
6. Gazis, V., Goertz, M., Huber, M., Leonardi, A., Mathioudakis, K., Wiesmaier, A., et al. (2015). *Short paper : IoT : Challenges, projects, architectures* (pp. 145–147).
7. Li, H., & Parlikad, A. K. (2016). Social internet of industrial things for industrial and manufacturing assets. *IFAC-PapersOnLine, 49*(28), 208–213.
8. Mezei, I., Malbasa, V., & Stojmenovic, I. (2010). Robot to robot. *IEEE Robotics and Automation Magazine, 17*(4), 63–69.
9. Wood, S. M., & Goodman, D. L. (2006). *Prognostics in high reliability telecom applications* (pp. 1–3).
10. Goyal, D., & Pabla, B. S. (2015). Condition based maintenance of machine tools—A review. *CIRP Journal of Manufacturing Science and Technology, 10,* 24–35.
11. Alaswad, S., & Xiang, Y. (2017). A review on condition-based maintenance optimization models for stochastically deteriorating system. *Reliability Engineering and System Safety, 157,* 54–63.
12. Zhao, G. (2011). Wireless sensor networks for industrial process monitoring and control: A survey. *Network Protocols and Algorithms, 3*(1), 46–63.
13. Grall, A., Dieulle, L., Bérenguer, C., & Roussignol, M. (2002). Continuous-time predictive maintenance scheduling for a deteriorating system. *IEEE Transactions on Reliability, 51*(2), 141–150.
14. Sun, J., Zuo, H., Wang, W., & Pecht, M. G. (2012). Application of a state space modeling technique to system prognostics based on a health index for condition-based maintenance. *Mechanical Systems and Signal Processing, 28,* 585–596.
15. Qian, F., & Niu, G. (2015). Remaining useful life prediction using ranking mutual information based monotonic health indicator. In *2015 Prognostics and System Health Management Conference, PHM* (pp. 1–5).

16. Wang, T., Yu, J., Siegel, D., & Lee, J. (2008, November) A similarity-based prognostics approach for remaining useful life estimation of engineered systems. In *International Conference on Prognostics and Health Management, PHM*.
17. Zhou, Y., Chioua, M., & Ni, W. (2016). Data-driven multi-unit monitoring scheme with hierarchical fault detection and diagnosis. In *24th Mediterranean Conference on Control and Automation, MED*.
18. Srinivasan, B. (2007). Real-time optimization of dynamic systems using multiple units. *International Journal of Robust Nonlinear Control, 17,* 1183–1193.
19. Ning, H., Liu, H., Ma, J., Yang, L. T., & Huang, R. (2016). Cybermatics: Cyber–physical–social–thinking hyperspace based science and technology. *Future Generation Computer Systems, 56,* 504–522.
20. Lapira, E. R. (2012). Fault detection in a network of similar machines using clustering approach (Ph.D. thesis). University of Cincinnati.
21. Cannarile, F., Compare, M., Di Maio, F., & Zio, E. (2015). Handling reliability big data: A similarity-based approach for clustering a large fleet of assets. In *Safety and Reliability of Complex Engineered Systems* (pp. 891–896).
22. Bleakie, A., & Djurdjanovic, D. (2013). Analytical approach to similarity-based prediction of manufacturing system performance. *Computers in Industry, 64*(6), 625–633.
23. Ramasso, E. (2012). *Joint prediction of observations and states in time-series: A partially supervised prognostics approach based on belief functions and KNN* (pp. 1–13).
24. Yang, C., Letourneau, S., Liu, J., Cheng, Q., & Yang, Y. (2017). Machine learning-based methods for TTF estimation with application to APU prognostics. *Applied Intelligence, 46*(1), 227–239.
25. Schneeweiss, C. (2003). Distributed decision making—A unified approach. *European Journal of Operational Research, 150*(2), 237–252.
26. Yu, R., Iung, B., & Panetto, H. (2003). A multi-agents based E-maintenance system with case-based reasoning decision support. *Engineering Applications of Artificial Intelligence, 16*(4), 321–333.
27. Bayar, N., Darmoul, S., Hajri-Gabouj, S., & Pierreval, H. (2015). Fault detection, diagnosis and recovery using Artificial Immune Systems: A review. *Engineering Applications of Artificial Intelligence, 46,* 43–57.
28. Ray, P. P. (2018). A survey on Internet of Things architectures. *Journal of King Saud University—Computer and Information Sciences*.
29. Yan, J., Koç, M., & Lee, J. (2004). A prognostic algorithm for machine performance assessment and its application. *Production Planning & Control, 15*(8), 796–801.
30. Jardine, A. K. S. (2013). *Maintenance, replacement and reliability* (Vol. 1542).
31. Wilson, D. R., & Martinez, T. R. (1997). Improved heterogeneous distance functions. *Journal of Artificial Intelligence Research, 6,* 1–34.
32. MATLAB, version 9.0.0 (R2016a), The MathWorks Inc., Natick, Massachusetts, 2016.
33. Iec Bipm, Ilac Ifcc, Iupac Iso, & Oiml Iupap. (2008). *Evaluation of measurement data Supplement 1 to the Guide to the expression of uncertainty in measurement Propagation of distributions using a Monte Carlo method*. Technical report.
34. Wilensky, U. (1999). *NetLogo*. http://ccl.northwestern.edu/netlogo/. Evanston, IL: Center for Connected Learning and Computer Based Modeling, Northwestern University Evanston, IL 2009 (February 26, 2009).
35. Biggs, M. B., & Papin, J. A. (2013). Novel multiscale modeling tool applied to pseudomonas aeruginosa biofilm formation. *PLoS ONE, 8*(10).
36. You, M.-Y., & Meng, G. (2011). A generalized similarity measure for similarity-based residual life prediction. *Proceedings of the Institution of Mechanical Engineers, Part E: Journal of Process Mechanical Engineering, 225*(3), 151–160.
37. Puuronen, S., & Terziyan, V. (1999). A similarity evaluation technique for cooperative problem solving with a group of agents. *Lecture Notes in Computer Science (including subseries Lecture Notes in Artificial Intelligence and Lecture Notes in Bioinformatics), 1652,* 163–174.

Chapter 16
The Perceived Value of Additively Manufactured Digital Spare Parts in the Industry: An Empirical Investigation

Sergei Chekurov, Sini Metsä-Kortelainen, Mika Salmi, Irene Roda and Ari Jussila

Abstract The purpose of this chapter is to verify the conceptual benefits of the implementation of additive manufacturing (AM) in spare part supply chains from the point of view of the industry. Focus group interviews consisting of five sessions and 46 experts in manufacturing were conducted for this study. The focus group interviews served to identify the issues in the adoption of digital spare parts (DSP) and to expand on the available literature. The benefits found in the reviewed literature were partially verified by the participants but certain limitations, such as the excessive need of post processing, supplier quality parity, and ICT inadequacies, were presented that were absent or not highlighted in literature. The information gathered from the participants made it possible to create a realistic model of a digital spare part distribution network. According to the focus group interviews, digital spare parts could be deployed immediately for a specific type of product in the long tails of company spare part catalogues. However, improvements in AM, company ICT infrastructure, and 3D model file formats need to be achieved for a larger deployment of DSP.

Keywords Spare parts · Digital manufacturing · Additive manufacturing · Supply network management · Distributed manufacturing · 3D printing

---

S. Chekurov · M. Salmi
Department of Mechanical Engineering, School of Engineering, Production Technology, Aalto University, Otakaari 1, 02150 Espoo, Finland

S. Metsä-Kortelainen · A. Jussila
VTT Technical Research Centre of Finland Ltd., P.O. Box 1000, 02044 Espoo, Finland

I. Roda (✉)
Department of Management, Economics and Industrial Engineering, Politecnico di Milano, Piazza Leonardo da Vinci 32, 20133 Milan, Italy
e-mail: irene.roda@polimi.it

A. Crespo Márquez et al. (eds.), *Value Based and Intelligent Asset Management*,
https://doi.org/10.1007/978-3-030-20704-5_16

1 Introduction

Additive manufacturing (AM) has certain advantages over traditional manufacturing in that it allows for viable lot-one-size production because no tooling and minimal set-up time are needed [1]. Therefore, the additional costs of part complexity and variability are significantly lower than in traditional manufacturing and the total delivery time for AM is typically shorter [2, 3]. Another major benefit of AM is that, compared to subtractive manufacturing, the expertise needed to operate AM machinery is more easily transferable between product types [4].

While AM was originally used to produce prototypes, it is now used increasingly for industrial end-use applications [5]. The increase in end-use applications is also observed in the annual AM global report produced by Wohlers Associates, in which they report that, in 2012, 28.1% of all additively manufactured objects were functional parts, while in 2016 the respective percentage was 33.8% [6, 7]. In addition to the growth of the relative percentage, the absolute number of produced end-use parts has grown greatly, because the worldwide revenue of AM products and services grew from $2.25 billion in 2012 to $6.05 billion in 2016 [6]. However, [8] reported that only 10% of the companies they interviewed use AM to produce end-use parts. Since the percentage given by Wohlers is calculated from the total amount of additively manufactured parts, it follows that although few companies use AM for end-use production, they use it to produce parts in large quantities.

With the fact that end-use components can now be created with AM in mind, several researchers have presented that spare parts could be digitized and additively manufactured. This article aims to validate the results of the researchers by analysing the results of focus groups and to define the concept of AM of spare parts.

1.1 Advantages of Additive Manufacturing in the Supply Chain

It has long been hypothesized that implementing AM in supply chains would lead to significant benefits, to the point that AM has been likened to the internet in its ability to cause a paradigm shift [9]. Pérès and Noyes were some of the first researchers to suggest implementing AM in supply chains in order to manufacture spare parts closer to the point of need [10]. They presented several avenues of research to verify the potential of the concept: research into technical features of spare parts (materials, tolerance, life cycle etc.), organisational features (logistics, storage, recycling etc.), and economical features (total cost of spare part). These research questions have been investigated extensively in recent literature [1, 10–13].

Some of the benefits of introducing AM in a supply network include reduced changeover time [14], decreased energy costs [15], increased sustainability [16] and cost reduction in long tail production [17]. The long tail of products is concept which Anderson [18] describes as containing products in low demand and that have a low sales volume. Another trend in the research concerning AM in supply chains is exploring what the role of the customer could be in the process. With the emergence of AM, the customer could potentially have control over the design and production of components alone or in collaboration with the company whose product is being redesigned. This type of consumers are referred to in literature as prosumers [10, 19, 20].

Pérès and Noyes noted in 2006 that AM was too limited to produce end-use components. While there has been significant improvement in the technologies since then, as is implied also by the growth of end-use part manufacturing according to Wohlers associates, there are still severe limitations. For example, the viable maximum part size in AM is smaller than is desirable when manufacturing precise components [21], the price of material for AM is higher than in conventional manufacturing [22], the cost of AM systems is high [23], and the material selection in AM is quite limited when taking into account the needs of companies from different sectors [24]. Additionally, materials must be approved for certain applications, which must be done to every single material [25]. For the strictest applications, each manufactured material batch requires verification of quality [26]. A thorough list of barriers to progression of AM for end-use products as perceived by industry has been collected and published by Thomas-Seale et al. [27].

Although there are still clear limitations, AM technology has advanced enough to be a viable manufacturing method for end-use components in certain applications [13]. With the notion that components could be moved digitally and produced locally, the question of piracy has been brought up by Lindemann et al. [13] and Appleyard [28], who present two opposing views. While Lindemann et al. present that protecting the 3D designs to deter their copying is the right approach Appleyard argues that companies might need to change their approach to spare part distribution to be more open.

1.2 Potential of AM in Spare Parts Management

Spare part management is a vital part of many capital-intensive businesses having direct impact on the availability of high-value capital assets, essential to the operational processes [29, 30]. In fact, unavailability of a spare part item when needed may lead to long unproductive downtimes affecting the operating company's profit [31]. Given the general function of spare parts to support maintenance activities, the policies that govern the spare parts inventories are different from those that govern other types of inventory such as raw material, work-in-process and finished goods inventories [32–34]. Specifically, two of the main critical issues when managing spare parts are the high uncertainty about when a part is required and about the quantity of its requirement that derives from the unpredictability of failures occurrence. These are also the reasons explaining why the level of spare parts inventory kept by companies is usually very high in order to try to avoid risk of unavailability, leading to high inventory holding cost [35]. Moreover, sourcing of spare parts is often limited to one or a few suppliers, causing constraints for the procurement lead time and the costs; or in the opposite case of multiple sourcing, the related risk of the variations of the quality of supplied materials can incur. Obsolescence may also be a problem; indeed, it is difficult to determine how many units of a spare part item to stock for an obsolescent machine [29, 32, 34]. All these challenges opened the path in the scientific literature to the investigation on spare parts integrated inventory management [32, 36] and on spare parts supply chain management [29, 32, 37–39]. Moreover, the relevance of studying the AM technology's impact on these topics is evident because the key challenge in spare part management is to maintain high spare part availability with low cost. AM can be of aid in this issue as producing spare parts by AM can lower inventory stock while maintaining a good level of stock out avoidance [40]. Sasson and Johnson [17] reported that AM can be beneficial in the production of spare parts belonging to the long tail, which Anderson [18] describes as the majority part of a company's inventory that consists of items with a low demand. Such spare parts include, for example, products that are in a purely after sales stage of their life cycle such as discontinued consumer products, retired machinery, and antique elevators [41].

2 Development of the Digital Spare Parts Concept

Several studies have specifically paid attention to the potentialities of AM technology in the context of spare parts supply chain to reduce the size of central and local storages, eliminate the need to locate uncommon spare parts in the distribution network, and diminish the duration and cost of logistics [42]. The concept of spare part production using AM has been investigated by numerous researchers. The main studies specifically focusing on AM and spare part management are collected in Table 1.

The studies on AM in supply chain and the digital spare part (DSP) concept shown in Table 1 present an increased value to practitioners in industries that produce or use spare parts. However, although the findings in the body of research of spare part production by AM are generally positive, in reality very few companies implement digital spare parts in their supply chain operations. There is an evident research gap between why the results of the research are so positive and why it is yet to achieve a notable status in the field of maintenance management.

This study attempts to investigate how industrial practitioners perceive the value of DSP proposed by the researchers and to verify whether the results of the body of research are realistic or attractive to the companies that would benefit from the application of this technology. To reach the research objective, the interest of companies needed to be gauged together with understanding if there are relevant limitations for AM of spare parts diffusion in industry. The research was guided by the following research questions:

RQ1: How do industrial practitioners perceive the value of the DSP concept?
RQ2: What are the main advantages and criticalities of the DSP implementation in the perception of industrial partners?

Table 1 Description of studies connecting AM and spare part management and their findings and identified obstacles

Authors	Description of study	Key findings	Obstacles of DSP deployment
Walter et al. [42]	Introduced the concepts of distributed manufacturing of spare parts through AM	Concept of distributed manufacturing	The general cost of AM is high
Pérès and Noyes [43]	Described on a conceptual level the use of AM in delivering spare parts to isolated locations	Concept of delivering spare parts to isolated locations digitally	AM is not suitable for end-use applications because of technical limitations
Holmström et al. [44]	Compares on a conceptual level centralized and distributed AM spare part manufacturing	Centralized AM more likely to be used than distributed at the beginning	Quality of parts cannot be guaranteed
Liu et al. [41]	Simulated supply chains with and without AM for six different aircraft parts	Lower safety inventory	No obstacles identified
Khajavi et al. [40]	An aircraft spare part scenario analysis study that demonstrated that introducing AM in the supply chain reduced transportation costs, inventory costs, and part obsolescence costs	Reduced transportation costs Reduced inventory costs Reduced part obsolescence cost	Level of automation of AM is low; cost of AM machinery is high; AM machinery is slow
Knofius et al. [45]	Developed a methodology to identify which spare parts would benefit from being additively manufactured	Identified more than 1000 positive business cases of AM in service logistics	Quality and availability of spare parts product data varies
Li et al. (2016) [46]	Theoretical comparison between simulated AM-based and conventional supply chains	Reduced transportation costs	The cost of AM equipment is high
Sasson and Johnson [17]	Explores a rollout scenario of AM in a supply chain and qualitatively evaluates supply chain reconfigurations	Increased costs of AM spare parts can be justified by downtime reduction	AM material costs are high

(continued)

Table 1 (continued)

Authors	Description of study	Key findings	Obstacles of DSP deployment
Sirichakwal and Conner [47]	Analysed inventory-related benefits with a simulation	Reduced lead time Reduced holding cost Reduced stock-out risk	Qualification and certification is needed for spare part production
Chekurov and Salmi [48]	Conducted a case study by simulating inserting AM in a consumer electronics warranty repair network	Reduced total repair time	AM machinery is slow

To this end, focus group interviews were conducted with the representatives of the industry to verify if their opinions line up with the implementation of AM in spare part supply chains brought up in the literature in Sects. 1.1 and 1.2.

Another goal of this study, in addition to answering the research questions, was to refine the concept of DSP. From a methodological point of view, the focal concept was created according to Podaskoff's [49] stages for developing good conceptual definitions. According to the first stage of the methodology, potential attributes have to be identified by collecting a representative set of definitions. Therefore, relevant literature was surveyed for what other researchers consider critical to the definition of DSP. The potential attributes of other researchers' works have been collected in Table 2 in the manner demonstrated by Maynes and Podsakoff [50]. To explore further the concept of DSP, the attributes of Table 2 were also subjected to a focus group study, as suggested by the first stage of the Podsakoff methodology [49].

In the second stage of the methodology, the attributes need to be condensed into a reduced list and the sufficiency and necessity of the attributes needs to be established. The necessity of an attribute for a concept in this case means that the attribute is necessary to define the concept, and the sufficiency means that a combination of attributes is sufficient to distinguish a concept from another. The key attributes of DSP from the literature were compared to the ones discovered in the focus groups to get the overlapping attributes. The remaining attributes were given a designation and evaluated for necessity and sufficiency for DSP, traditionally produced spare parts, and additively manufactured parts.

3 Methodology

A focus group is a form of qualitative research in which groups of individuals composed of selected backgrounds are interviewed simultaneously. A focus group

Table 2 Summary of attributes of digital spare parts in literature

Study	Conceptualization of digital spare parts	Key attributes
Pérès and Noyes [43]	• "In order to shorten the time of immobilisation of a system having to be repaired, it is essential to have whatever the place and at any time the part needed to replace the one, which has failed"	• Delivered to replace a defective component
	• "… to be able to create, on demand and in situ, the part required to proceed to the maintenance intervention"	• Manufactured on site
	• "… a distant preparation of the digital files built from the CAD data for the optimisation of the part positioning …"	• Parts built from CAD data
	• "…a transfer of digital data through adapted networks …"	• Production data transferred by network
Holmström et al. [51]	• "…on demand and centralized production of spare parts is proposed as the most likely approach to succeed"	• Can use centralized manufacturing
	• "… if RM [rapid manufacturing] technology develops into a general purpose technology the distributed approach becomes more feasible"	• Can use distributed manufacturing

approach was chosen to conduct this research because it allows to probe deeper than quantitative methods and allows for the pursuit of new and emerging ideas [52].

Conducting focus group interviews is an accepted tool in management research to gauge the interest of company leaders in future possibilities [53]. As focus group interviews are designed to challenge assumptions, they are a good instrument in determining the perceived value of research concepts in real life applications [54–57].

A series of semi-structured focus group interviews were organized in Espoo, Finland in April 2016, in which 46 individuals from 34 companies and institutions participated. The recruitment of the participants was carried out by public invitation on the websites of the hosting organizations and national industrial mailing lists. Heterogeneity in the length of work experience, company types and sectors was sought and therefore no restrictions were set on who could participate, as long as they were professionals working in the field related to manufacturing. The participants were not compensated for their participation.

3.1 Demographic Information About Participants

The participants of the focus group interviews were experts in the field of manufacturing and familiar with the principles of AM. The companies involved were original equipment manufacturers (OEMs), manufacturing subcontractors (MSCs), AM service providers (AMSPs) and software developers (AMSDs), and national institutes (NIs). The number of participants ranged between 8 and 11 in each focus group. The work experience and position in company of participants is shown in Table 3 and the complete demographic information is given in Appendix.

All AM service providers that participated in the focus groups have B2B and B2C approaches and the AM software developers support the interaction between OEMs and AM service providers. The manufacturing subcontractors were traditional engineering companies mostly specializing in metal components. The participating national institutes were the national standardization agency and the national innovation agency. Due to the prevalent sectors in the Finnish industry, the present OEMs were mostly manufacturers of large industrial machinery. These OEMs are perfectly positioned to leverage the benefits of DSP because they not only manufacture equipment but also have extensive service operations for the maintenance of their own and their competitors' products.

> We divide our business in two – do we need them (spare parts) ourselves or do we sell them to our customers – these are two very different business models. (G5, OEM)

The distribution of the participants' years of experience is wide and the median is 15 years. The distribution is shown in five-year increments in experience in Table 3. The average work experience did not differ significantly between the types of company. There was a much bigger discrepancy in positions within the company when comparing the different company types. The participants from AM service providers and software developers were mostly from high in the company hierarchy, while the participants from national institutes were specialists or first-line managers. The participants from the other company types were quite evenly distributed. The reason why AM companies had an uneven distribution in the company positions is that they are limited in size, and therefore the top management tends to the operational side of business.

3.2 Organization of the Focus Groups

In a manner suggested by the focus group interviews of [58], the participants were collectively given a short introduction to the general concept of DSP before being divided into groups. The participants were divided into five groups with the consideration that no group should have two individuals from the same company. The participants were divided into five groups with the consideration that no group should have two individuals from the same company. Two researchers with extensive knowledge of the topic moderated a session of each group. In total, ten researchers participated

Table 3 Focus group participant characteristics

	AM service provider (n = 7)	AM software developer (n = 2)	Manufacturing subcontractor (n = 9)	National institute (n = 3)	OEM (n = 22)	Other (n = 3)	Total (n = 46)
Work experience in years							
<5	1	0	1	1	4	1	**8**
5–10	2	0	2	0	5	0	**9**
10–15	2	2	1	1	4	0	**10**
15–20	0	0	3	0	4	2	**9**
20–25	0	0	2	0	3	0	**5**
>25	2	0	0	1	2	0	**5**
Position in company							
Specialist	0	0	2	2	9	1	**14**
First-line management	0	0	3	1	4	0	**8**
Middle management	1	1	0	0	6	1	**9**
Top management	6	1	4	0	3	1	**15**

Table 4 Semi-structured questioning routine used in the focus group interviews

Theme	Question	Relevant literature
Theme 1: current status of spare parts management	1. What are the biggest problems with spare part production and logistics?	Kennedy et al. [32], Driessen et al. [29]
Theme 2: digitization of spare parts, new possibilities	2. What type of products will benefit from being additively manufactured?	Knofius et al. [45]
	3. What percentage of spare parts can be additively manufactured?	Knofius et al. [45]
	4. What do you think of the possibilities of digitizing existing spare parts?	Khajavi et al. [59]
	5. What are the obstacles in digitizing spare parts?	Knofius et al. [45]
	6. What kind of changes will digital spare parts bring to the service functions of a company?	Liu et al. [41], Li et al. [46]
	7. Should spare part production become more open or should it stay locked to the OEM?	Appleyard [28], Lindemann et al. [13]
Theme 3: digital spare part network and its requirements	8. What type of actors and skills are needed?	Holmström et al. [44]
	9. What could be the position of your company in the new paradigm?	Holmström et al. [44]
	10. What action is required from the companies and governmental institutions?	Holmström et al. [44]
	11. Is it realistic to create a DSP network?	Holmström et al. [44]
	12. What kind of data issues might arise?	Appleyard [28], Lindemann et al. [13]

in the focus group moderation. One of the researchers in each group was chosen to be the facilitator, while the second researcher was left in charge of taking continuous notes. The sessions included extensive written tasks that could not be recorded via audio and therefore the second researcher was strictly relegated as a secretary.

Each group received the same pre-defined questioning routine to facilitate the direction of discussions and to make the results of different group interviews comparable with one another. The questions were based on literature and composed jointly during a meeting of the ten moderators. The semi-structured question routine along with references that inspired each question can be seen in Table 4.

The data were collected through a four-step process. The participants were first given a task (e.g. "Write down the problems with digitization of spare parts from the point of view of your company."), which they were to complete on their own and for which they were allocated a specific amount of time. In the second step, time was given to compare the notes internally in small groups that were created spontaneously. The third step consisted of synthesizing the acquired data in smaller groups. Finally, in the fourth step, the groups presented their views, which were then open to discussion. These views were then recorded without critical evaluation. The focus group sessions were conducted simultaneously and the duration of each session was two hours.

3.3 *Data Parsing*

The data collected from the focus groups consisted of notes written down by the secretaries and the written material produced by the participants. The moderator and secretary made a summary of the interviews shortly after their end to ensure that the text were representative of the actual views and opinions of the participants. The content was codified by one of the moderators using what [60] call the "free coding approach" according to the codes that emerged inductively when analysing the collected data. The final codes were:

- Problems in spare parts management
- New possibilities of digitization of spare parts
- Obstacles in digitization of spare parts
- Description of required properties of digital spare parts
- Description of a digital spare parts network
- Factors that facilitate the development of a digital spare parts network

Because the participants were professionals discussing the subject of their work, and because the subject field was narrow, the codes were limited and clearly identifiable. To synthesize, the results were cross-referenced and summarized as one. The summaries of the interviews were compared and analysed by all ten moderators together.

4 Results

In this section, the data collected from the focus group interviews are introduced systematically. The points of view of the participants are presented according to the codes and accompanied by particularly relevant quotations.

4.1 Problems in Spare Parts Management

During the interviews, three aspects of spare parts management were mentioned most often. First, the participants felt that the availability of spare parts is a crucial problem because getting a replacement can take long in unexpected cases and rare components have long delivery times.

> In the worst case getting components can take months. Just getting the raw material for the component can take up to twelve months if it is a special material. (G1, OEM)
>
> Certain sizes of bearings are only manufactured once per year. If the batch is sold out, you have to wait a year. (G1, OEM)

Second, the participants emphasized that they would be ready to pay a lot of money for a spare part to avoid unexpected and lengthy process inactivity caused by part failure. In addition to the loss of productivity due to stock-outs, the risk of downtime causes large spare parts inventories, high storage expenditure and idleness of assets and the most critical parts are kept in smaller inventories around the world.

> The cost is secondary because the downtime of a process is the worst case scenario. We need large warehouses, which makes our assets idle. The customers have their own storages for critical components and would be willing to pay to eliminate them. (G1, OEM)

Third, the propensity of employees to commit costly mistakes when hurrying while attempting to fix machinery was mentioned often. Employees try to compensate for long delivery times by hurrying, which leads to mistakes or to ordering entire new products instead of spare parts.

> Rushing and scrambling inevitably leads to mistakes. (G3, OEM)

Other often-mentioned issues included that the inventories are especially big when considering long tail products and that minimum order quantities are too big when only one part is needed.

4.2 New Possibilities of Digitization of Spare Parts

Data from the focus groups suggest that there has been enough previous interest among the participants concerning the concept of digitization of spare parts to generate their own ideas of new delivery models and sum them up concisely.

> Someone presses a button here, the printer starts over there and then the data disappears from there (G3, OEM)

The concept of customers participating in the creation and manufacturing of spare parts was brought up often in groups 2 and 4 but not in the others. The participants in these groups theorized that industrial design could benefit from becoming open source to leverage the community of designers. The digitization of spare parts also raised questions about how the laws and rules of the physical world will apply for example regarding customs clearance.

Table 5 The views of the focus group participants on new possibilities and their enablers and implications of digitization of spare parts for maintenance operations

Possibilities of digitization of spare parts	Enabler of possibilities	Implications for maintenance operations
Reduction of delivery time	Spare parts can be printed anywhere where the machinery is available	Lower delivery time and costs Improved OEM service speed and flexibility
Spare parts do not have to pass through customs	Parts can be produced directly from digital files	Spare parts can be delivered at a lower cost
Data can travel where physical parts cannot	Parts do not require specific know-how to be printed in different locations	Improved possibility to deliver spare parts in hard to reach locations
Positive environmental impact	AM uses only the material needed to produce components.	Lower emissions for the supply network
Upgraded spare parts	AM requires little set-up	Components can be improved based on field maintenance data each time they have to be replaced
Emergency spare parts		Temporary spare parts can be installed quickly until the actual spare parts are delivered

> You will be able to make a lot of money by 3D modelling from home (G5, AMSP)
>
> If a customer has a 3D scanner and a printer he does not need a spare part supply chain (G2, AMSP)
>
> How will the customs work in the future? If we send a digital spare part to Russia will we need to declare it? (G2, OEM)

The participants listed possibilities of digitizing spare part manufacturing and their enablers and implications were brought up during the discussion. The enablers of the possibilities are the reasoning of the participants behind the possibilities and the implications are the continuations of those ideas. The possibilities along with their enablers and implications are collected in Table 5.

4.3 *Obstacles in Digitization of Spare Parts*

The participants were asked to list the obstacles regarding digitization of spare parts, which were then discussed. The causes of the obstacles and the real-world implications for maintenance operations were discovered during these discussions. Intellectual property rights (IPR) issues of spare parts was clearly an issue about which most participants were concerned and wanted means of protection to be developed.

Table 6 The answers collected from the participants of the focus groups relating to the obstacles and their causes and implications of digitization of spare parts for maintenance operations

Obstacle in digitization of spare parts	Cause of obstacle	Implication for maintenance operations
High cost of AM	High cost of AM machine hour price and materials	Increased price of producing components
Limited size of possible components	Limited build envelopes of AM machinery	Limited choice of components can be produced with AM
Inadequate quality of spare parts	AM accuracy limitations	
Variable quality between AM machines	Lacking control of process parameters in AM	Only parts that fit in the quality variability can be produced with AM
Variable quality between shipments of AM materials	Lack of standardization of AM material production processes	
Piracy	Data leaks in the supply network and reverse engineering of components	Loss of sales
File version management	Poor ICT	Lack of conviction in correct spare parts
3D model unavailability	Component manufactured based on 2D drawings or design subcontracted	More work in 3D modelling and increased labour cost
Difficulty in making 3D models from obsolete components	Imprecision of 3D scanning	
Parts are not ready after 3D printing	Parts need to be post processed and 3D scanning does not provide this information	

With 3D printing you can create IDs for the spare parts, with which you could see if the spare part is licensed or not" (G2, AMSP)

The obstacles along with their causes and implications are presented in Table 6.

4.4 Descriptions of Required Properties of Digital Spare Parts

In the discussions regarding the ideal properties of DSP components, the participants made it clear that a part has to be technically and economically viable to be considered for DSP distribution. This means that although a part could technically be manufactured with AM, it should only be used if the distribution advantages outweigh the increased costs of AM. Because standard parts are easily manufactured with conventional manufacturing methods, there is no need to involve DSP.

> If the part is simple, a handy machinist will create it in the same time it would take another to 3D model it. (G1, MSC)

The OEM participants were asked to write down the estimated percentage of parts in their companies' spare parts libraries that can be acceptably manufactured with AM. The estimation varied between 2 and 75% by the participants, but most answers stayed between 5 and 10%.

The participants were asked to write down the properties of spare parts that could affect their applicability in the DSP concept. Afterwards, the reasons behind why the participants chose the properties were discussed. The essential technological and control properties along with the causes for the preference for DSP from the perspective of the participants are presented in Table 7 along with their preferences and their causes.

4.5 DSP Network and Its Requirements

To investigate what the DSP network should be composed of, the participants were asked to list the necessary actors and expertise that should be found in a DSP network. The participants were tasked with writing down the actors of a digital spare part distribution network and placing them in their respective places in the network. The actors that the participants proposed were:

- Network administrator
- AM service provider
- Digitization department of the service provider
- OEM
- Spare part ordering personnel of OEM
- OEM's model database
- AM plant
- Maintenance site
- Material supplier
- Software developers

The participants had a strong preference for an international service bureau network so that spare parts could be manufactured close to the customer but the 3D file of the part could be stored in one location. In particular, the OEMs would prefer to order their DSPs from one place that handles the entire spare part production and delivery process.

> We need a joint enterprise where the 3D machinery is centralized for better quality control, IPR safety and no profit margin of external service providers … big companies have the readiness to invest in this within five years. Big companies can then sell machine time to smaller companies later on. (G3, OEM)

The role of service bureaus was emphasized because they need to be reliable and supply a choice of different service contracts.

Table 7 The main properties of spare parts affecting the viability of digitization according to the participants

Property of spare parts	Property type	Preference for DSP	Cause for preference
Complexity	Technological	High	The preference for high complexity comes from the fact that AM can manufacture complex products as easily as parts with simpler features
Size	Technological	Low	The preference for smaller size comes from AM technology's current limitation of being able to produce parts only up to a certain size
Criticality	Control	High	Highly critical spare parts cause costly periods of downtime if they are not immediately available. Therefore, the decreased lead-time enabled by distributed manufacturing of DSPs is a major advantage
Demand pattern	Control	High variation Low volume	DSP is a benefit for items with high demand variation because additive manufacturing can meet demand very flexibly. Moreover, the preference for low volume comes from the fact that while AM is beneficial for lower volumes; at higher volumes, it usually becomes more expensive than manufacturing components by conventional means
Value	Control	High	High value makes stocking an unattractive solution. With the use of AM, as there is no need to maintain physical inventories and the inventories are stored on a server, the cost of inventory is reduced dramatically. The small cost that remains is that of server hosting
Delivery time predictability	Control	Low	Low delivery time predictability is also a benefit to DSPs because the delivery time predictability for AM is very reliable
Specificity	Control	High	The lower the number of suppliers the better the part is suited for DSP. Because additive manufacturing requires no moulds and few instructions, parts can be manufactured without extensive expertise. Therefore, the number of suppliers for DSPs can be much higher than that of traditional suppliers
Life cycle stage	Control	Very early or late	At the beginning of the life cycle, the spare parts can be created in case the main production is not ramped up yet. At the end of the life cycle, DSP can be beneficial because the tools or the supplier might be gone
Update rate	Control	Frequent	Parts that become obsolete (through either upgrades or deterioration) have to be disposed of. If they are produced only when necessary, the stocking and disposal costs can be avoided

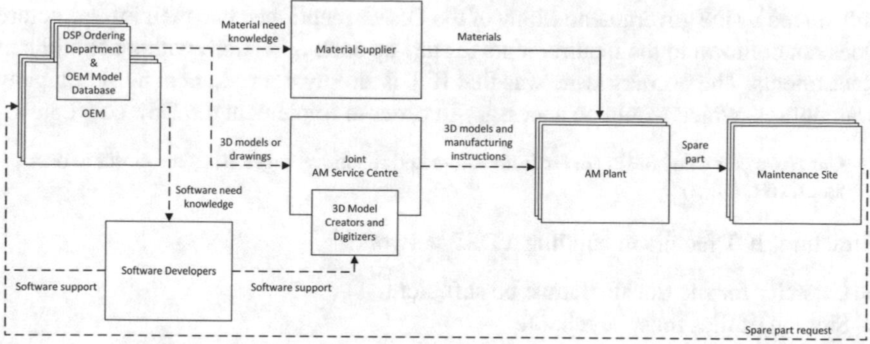

Fig. 1 A model of the optimal digital spare part network according to the participants. In this model, the AM service centre is the network administrator. Solid arrows signify the transfer of objects and dotted arrows signify the transfer of information

> We need a platform with which we can deliver them (3D files of components) safely … and an actual manufacturer close to the customer (G5, OEM)

The model of a DSP network generated by the authors based on the answers by the participants is presented in Fig. 1. The network administrator of the DSP network model is a service centre jointly owned by the OEMs. The 3D models and drawings are stored in each OEM's model database, as the participants showed strong preference to host the files on their own servers.

> It is better and safer if the 3D model is acquired from the company every time it is printed. We must know when a spare part is produced (G2, OEM)

In the process flow of the model, an OEM sends a request to the AM service centre to manufacture the spare part along with its 3D models or drawings. The digitization department of the service centre transforms into files fit for manufacturing and sends them onwards to the AM facilities close to the maintenance site. Once the spare part is manufactured, it is transported to the maintenance site to be installed.

The network model features the material supplier and software developers as the two external parties. The software developers offer software support to the OEM and the service provider and receive input on how they should develop their software to suit their needs better. The material supplier identifies potential materials from the OEM model database and supplies the AM plants with materials that suit the needs of the OEMs.

4.6 *Facilitating Factors in the Development of a Digital Spare Parts Network*

The company representatives shared their opinion on factors that can affect the creation of DSP networks. The factors were divided into ICT and issues that could be influenced by the government. Many of the OEMs mentioned that their infrastructure does not conform to the requirements to start up such operations without significant investments. The primary issue was that ICT is simply not present at most company warehouses, which would be necessary in order to implement the DSP concept.

> Our spare part warehouses are far from automated robotic storages. They are closer to dark attics. (G1, OEM)

Structural ICT factors in building a DSP network:

- Capacity for file transfers must be sufficient
- Storage of files must be reliable
- File transfer costs must be low
- Transfer speeds must be sufficiently fast
- Internet network conditions must be stable and fast
- Data formats must be standardized

Government factors in building a DSP network:

- Existing workforce and upcoming engineers need to be educated in AM and DSP
- Design for AM and software development for manufacturing needs to be taught more extensively in higher education institutions
- Upper-level management needs to be educated in the concepts of strategic DSP implementation
- Legal services need to become involved in the DSP concept to investigate the possibly emerging IPR issues
- Investment support for AM infrastructure development should be offered

5 Discussion of Focus Group Results

The focus group interviews verified that companies find implementing AM in spare part supply chains a valuable concept for spare part management and provided information to form a more comprehensive picture of the possibilities of the digital spare part concept.

The problems with spare part management that are listed in Sect. 3.1 are very similar to those found in the work of [32, 44]. The only exception is for the problem of rushing of maintenance operators for compensating long delivery times, which was clearly represented in the focus groups more than in the literature.

All the possibilities of digitization of spare parts listed in Table 5 were also found in the literature findings presented in Table 1 except for the high interest of the focus group participants in bypassing customs by manufacturing spare parts in the country where they are needed and the concepts of improved and emergency spare parts.

The obstacles of digitization of spare parts in Table 6 were largely similar to the ones found in the literature overview in the introduction. The high cost of AM compared to subtractive manufacturing was often cited as a large obstacle by the participants as well as by Sasson and Johnson [17], Holmström et al. [44], Khajavi et al. [59] and Li et al. [46]. The limited size of possible components due to AM restrictions was indicated as a major obstacle by the participants because a many of their spare parts are too large for current AM build envelopes. Although the possible size of components was considered by Knofius et al. [45] in identifying business cases of DSP, it has only been explicitly stated as an obstacle by Pérès and Noyes [43]. The obstacle of insufficient and varying quality of AM components was likewise brought up often by the participants but was only raised as a clear issue by Pérès and Noyes [43] and Holmström et al. 44]. On the other hand, the relative slow speed of AM referenced as obstacles by Khajavi et al. [40] and Chekurov and Salmi [48] were not mentioned often by the participants. File version management and 3D model unavailability were listed by some participants as significant obstacles and similar views were presented by Knofius et al. [45]. The potential of losing revenue to piracy was mentioned often in the focus group results although they were only present in two articles in the referenced literature. In their article, Appleyard [28] discusses the implications of piracy in AM and concludes that companies need to change their business models to fully leverage the AM paradigm or suffer financial losses due to piracy. Lindemann et al. [13] present an opposing view by demonstrating copy protection of AM parts. From the points of view of the focus group participants, the participants would prefer to use the approach presented by Lindemann et al. when making their spare parts available as DSPs.

In addition to the issues of ICT nature, the obstacle of post processing components was more prominent in the focus groups than in the mentioned literature.

According to Table 7, the ideal parts for the DSP concept from the technical perspective are physically small and complex. These aspects are linked with the technological requirements of AM, and in this regard, the [13] publication discusses ideal part selection for AM. All of the control properties in Table 7 are traditional properties of spare part classification [34].

The description of the ideal DSP in Table 7 is very close to that of a part belonging to the long tail of a company's product catalogue. This is in line with previous findings of [17, 51] who noted that long tail products are good candidates for AM.

A traditional spare part supply network in the companies that were interviewed can take on a multitude of configurations, but they all have in common the fact that spare parts are sourced from a handful of suppliers, stored in central and secondary warehouses, and transported to the repair site when the need arises. In a scenario in which a company implements DSP, the transportation of data becomes more important than physical part movement. To mirror this, the role of the material supplier and the software developers is significantly larger in the DSP network especially in the early stages. Other differences between the traditional and DSP network models are the replacement of a central storage with the model database, removal of regional storage spaces, and the introduction of distributed manufacturing. These changes make it possible for spare parts to be delivered to the location much faster than in the traditional model. The DSP network model acquired by the means of this study shares notable aspects with the models developed by Khajavi et al. [40] and Holmström et al. [44] whose models demonstrate the concept of distributed manufacturing and the possibility of multiple OEMs sharing the supply network. The model constructed based on the priorities of this focus group study also includes these aspects but in addition emphasizes the roles of software developers and material suppliers..

6 Further Definition of the Digital Spare Parts Concept

The second stage of the Podsakoff methodology consists of condensing the list of identified attributes of the concept to a smaller list and evaluating if they are indeed necessary to the concept and if they can be used to sufficiently describe the concept without being confused with another [49]. In this study, the condensation of the attribute list is done by comparing the key attributes found in the literature with the ones acquired from the focus group study. The overlapping attributes that emerge from both sources are then compared with the concepts of "Traditionally produced spare part" and "Additively manufactured part" and their sufficiency is evaluated. In comparison with Table 2, the missing attributes are "manufactured on site" and "can use centralized manufacturing" because the participants favoured the distributed manufacturing model. The attributes present in both the literature and the focus group reviews are presented in Table 8.

In the third stage of the Podsakoff methodology, the preliminary definition of the concept is developed. This is done by describing the necessary and sufficient attributes of the concept compared to related concepts, specifying the entity, general property, dimensionality and stability of the concept and by identifying the consequences of the concept [49]. The entity of the DSP concept is a component and its general property is the means of its delivery. At this level of abstraction, the concept of DSP is

Table 8 Identifying necessary and sufficient attributes of the concept of a "digital spare part"

Attributes	Digital spare part	Traditionally produced spare part	Additively manufactured part	Conclusions
A1: delivered to replace a defective component	Present	Present	Absent	Necessary but not sufficient
A2: parts built from 3D model data	Present	Absent	Present	Necessary but not sufficient
A3: production documents transferred by network	Present	Absent	Present	Necessary but not sufficient
A4: can use distributed manufacturing	Present	Absent	Present	Necessary but not sufficient

unidimensional, as all of the attributes exist in the same conceptual domain of system operations. Although the technical and economic performance of components in the DSP concept varies, all conceivable components share the key attributes. The concept is therefore stable among all variations of the entity.

The concept of DSP has multiple necessary attributes that sufficiently define it. Any combination of A1 and the other attributes is necessary and jointly sufficient to define a DSP. While digital spare parts and traditionally produced spare parts have one attribute, their main function, in common, the rest of the attributes apply only to digital spare parts, as traditionally produced spare parts cannot be digitally distributed. Similarly, DSP and additively manufactured parts have the possibility of digital distribution (A2–A4) in common but do not share the main function A1.

The consequences that were found both in literature and in the focus group interviews were the reduction of repair time, delivery time and costs, emissions and material waste, and inventory.

By combining the defining attributes in Table 8 and the overlapping consequences, the DSP concept could be described as "A concept in which defective components are replaced by manufacturing spare parts close to the location of need from 3D model data that are transferred by network with the main consequences of reducing repair time, delivery time and costs, emissions, and inventory".

7 Conclusions

The perceived value of implementing AM in supply networks of digital spare parts was measured and the concept of DSP was expanded by focus group interviews with manufacturing professionals. The interviews were conducted to investigate the possible benefits of DSPs and their importance to industrial companies. The interest in producing spare parts digitally was very high among the participants, but there are several issues to overcome before large-scale implementation can take place. It was noted that neither the AM technology, nor the ICT infrastructure of the companies that would make DSP distribution possible are robust enough for the digital distribution method to be deployed fully in the interviewed companies. However, the DSP distribution method could already be applied to specific types of components. Through the participants' description of the ideal DSP, it was found that long tail products make excellent candidates for digital distribution. It is therefore likely that the first implementations of DSP networks will be based on such products.

The cost of AM was not seen as a prohibitive issue for long tail products by the participants. Indeed, it was well understood that even though AM parts are generally more expensive to produce, the cost savings come from other areas of implementation. In the case of long tail spare parts; these could be related to lower warehousing and transportation costs as well as faster lead times and delivery times.

Another potential DSP subgroup is the improved spare parts that have better qualities than the original ones. This can mean, for instance, better design by topology optimization or reduced number of joints to decrease the risk of failure. In order for

DSP to become a more viable approach, AM technology has to be developed to correspond with more spare parts in companies' catalogues. This could be achieved, for example, by faster machines that produce parts that require less post treatment, or by expanding the limited material library of AM. In addition, old designs have to first be digitized for production with AM. To improve the possibility of manufacturing future spare parts digitally, any new designs have to be recorded systematically according to the attributes described by the participants.

While it will take some time for AM technology development to catch up with company needs, there are already steps that companies can take towards its implementation. The companies that view digital spare parts as advantageous and plan to utilize the concept in the future should start the preparation related to part documentation and ICT system capabilities, developing required competences and finding the right partners. Ultimately, the actual structure of the DSP network will depend on decisions of the current operators of the spare parts supply chain as well as on the actions of the new network members.

8 Limitations and Future Work

The study was conducted in Finland and as such, the views presented in the focus groups represent the bias of the Finnish industry. Further focus group research aimed at different sectors in different countries could expand on the results. In addition, a quantitative study regarding the readiness levels of the DSP concept and a more in-depth interview with single individuals from companies that already use AM in their supply chains using the same question routine as in this study could yield data that are more detailed. Such a study could be focused on developing mathematical models for DPS enabled supply chains that takes the findings of this study into account. Furthermore, this study can be used for initial implementation of DSP in companies.

The coding strategy in parsing the data of the focus group interviews was developed inductively. While this open-ended approach to coding yielded valuable results, further studies on the same topic should use predefined coding strategies to achieve a higher level of accuracy of data.

Regarding the concept definition of DSP, this study works in the domain of the first three stages and ends the process by providing an initial definition of the concept. The concept of DSP needs to be developed further by following the fourth stage of Podsakoff's concept definition, in which the conceptual definition of the concept is further refined by reducing the complexity of the language and by soliciting the feedback from peers [49].

Acknowledgements The research on which this chapter is based was conducted as a part of the Digital Spare Parts (DIVA) project and funded by a consortium of companies and Tekes—the Finnish Funding Agency for Innovation.

Appendix

See Table 9.

Table 9 Complete demographic information of the focus group participants

Group	Company type	Position in company	Years of experience
1	AM service provider	Top management	10–15
1	AM service provider	Top management	>25
1	Manufacturing subcontractor	Top management	20–25
1	Manufacturing subcontractor	Top management	15–20
1	Manufacturing subcontractor	Top management	15–20
1	OEM	Middle management	10–15
1	OEM	Middle management	15–20
1	OEM	Middle management	10–15
2	AM service provider	Top management	10–15
2	AM service provider	Top management	5–10
2	AM software developer	Middle management	10–15
2	Manufacturing subcontractor	Top management	10–15
2	National institute	Specialist	<5
2	National institute	First-line management	>25
2	OEM	Specialist	5–10
2	OEM	Specialist	10–15
2	OEM	Specialist	<5
2	OEM	Middle management	>25
2	OEM	Top management	5–10
3	Manufacturing subcontractor	First-line management	<5
3	Manufacturing subcontractor	First-line management	15–20
3	National institute	Specialist	10–15
3	OEM	First-line management	10–15
3	OEM	Specialist	<5
3	OEM	Specialist	15–20
3	OEM	Middle management	20–25
3	OEM	Specialist	<5
3	Other	Specialist	<5

(continued)

Table 9 (continued)

Group	Company type	Position in company	Years of experience
4	AM service provider	Middle management	>25
4	AM software developer	Top management	10–15
4	Manufacturing subcontractor	First-line management	5–10
4	OEM	First-line management	5–10
4	OEM	First-line management	5–10
4	OEM	Top management	>25
4	OEM	Specialist	<5
4	OEM	Middle management	20–25
4	OEM	Specialist	20–25
4	Other	Top management	15–20
5	AM service provider	Top management	<5
5	AM service provider	Top management	5–10
5	Manufacturing subcontractor	Specialist	20–25
5	Manufacturing subcontractor	Specialist	5–10
5	OEM	First-line management	15–20
5	OEM	Top management	15–20
5	OEM	Specialist	5–10
5	Other	Middle management	15–20

References

1. Mellor, S., Hao, L., & Zhang, D. (2014). Additive manufacturing: A framework for implementation. *International Journal of Production Economics, 149,* 194–201.
2. Weller, C., Kleer, R., & Piller, F. T. (2015). Economic implications of 3D printing: Market structure models in light of additive manufacturing revisited. *International Journal of Production Economics, 164,* 43–56.
3. Simhambhatla, S., & Karunakaran, K. P. (2015). Build strategies for rapid manufacturing of components of varying complexity. *Rapid Prototyping Journal, 21*(3), 340–350.
4. Jonsson, P., & Holmström, J. (2016). Future of supply chain planning: Closing the gaps between practice and promise. *International Journal of Physical Distribution & Logistics Management, 46*(1), 62–81.
5. Kellens, K., Baumers, M., Gutowski, T. G., Flanagan, W., Lifset, R., & Duflou, J. R. (2017). Environmental dimensions of additive manufacturing: Mapping application domains and their environmental implications. *Journal of Industrial Ecology, 21,* S49–S68.
6. Wohlers, T., & Report, Wohlers. (2017). *Additive manufacturing and 3D printing state of the industry: Annual worldwide progress report* (p. 2017). Fort Collins, CO: Wohlers Associates Inc.
7. Wohlers, T., & Report, W. (2013). *Additive manufacturing and 3D printing state of the industry: Annual worldwide progress report* (p. 2013). Fort Collins, CO: Wohlers Associates Inc.

8. Schniederjans, D. G. (2017). Adoption of 3D-printing technologies in manufacturing: A survey analysis. *International Journal of Production Economics, 183*, 287–298.
9. Holmström, J., & Partanen, J. (2014). Digital manufacturing-driven transformations of service supply chains for complex products. *Supply Chain Management: An International Journal, 19*(4), 421–430.
10. Rylands, B., Böhme, T., Gorkin, R. I., & Fan, T. B. (2016). The adoption process and impact of additive manufacturing on manufacturing systems. *Journal of Manufacturing Technology and Management, 27*(7), 969–989.
11. Oettmeier, K., & Hofmann, E. (2016). Impact of additive manufacturing technology adoption on supply chain management processes and components. *Journal of Manufacturing Technology and Management, 27*(7), 944–968.
12. Lyly-yrjänäinen, J., Holmström, J., Johansson, M. I., & Suomala, P. (2016). Effects of combining product-centric control and direct digital manufacturing: The case of preparing customized hose assembly kits. *Computers in Industry, 82*, 82–94.
13. Lindemann, C., Reiher, T., Jahnke, U., & Koch, R. (2015). Towards a sustainable and economic selection of part candidates for additive manufacturing. *Rapid Prototyping Journal, 21*(January), 216–227.
14. Tuck, C., Hague, R., & Burns, N. (2007). Rapid manufacturing: impact on supply chain methodologies and practice. *International Journal of Services and Operations Management, 3*(1), 1–22.
15. Gebler, M., Schoot Uiterkamp, A. J. M., & Visser, C. (2014). A global sustainability perspective on 3D printing technologies. *Energy Policy, 74*(C), 158–167.
16. Kothman, I., & Faber, N. (2016). How 3D printing technology changes the rules of the game: Insights from the construction sector. *Journal of Manufacturing Technology and Management, 27*(7), 932–943.
17. Sasson, A., & Johnson, J. C. (2016). The 3D printing order: variability, supercenters and supply chain reconfigurations. *International Journal of Physical Distribution & Logistics Management, 46*(1), 82–94.
18. Anderson, C. (2006). *The long tail: Why the future of business is selling less of more*. Hachette Books.
19. Rayna, T., & Striukova, L. (2016). From rapid prototyping to home fabrication: How 3D printing is changing business model innovation. *Technological Forecasting and Social Change, 102*, 214–224.
20. Fox, S., & Li, L. (2012). Expanding the scope of prosumption: A framework for analysing potential contributions from advances in materials technologies. *Technological Forecasting and Social Change, 79*(4), 721–733.
21. Gausemeier, J., Echterhoff, N., Kokoschka, M., & Wall, M. (2012). *Thinking ahead the future of additive manufacturing—Future applications* (Study Part 2).
22. Scott, A., & Harrison, T. (2015). Additive manufacturing in an end-to-end supply chain setting. *3D Printing and Additive Manufacturing, 2*(2), 65–77.
23. Thomas, D. (2016). Costs, benefits, and adoption of additive manufacturing: a supply chain perspective. *International Journal of Advanced Manufacturing Technology, 85*, 1857–1876.
24. Singh, S., Ramakrishna, S., & Singh, R. (2017). Material issues in additive manufacturing: A review. *Journal of Manufacturing Processes, 25*, 185–200.
25. Gray, G. T. I., et al. (2017). Structure/property (constitutive and spallation response) of additively manufactured 316L stainless steel. *Acta Materialia, 138*, 140–149.
26. Portolés, L., Jordá, O., Jordá, L., Uriondo, A., & Esperon-miguez, M. (2016). A qualification procedure to manufacture and repair aerospace parts with electron beam melting. *Journal of Manufacturing Systems, 41*, 65–75.
27. Thomas-Seale, L. E. J., Kirkman-Brown, J. C., Attallah, M. M., Espino, D. M., & Shepherd, D. E. T. (2018). The barriers to the progression of additive manufacture: Perspectives from UK industry. *International Journal of Production Economics, 198*(February 2017), 104–118.
28. Appleyard, M. (2015). Corporate responses to online music piracy: Strategic lessons for the challenge of additive manufacturing. Business Horizons.

29. Driessen, M., Arts, J., Van Houtum, G.-J., Rustenburg, J. W., & Huisman, B. (2015). The management of operations maintenance spare parts planning and control: A framework for control and agenda for future research. *Production Planning & Control, 26*(5), 407–426.
30. Behfard, S., Al Hanbali, A., van der Heijden, M. C., & Zijm, W. H. M. (2018). Last time buy and repair decisions for fast moving parts. *International Journal* of *Production Economics, 197*(August 2017), 158–173.
31. Sarker, R., & Haque, A. (2000). Optimization of maintenance and spare provisioning policy using simulation. *Applied Mathematical Modelling, 24*(10), 751–760.
32. Kennedy, W. J., Wayne Patterson, J., & Fredendall, L. D. (2002). An overview of recent literature on spare parts inventories. *International Journal* of *Production Economics, 76*(2), 201–215.
33. Molenaers, A., Baets, H., Pintelon, L., & Waeyenbergh, G. (2012). Criticality classification of spare parts: A case study. *International Journal of Production Economics, 140*(2), 570–578.
34. Roda, I., Macchi, M., Fumagalli, L., & Viveros, P. (2014). A review of multi-criteria classification of spare parts: From literature analysis to industrial evidences. *Journal of Manufacturing Technology and Management, 25*(4), 528–549.
35. Ledwoch, A., Yasarcan, H., & Brintrup, A. (2018). The moderating impact of supply network topology on the effectiveness of risk management. *International Journal* of *Production Economics, 197*(January 2017), 13–26.
36. Cavalieri, S., Garetti, M., Macchi, M., & Pinto, R. (2008). The management of operations a decision-making framework for managing maintenance spare parts. *Production Planning & Control, 19*(4), 379–396.
37. Huiskonen, J. (2001). Maintenance spare parts logistics: Special characteristics and strategic choices. *International Journal of Production Economics, 71*(1–3), 125–133.
38. Martin, H., Syntetos, A., Parodi, A., Polychronakis, Y., & Pintelon, L. (2010). Integrating the spare parts supply chain: An inter-disciplinary account. *Journal of Manufacturing Technology and Management, 21*(2), 226–245.
39. Kazemi Zanjani, M., & Nourelfath, M. (2014). Integrated spare parts logistics and operations planning for maintenance service providers. *International Journal* of *Production Economics, 158*, 44–53.
40. Khajavi, S. H., Partanen, J., & Holmström, J. (2014). Additive manufacturing in the spare parts supply chain. *Computers & Industrial*, *65*(February 2016), 50–63.
41. Liu, P., Huang, S. H., Mokasdar, A., Zhou, H., & Hou, L. (2014). The impact of additive manufacturing in the aircraft spare parts supply chain: Supply chain operation reference (scor) model based analysis. *Production Planning & Control, 25*(13–14), 1169–1181.
42. Walter, M., Holmström, J., & Yrjölä, H. (2004). Rapid manufacturing and its impact on supply chain management. In *Proceedings of the Logistics Research Network Annual Conference*.
43. Pérès, F., & Noyes, D. (2006). Envisioning e-logistics developments: Making spare parts in situ and on demand. State of the art and guidelines for future developments. *Computers in Industry, 57*(6), 490–503.
44. Holmström, J., Partanen, J., Tuomi, J., & Walter, M. (2010). Rapid manufacturing in the spare parts supply chain. *Journal of Manufacturing Technology and Management, 21*(6), 687–697.
45. Knofius, N., Van Der Heijden, M. C., & Zijm, W. H. M. (2016). Selecting parts for additive manufacturing in service logistics. *Journal of Manufacturing Technology and Management*.
46. Li, Y., Jia, G., Cheng, Y., & Hu, Y. (2017). Additive manufacturing technology in spare parts supply chain: A comparative study. *International Journal of Production Research, 55*(5), 1498–1515.
47. Sirichakwal, I., Conner, B. (2016). Implications of additive manufacturing for spare parts inventory. *3D Printing* and *Additive Manufacturing*.
48. Chekurov, S., & Salmi, M. (2017). Additive manufacturing in offsite repair of consumer electronics. *Physics Procedia*.
49. Podsakoff, P. M., MacKenzie, S. B., & Podsakoff, N. P. (2016). Recommendations for creating better concept definitions in the organizational, behavioral, and social sciences. *Organizational Research Methods, 19*(2), 159–203.

50. Maynes, T. D., & Podsakoff, P. M. (2014). Speaking more broadly: An examination of the nature, antecedents, and consequences of an expanded set of employee voice behaviors. *Journal of Applied Psychology, 99*(1), 87.
51. Holmström, J., & Partanen, J. (2014). Digital manufacturing-driven transformations of service supply chains for complex products. *Supply Chain Management*.
52. Morgan, D. L. (1988). *Focus groups as qualitative research*, First. Newbury Park: SAGE Publications, Inc.
53. Flynn, B. (1990). Empirical research methods in operations management. *Journal of Operations Management, 9*(2), 250–284.
54. Barbour, R. (2007). *Doing focus groups*, First. London: SAGE Publications, Inc.
55. Fern, E. F. (2001). *Advanced focus group research*, First. Thousand Oaks: SAGE Publications, Inc.
56. Greenbaum, T. L. (1998). *The handbook for focus group research*, Second. Thousand Oaks: SAGE Publications, Inc.
57. Puchta, C., & Potter, J. (2005). *Focus group practice*, First. London: SAGE Publications, Inc.
58. Angell, L. C., & Klassen, R. D. (1999). Technical note: integrating environmental issues into the mainstream: An agenda for research in operations management. *Journal of Operations Management, 17*(5), 575–598.
59. Khajavi, S. H., Partanen, J., & Holmström, J. (2014). Additive manufacturing in the spare parts supply chain. *Computers in Industry, 65*(1), 50–63.
60. Campbell, J. L., Quincy, C., Osserman, J., & Pedersen, O. K. (2013). Coding in-depth semistructured interviews: Problems of unitization and intercoder reliability and agreement. *Sociological Methods & Research, 42*(3), 294–320.

Chapter 17
Summary of Book Findings. Mapping Chapters to Framework Dimensions

Adolfo Crespo Márquez, Marco Macchi and Ajith Kumar Parlikad

Abstract This book explored the concept of 'value-based asset management' in depth and breadth. We recognise that the reader might find the range of issues covered by the book too daunting, and might wish to focus only on those parts and chapters that seem most relevant to their research or practice. Therefore, the aims of this final chapter are two-fold. Firstly, it provides an easily digestible summary of each chapter and the key learning outcomes of the research carried out therein. Secondly, we knit the different chapters together by aligning them with the value-based asset management framework, and its dimensions as presented in the introductory chapter. This will hopefully enable the reader to not-only rapidly understand the outcomes of the book, but to appreciate the applicability of the ideas addressed by the generalized framework in different industry sectors, asset types and at various levels of an organisation.

Keywords Asset management · Framework · Value model · Asset management system

A. Crespo Márquez (✉)
Department of Industrial Management, School of Engineering, University of Seville, Camino de los Descubrimientos s/n, 41092 Seville, Spain
e-mail: adolfo@us.es

M. Macchi
Department of Management, Economics and Industrial Engineering (DIG), Politecnico di Milano, Via Lambruschini 4/b, 20156 Milan, Italy
e-mail: marco.macchi@polimi.it

A. K. Parlikad
Department of Engineering, Institute for Manufacturing, University of Cambridge, Cambridge CB3 0FS, UK
e-mail: aknp2@cam.ac.uk

A. Crespo Márquez et al. (eds.), *Value Based and Intelligent Asset Management*,
https://doi.org/10.1007/978-3-030-20704-5_17

1 Summary of Book Parts, Chapters and Findings

A general rationale of each Part and Chapter is summarized in the following Sections of this Chapter; this should be helpful to summarize and get a final outlook of what could be found in the book. This summary will also motivate a subsequent set of maps that will be introduced later, in order to see the different Chapters from the perspective of the generalized framework defined in the introductory chapter of this book.

Part 1. Long-Term Vision for Proper Asset Management
This Part aims at illustrating a long-term vision in order to properly implement Asset Management in a company.

The introductory chapter (Chap. 1) initially discusses the fundamental concepts at the background of the entire book, and introduces a generalized framework providing the relevant dimensions of value-based and intelligent asset management. Afterwards, this Part aims at proving that Asset Management requires to model value delivered by the assets, being capable of evaluating the long-term value proposition and the proneness towards value delivery in terms of multiple impacts. To this end, it includes three other contributions providing a focus on multiple impacts considered when an organization (a private company or a public institution) aims to manage the long lives of physical assets driven by a long-term vision.

An overarching framework is presented, focusing on infrastructure assets used to deliver the services and thus to generate, as long-term objective, the societal outcomes required by citizens in smart cities (Chap. 2). Besides, in view of life management of assets in manufacturing plants, technical performances and costs are jointly considered within a value assessment that joins technical concerns with profitability, to support asset-related decision-making (Chap. 3). Eventually, complex asset condition is measured through a composite metric in order to assess the long-term deterioration of assets, quantifying the growth of likelihood of generalized failures requiring a capital expenditure for assets refurbishment or replacement, and the proximity to the end of useful life (Chap. 4). A short summary is now provided for each chapter.

Chapter 2—*A Conceptual Framework for the Alignment of Infrastructure Assets to Citizen Requirements in Smart Cities*.

This Chapter sets the tone for understanding the concept of 'value' at the broadest scale—the society—and provides a framework to create a link between societal outcomes and the assets that are used to generate those outcomes. It is accepted that the built environment including infrastructure within a city has a direct impact on the quality of life for citizens that live, work and visit the city. This relationship is generally understood at a high-level but not when considering the performance of individual assets to the citizen requirements. The key conceptual contribution of this chapter is the need for explicit definition of the functional output of infrastructure assets and systems to create an understanding of how a city or nation's infrastructure comes together to deliver the services required by its citizens and meet the citizen requirements.

The framework provided in the chapter can not only be used at this scale, but also can be used to identify the links between an organization or a system and the assets that support the delivery of value to the organization or system. The framework offers much needed clarity to asset managers to translate the key objectives of an organization to design and management of its physical assets. Clear line of sight between asset performance and organizational objectives not only helps in designing the system and assets to generate value, but also to ensure—through proper data collection and analysis—that decisions made throughout the life of those assets actually deliver against the objectives that informed the design of the assets.

Chapter 3—*Application of a Performance-Driven TCO Evaluation Model for Physical Asset Management*.

This Chapter studies the adoption of Total Cost of Ownership (TCO), starting from a literature-based framework of its potential applications to support decisions at different stages of the life cycle of assets in manufacturing plants. Given this scope, it argues for the relevance of integration of technical performances evaluation into the cost models. Therefore, it proposes a modeling approach to build a performance-driven TCO evaluation model for asset management.

The modelling approach aims at improving the quantification process of costs, particularly the operational expenditure (OPEX) based on Reliability, Availability and Maintainability (RAM) analysis. It consists of the joint use of: (i) a cost break down structure (CBS), designed to perform the necessary TCO analysis and trade-offs; (ii) a stochastic simulation for modeling the casual nature of stochastic phenomena such as asset failures; (iii) the Reliability Block Diagram (RBD) logic, used to express the interdependencies of performances of individual assets and their impacts over the whole production system. The adoption of well-known performance measures in manufacturing as overall equipment effectiveness (OEE) is also proposed in order to assess the cost of performance losses (availability, performance and quality losses).

Overall, theories of performance measurement and modelling for reliability engineering are backing up the financial and economic evaluations, with the intention to provide a decision support for harmonizing the trade-offs of performances and costs. Indeed, the envisioned model is proposed to guide asset managers in a decision-making process driven by a concept of lifecycle value that explicitly relates economics to the technical performance evaluation. This would allow combining the technical, economic and financial evaluations required by asset-related decision-making, thus helping to translate the technical concerns into profitability.

Chapter 4—*Defining Asset Health Indicators (AHI) to Support Complex Assets Maintenance and Replacement Strategies. A Generic Procedure to Assess Assets Deterioration*.

This Chapter proposes a generic procedure to compute an asset health index or indicator (AHI) to measure the current degradation of a complex asset. Through the procedure, the contribution aims at generalizing the practice induced by methodologies selected by means of a review mainly from electrical infrastructures sector.

Accordingly, the AHI is a composite metric intended to provide information of asset capability over time; in particular, it is a measure of the condition of the asset and the proximity to the end of its useful life. Besides the age, the AHI integrates, as weighted factors, operational observations, field inspections and laboratory tests, and factors linked to the functional/geographic location enabling to consider also the environmental conditions.

As the asset health deteriorates, AHI is a measure of the growth of likelihood of generalized failures requiring a capital expenditure to either refurbish or replace the assets. Before this happens, the aim is to proactively act against it: AHI is a model developed to this end, leading to the quantification of the remaining useful life, and to set correspondent recommendations for decisions in regard to assets replacement investment and maintenance programs.

Overall, the AHI model extends theories of condition monitoring to quantify the general health, and health deterioration over time, of a complex asset. The extension is firstly related to the fact that the AHI composes different information used to track the degradation of the subsystems of the asset; afterwards, the extension regards the application scope, as a good understanding of how assets degrade over the long-term can be incorporated in long-term planning. Therefore, an organization has a model aiding to improve the decision-making process of the asset manager when prioritizing maintenance actions, overhauls, and replacements of assets through capital investments. This would enable to assure and optimize the lifecycle value of assets as they approach the end of their useful life.

Part 2. Focusing on Value-Based Asset Management

Part 2 focuses on value-based decision-making approach, remarking its relevance as an overall perspective for management of the assets over their life cycle. Indeed, the concept of value is now *operationalized* to drive the day to day management decisions and activities.

To this end, it is firstly recommendable to develop the AM business process, embedding it within the organization of extant companies. It is the primary concern of Chap. 5, whose main focus is the provision of a framework to guide the implementation of AM based on the asset-related decisions at different control levels (i.e., strategic, tactical or operational) and asset lifecycle stages (Beginning of Life, Middle of Life, End of Life). Embedding the AM as business process, creates the context to ensure a sustainable value creation from assets. However, this is not enough.

Value modelling is also required, according to the intended use discussed in Chap. 1. Given this as a background, the two chapters (Chaps. 6 and 7) included in this part, are *operationalizing* the creation of value models, by relying on the adoption of specific engineering methodologies, integrated to provide a decision support. In fact, such methodologies serve with the purpose to set the value model and to elaborate the required information for asset-related decision-making.

Chapter 6 introduces a systematic approach to make value-based asset management decisions. Herein, the aim is to identify and quantify the different ways in which an asset provides value to its different stakeholders: the 'value-mapping' process proposed in the Chapter provides a method to map the stakeholder's requirements to the value provided by the asset and the actions that can influence value, thus relevant for effective decisions.

Finally, considering the importance of criticality analysis in understanding the value delivered by assets, Chap. 7 focuses on the adoption of criticality analysis as a methodology to define the assets hierarchy. This concept is illustrated in the context of guiding the felling and pruning process within the context of power lines. Criticality analysis demonstrates its capability, as a value model, to effectively align the expenditures to criticalities due to value for business.

A short summary is now provided for each chapter.

Chapter 5—A framework to embed asset management in production companies.

This Chapter relies on the acknowledgment of Asset Management as a holistic business process involving different functions within an organization to manage the assets along their lifecycle. The main recognition regards the fact that, with Asset Management, assets are brought at the center of the management process to generate and/or preserve the value, while several activities and enterprise functions—amongst them, maintenance—are required along the lifecycle of the assets.

Thus, the Chapter presents a framework to guide embedding AM in an organization. The framework addresses particularly production companies, as it is the result of a focused literature review and empirical investigation in this specific scope; nevertheless, it also adopts general concepts backed up by the current standardization, specifically the ISO 5500X series.

The framework is built upon two dimensions—the asset life cycle and the hierarchical level of the asset-control activities—recommended to position any asset-related decision—namely the Beginning of Life (BoL), Middle of Life (MoL), and End of Life (EoL) stages, and the strategic, tactical or operational control. Besides, the framework assumes four principles—i.e., life cycle, system, risk and asset-centric orientation—to define and guide the approach, methods, and systems to be adopted in order to integrate AM within a company.

Overall, the framework reflects the importance of the four founding principles in order to structure an Asset Management system; in particular, it remarks their role as required background upon which the AM decision-making process should be developed in order to ensure a sustainable value creation from assets.

Chapter 6—An approach to value-based infrastructure asset management.

This Chapter presents one of the fundamental methodological contributions of this research. The chapter describes a systematic approach to make value-based asset management decisions. The approach aims to tackle the fundamental issues of identifying and quantifying the different ways in which an asset provides value to its different stakeholders. Further the approach expands on this to explain how this understanding can be used to make effective decisions that aims to maximise this value. A so-called 'value-mapping' process provides an efficient method to map the stakeholder's requirements to the value provided by the asset and the actions that can influence value. This value-map can then be used to assess the value and make effective decisions.

This approach for understanding and applying value to decision-making is demonstrated in the chapter through a case study involving metro transportation tunnels. The essential consideration of value is expected to allow organisations in evaluating the balance between cost, risk and performance and thereby allowing better informed decisions. The head of profession at the city metro where the methodology was applied to develop a repair strategy for tunnels commented: "*[This methodology] offers an opportunity to assess systematically what we should value in each case and guide the decision making accordingly.*"

This methodology was further applied in a local county council for prioritizing maintenance across their portfolio of bridges. The Area Bridge Manager at the council commented on the benefit of this approach as follows: "*[This approach] provides us with a degree of confidence to justify the expenditure of our highway structures to target our limited resources to the benefit of the local communities.*"

Chapter 7—Exploiting EAMS, GIS and Dispatching Systems Data for Criticality Analysis. Improving Maintenance and Felling and Pruning Management in Power Lines.

This Chapter presents two very important aspects that must be highlighted: the first one is related to the utilization of digital information, already existing in the company, to speed up the process of assets hierarchy definition; a second and probably more important point is the introduction of asset value, through the criticality analysis of the power lines, to define this assets hierarchy and to guide the felling and pruning process.

The felling and pruning process has been therefore re-engineered, in fact the company has accomplished an entire digital transformation of the process, that now has been centralised and supported, entirely, by a new software application named "*Felling and Pruning Management System*", that can be considered a real value-based management system.

Value provided by the different lines to the business can change over time with the new network developments, configuration or topology modifications. Nevertheless, this information is permanently updated by the system and taking into account in the maintenance planning decision-making process. According to actual value, appreciated in asset criticality of the power lines, their maintenance plans are optimized, minimising risks but also complying with the applicable Spanish and autonomous regional legislation.

Last news received from the company refer to important achievements with more than 30% savings in the felling and pruning process (representing close to 2 Million € net savings per year). Beside this achievement, the company and authorities are currently studying to change existing regulations concerning power lines maintenance, linking maintenance activities frequencies to criticality.

Part 3. Advances in Operational Decision Making—CBM/PHM

Part 3 focuses on the advanced developments around asset-related decision-making at the operational level. The tools herein adopted aim to support the failure prediction and/or the determination of assets conditions, with the ultimate goal to lead to the release and execution of optimal or improved maintenance activities.

The primary feature, common to such advances, is the integration of physical processes and cyber space with intelligent computation, implemented by exploiting the scalability of machine learning models, built both on Artificial Intelligence and statistical modelling techniques. Furthermore, considering the industrial practice, it is worth remarking that this innovation can be developed not only from green fields but also from brown fields, integrating the capabilities of extant plant automation (including all levels from SCADA systems to PLCs, CNCs and RTUs) with new potentials represented by asset condition monitoring, diagnostic and prognostic systems developed under the framework of CPS-based/CPS-like integration. Overall, it enables an approach that aims at implementing the most adequate solutions, finding a compromise between accuracy and computational power of the advanced tools. The chapters within this Part 3 illustrate different tools, built based on such common traits while adapting to the specific needs of the operating scenarios wherein they are implemented.

Chapter 8 illustrates the idea of using the artificial neural networks on top of the legacy supervisory control and data acquisition system (SCADA), to implement an innovative failure detection and power generation forecasting method. The effectiveness of the model and method are demonstrated in inverter failure prediction in Grid Connected Photovoltaic systems. Chapter 9 fosters to apply artificial neural networks per failure mode and their practical implementation in SCADA systems for different Photovoltaic plants. In this case, failure mode prediction is built on process-control variables as an approximation to real production in the Photovoltaic plant in absence of failures. Based on it, both early detection and diagnostics can be applied, with scalability, to all power inverters in the plant. Chapter 10 considers reliability modelling to optimise/improve the condition-based maintenance (CBM) policy for systems subject to both aging and cumulative damage: it exploits Proportional Hazards Model (PHM) in two cases—when distribution parameters of the degradation process are known in advance or unknown, and a case study of Asphalt Plug Joint in a bridge.

Chapter 11 presents the implementation of a condition monitoring system on an Electric Arc Furnace (EAF) of a steel making, developed to support risk-informed decisions of maintenance planners and engineers when the EAF is running. Condition monitoring aims to enrich plant automation, with state detection and health assessment, to interpret data manipulated from PLC and SCADA levels; to this end, statistical modelling is used combined with knowledge gathered from Hazard and Operability Analysis (HAZOP), to detect incipient failures through monitoring of the burning system, before causing a downtime, that may lead to disastrous effects on people safety, integrity of assets and production cycle.

The higher availability of on-line data at lower cost, leads to new potentials for dynamic reliability modelling to schedule maintenance activities based on the current condition of both assets and operations, thus providing an enhanced opportunistic maintenance capable to effectively take advantage of changing operating requirements. It is the case addressed by the proposed modelling for a real-time dynamic opportunistic maintenance of wind farms, significantly outperforming opportunistic maintenance strategies based on static thresholds (Chap. 12).

A short summary is now provided for each chapter.

Chapter 8—*A CPS for Condition Based Maintenance Based on a Multi-Agent System for Failure Modes Prediction in Grid Connected PV Systems.*

This Chapter is devoted to structure a management framework for Condition Based Maintenance (CBM). The CBM system monitors continuously the performance and condition (several state variables) of the different Photovoltaic (PV) plant elements, helping to determine the optimal time for specific maintenance activities, verifying at the same time the effectiveness of the actions. In fact, to increase the availability and performance of a PV plant, failure control must be based on a sustainable and well-structured condition monitoring, which relies on procedures fulfilling international standards, in order to keep and improve solutions over time.

Due to the huge amount of data that have to be accessed to during the operation of these PV plants, the systems supporting maintenance decision-making need to be carefully analyzed and the solution adopted must allow a very practical implementation. The idea presented in this contribution is to use a Cyber-Physical Systems (CPS) with Artificial Neural Networks (ANN) to that purpose.

Cyber-Physical Systems (CPS) approach easies the integrations of physical processes, network of assets and intelligent computation, while artificial intelligence supports this solution from the scalability point of view. An important aspect of the work is how to cope with the need of these solutions to adapt to locational and operational changes. The artificial neural networks (ANN) are developed on top of the legacy supervisory control and data acquisition system, to implement an innovative failure detection and power generation forecasting method. The effectiveness of the model and method are demonstrated in inverter failure prediction of these power plants.

Chapter 9—Failure Mode Prediction and Energy Forecasting of PV Plants to Assist Dynamic Maintenance Tasks by ANN Based Models.

This Chapter shows the importance to take into account dynamic changes in environmental parameters and asset operating conditions when developing reliability studies, modeling assets degradation and projecting solar energy production in Photovoltaic (PV) solar plants.

To that end, Artificial Neural Networks (ANN) models were used to find complex relations among model input variables, in this case with high tolerance to data uncertainty, and to provide predicted variable patterns on-real time. ANN models performed well sometimes with not enough data quality, not always accessible or available at a reasonable cost.

Prediction could be done for process-control variables (like accumulated active energy of the inverter) obtaining an approximation to real production in the PV plant in absence of failures. This model could be then easily implemented on the SCADA systems, and replicated for all power inverters in the plant. With this process early detection of incipient failures in inverters could be achieved.

Besides detection, ANN models may also offer help to identify and diagnose failure modes, like the tracker blocking in this Chapter, replacing other statistical models and tools or simply through integration within these tools.

All in all, the Chapter fosters to apply ANN models per failure mode and their practical implementation in SCADA systems for different plants. This is demonstrated to ease and condition-based maintenance and risk modelling (residual life until total equipment failure), enabling reductions of corrective maintenance direct and indirect costs.

Chapter 10—*Condition-based maintenance for systems with aging and cumulative damage based on proportional hazards model.*

The emergence of novel sensor technologies and the drop in their costs has led to the popularity in condition monitoring and condition-based maintenance (CBM). It is very clear that CBM has significant cost and risk advantages compared to planned or corrective maintenance. However, effective implementation of CBM depends on the robustness of the mathematical model used to represent degradation and failures, as well as the technique used for determining the optimal CBM policy. This chapter focusses on addressing one of the key challenges in reliability modelling—how to optimise the condition-based maintenance (CBM) policy for systems subject to both aging and cumulative damage.

The Chapter proposes Proportional Hazards Model (PHM) as a means to characterize the joint effect of aging and cumulative damage. CBM models are developed for two cases: one assumes that the distribution parameters of the degradation process are known in advance, while the other assumes that the parameters are unknown and need to be estimated during system operation. In the first case, an optimal maintenance policy is obtained by minimizing the long-run cost rate. For the case with unknown parameters, periodic inspection is adopted to monitor the degradation level of the system and update the distribution parameters. A case study of Asphalt Plug Joint in a bridge is employed to illustrate the practical utility of this model and to explore the benefits of the model in improving maintenance policies.

The results in this Chapter show that the degradation rate exerts a significant impact on system lifetime distribution. Engineers or managers are suggested to pay more attention to improving the accuracy of the degradation rate estimation.

Chapter 11—*Implementation of a Condition Monitoring System on an Electric Arc Furnace through a risk-based methodology.*

This Chapter presents the implementation of a condition monitoring system on an Electric Arc Furnace (EAF) of a steel making plant located in Italy. It illustrates how the industrial application of the system is engineered and implemented with the objective to convert data from the plant floor into information, to be displayed at a supervisor level to support risk-informed decisions of maintenance planners and engineers.

The methodology used to this end is a step-wise approach extended from a risk analysis to the deployment of the condition monitoring system. The approach firstly designs the system building on a process hazard analysis (PHA), done not only to map the asset behaviour, but also to identify the key variables, highly representative to monitor the asset. In particular, a Hazard and Operability Analysis (HAZOP) helps modelling the behaviour of the burning system of the furnace in normal and abnormal circumstances: it enables to synthesize the process knowledge sourced by a team of experts (process experts, maintenance experts, etc.). A follow-up is to identification of the key variables to control the deviation of normal and abnormal circumstances (i.e. variables already monitored for process control reasons, as process-control variables). Thus, the condition monitoring is defined and later implemented in a correspondent tool enriching the plant automation, with functionality of state detection and health assessment, to interpret data manipulated from PLC and SCADA levels. Specifically, statistical modelling is used combined with the knowledge gathered from the HAZOP, to detect incipient failures through monitoring of the burning system, before causing a downtime, that may lead to disastrous effects on people safety, integrity of assets and production cycle.

Chapter 12—*A Dynamic Opportunistic Maintenance Model to Maximize Energy-based Availability while Reducing the Life Cycle Cost of Wind Farms*.

This Chapter is a good example of dynamic models applied to optimize maintenance activities scheduling. The idea is to schedule these activities when weather conditions are more advantageous. This is reached through the design of dynamic reliability thresholds, which depend on wind speed and generated turbine power.

The availability of the wind turbines for energy production was maximized during their most profitable periods. That's why the proposed dynamic opportunistic maintenance policy significantly outperforms traditional (based on static thresholds) opportunistic maintenance strategies.

In a moment in which on-line data is becoming more available and less costly, the possibility to implement policies using dynamic interpretation rules for the release of maintenance preventive activities is really compelling. The level of intelligence, measured by the number and precision of the new interpretation rules to manage maintenance, can definitively increase. Beside this, rules can be established considering a more systemic approach for the complete wind farms, according to their configuration and economic dependencies.

It is also important to highlight the effort required to build the mathematical simulation model. The format of the model considers maintenance capacity constraints, multiple failure modes, precise indenture levels and entire life cycle cost calculation. All these features serve to offer a very outstanding market solution, to meet the challenge of real-time management of opportunistic maintenance.

Part 4. Emerging Value-based and Intelligent Asset Management Processes

This Part 4 aims at presenting new emerging processes reaping the benefits of new ideas that can be nowadays implemented as a consequence of new technologies, and that can certainly embed more intelligence and orientation to value in existing asset management systems.

Blending technologies and engineering methodologies is required to efficiently and effectively manage and process information, to support the asset-related decision-making process over the asset lifecycle. It is a primary reason why to consider the promises of digitization promoted by Industry 4.0. Therefore, emerging concepts in this field are relevant, as they contribute building the foundation of an Intelligent Asset Management system. Therefore, talking about the perspective of Digital Twins in CPS-based system (as done by means of a literature review in Chap. 13, focused on CPS-based production systems) is a relevant issue to build and implement value models in a digitized environment, with potentials of decision support along the whole life cycle of the assets.

The potential of integration due to IIoT can be also addressed, with the goal to develop new possibilities for collaborative approaches to problem solving, to improve system-level performance in industrial environments. It is the case that can be potentially built by means of a Social Network of Collaborating Industrial Assets, defined as networks of cooperating assets with sensing, communicating and computing capabilities (Chap. 14). One of the applications of these networks is collaborative prognostics: machines act as truly "smart" and social assets, capable to communicate and learn failure patterns from each other. In a conceptual proof of such collaborative prognostics (Chap. 15), Social Asset Networks show a faster convergence and lower cost of prognostics, if compared to typical self-learning and fleet-wide approaches.

Eventually, implementing Additive Manufacturing in supply networks of Digital Spare Parts is fostering new scenarios, for which is possible for spare parts to be delivered to the location where they are needed much faster than in the traditional model. Thus, leveraging on an innovative business process and technology support, Digital Spare Parts is expected to change, in the near future, both challenges and decisions in design and control of spare parts supply chain network (Chap. 16).

A short summary is now provided for each chapter.

Chapter 13—*A review of the roles of digital twin in CPS-based production systems*.

This Chapter focuses on the concept of Digital Twin (DT), considering its growing use in industry and research initiatives, and its potentials for decision-making in regard to engineered systems along their lifecycle.

As recently the concept is highly discussed due to the push of Industry 4.0 wave, and the scientific literature does not provide a unique definition, the analysis provided in the Chapter firstly aims at tracing the definitions starting from the initial conceptualization in the aerospace field, to the most recent interpretations in the manufacturing domain and, more specifically, in the scope of Industry 4.0 and smart manufacturing research. The Chapter analyses literature contributions about Digital Twin, released after 2012, in order to trace the evolution of the concept, related to its use, to connected simulation models, to diagnostics and prognostics features. As a result, the DT is understood as the virtual and computerized counterpart of a physical system that can be used to simulate it for various purposes (such as monitoring, optimization, prediction), while exploiting a real-time synchronization of the sensed data from field; such a synchronization is possible thanks to the enabling technologies of Industry 4.0 (e.g. Internet of Things) and, as such, the DT is deeply linked with it.

Research on the DT in manufacturing is an evolution of the stream of Virtual Factories. It is leading to the digitalization of the plant integrated with the real system, coming in help during all the lifecycle of its component assets. Nevertheless, it is evident that this lifecycle support is at its infancy: information generated at design phase of a production system is not yet properly exploited during the operational phases.

For the future, a full exploitation of advanced technologies, as the CPS (Cyber-Physical Systems), open the way to new and advanced uses of the DT, including diagnostics and prognostics purposes—already discussed in previous researches.

In particular, the Chapter fosters the adoption of a DT to mirror the status of the production system and to perform real-time optimizations, decision making and predictive maintenance according to the sensed conditions. Overall, DT will be a core concept within the transformative process due to digitalization, and the use of CPSs and DTs for asset management may be expected to support decisions in a digitized environment to assure value delivery at different asset-control levels and lifecycle stages.

Chapter 14—*A Social Network of Collaborating Industrial Assets*.

The Industrial Internet of Things (IIoT) concept, where data gathered through sensors embedded in the machines are leveraged to perform real-time failure detection and prediction for machines in a machine fleet, is widely regarded to have the potential to significantly reducing maintenance cost and machine downtime. However, the potential of this integration in improving system-level performance in industrial environments has rarely been explored.

Combining techniques in distributed intelligence and machine learning, this Chapter exploits the power of IIoT and presents the building blocks for an innovative paradigm of collaborative social networks of machines. This introduces the concept of Social Asset Networks, defined as networks of cooperating assets with sensing, communicating and computing capabilities.

The Chapter presents the building blocks of Social Asset Networks and characteristics of one of its important components—Social Assets. A sketch of the general architecture needed to enable a Social Network of Collaborating Industrial Assets is proposed and two illustrative application examples are given. One of the applications is the idea of collaborative prognosis, the fundamental mathematical principles of which is provided in next Chap. 15 of this book.

Chapter 15—*Collaborative Prognostics in Social Asset Networks*.

This Chapter provides the foundations for a novel approach to failure prognosis and predictive maintenance. This technique exploits the concept of social networks of machines, and enables machines to communicate and learn failure patterns from each other. This radically enhances their ability to predict failures, and improve reliability and performance. This provides an important stepping stone to the field of individualised asset prognostics.

In this approach, the information obtained from the medium by means of sensors is synthesised into a Health Indicator, which determines the state of the asset. The Health Indicator of each asset evolves according to an equation determined by a triplet of parameters. The Chapter shows how by exchanging their estimates of remaining useful life, networked assets are able to precisely determine their Health Indicator dynamics and reduce maintenance costs. This is done in real time, with no centralised library, and without the need for extensive historical data.

The Chapter also compares Social Asset Networks with the typical self-learning and fleet-wide approaches, and show that Social Asset Networks have a faster convergence and lower cost. This study simply serves as a conceptual proof for the potential of collaborative prognostics for solving maintenance problems, and further studies have been carried out to justify the implementation of such a system in a real industrial fleet. In essence, this Chapter shows the real potential of truly "smart" assets to revolutionise the world of predictive maintenance.

Chapter 16—*The perceived value of additively manufactured digital spare parts in the industry: an empirical investigation.*

This Chapter presents an empirical investigation conducted in Finland through focus groups involving different actors in terms of types of company—AM service provider, Manufacturing subcontractor, AM software developer, OEMs and others—and position in the company—Top management, middle management, Specialist and First-line management. The perceived value of implementing Additive Manufacturing (AM) in supply networks of Digital Spare Parts (DSP) is measured, and the concept of DSP is expanded by focus group interviews.

Amongst the evidences achieved during the study, it is worth remarking the high interest in producing spare parts digitally, with particular emphasis on the long tail products—products in low demand and that have a low sales volume—as excellent candidates. This corresponds, in spare parts management, to the main benefits of reducing repair time, delivery time and costs, emissions, and inventory. Indeed, looking at the trade-offs induced by the conflicting value drivers and metrics, the cost of AM was not seen as a prohibitive issue for long tail products by the participants of the focus group. Another potential DSP benefit is the improved spare parts that have better qualities than the original ones, meaning, for instance, better design by topology optimization or reduced number of joints to decrease the risk of failure.

However, several issues should be overcome before large-scale implementation of DSP can take place. From a technology's perspective, it was observed that neither the AM technology, nor the ICT infrastructure of the companies that would make DSP distribution possible, are robust enough for the digital distribution method to be deployed fully in the interviewed companies. Besides, other changes are required. AM technology has to be developed to correspond with more spare parts in companies' catalogues.

Overall, this Chapter enables describing DSP as "A concept in which defective components are replaced by manufacturing spare parts close to the location of need from 3D model data that are transferred by network with the main consequences of reducing repair time, delivery time and costs, emissions, and inventory". Accordingly, this envisions the scenario in which a company implements DSP: the transportation of data will become more important than physical part movement; the role of the material supplier and the software developers will be significantly larger in the DSP network; the traditional and DSP network models will be replaced by a central storage with the model database, regional storage spaces will be removed, and the introduction of distributed manufacturing will be occurring. These changes make it possible for spare parts to be delivered to the location much faster than in the traditional model. Therefore, leveraging on an innovative business process and technology support, DSP is expected to change, in the near future, both challenges and decisions in design and control of spare parts supply chain network, which is a valuable topic within the frame of Value-based and Intelligent Asset Management.

2 Mapping Chapters to Framework Dimensions

As mentioned in the first Chapter, this book aims to foster a future perspective, making a step forward with respect to the current understanding of Asset Management primarily geared with well-known standards.

To this end, according to the framework that has been established in the introductory chapter of the book, this final section aims to map the different Chapters insights to the framework dimensions; it enables to show our concrete understandings and expectations on what is a Value-based and Intelligent Asset Management.

It is worth pointing out that this book compiles information gathered from research and innovation efforts in projects relevant to this scope, especially considering the evidences from state of the art and current research trends of Physical Asset Management (PAM) and Operations and Maintenance (O&M) of industrial plants and infrastructures. These are summarized in Tables 1 and 2, as key learning outcomes of the researches carried out, knitted together based on the dimensions of the framework.

In Table 1 we look at the first framework dimension: in this Table the value model as illustrated by each Chapter is specified, specially detailing its characteristics according to what is argued by this dimension.

In Table 2, we map the Chapters with respect to the other framework dimensions, thus identifying the AM resources—i.e. the "tools"—constituting the Asset Management system, as specifically focused by each Chapters.

3 Conclusions

This chapter clearly shows that the idea of 'value-based' and 'intelligent' asset management can be applied across different industry sectors from manufacturing to infrastructure, and to a wide variety of decisions throughout the lifecycle of assets of different types. It is also clear that a value perspective is useful not only when looking at a particular asset or a component as the unit of analysis, but to complex systems and even at a city scale. Some of the overall conclusions and findings from the book as summarised by this Chapter are as follows.

- Value models to support decision-making are majorly oriented to value realization at system- or network-level. Fleet or city levels are, for different reasons, also pointing to a systemic value, as they (e.g., the fleets) comprise more individual assets spread across different locations or (e.g., the cities) are a case of a complex system of systems (e.g., different types of infrastructure systems providing value to citizens).
- The whole-life assessment of value realization requires the adoption of different value models related to different lifecycle stages. This variety is spread amongst models potentially applicable throughout the all lifecycle stages, and models majorly/solely applicable to a single stage. According to the compilation done

Table 1 Map of book chapters with respect to the value model characteristics

	Driven by stakeholders' requirements, value drivers/metrics	Value realization at asset system/network level	Whole-life assessment of value realization
Chapter 2—A Conceptual Framework for the Alignment of Infrastructure Assets to Citizen Requirements in Smart Cities	Society's requirements—health, education, quality of life	Cities level	Applicable to Beginning of Life, Middle of Life and End of life stages
Chapter 3—Application of a performance-driven TCO evaluation model for physical asset management	Production and maintenance requirements leading to related KPIs, especially dealing with technical performance losses	System level	Applicable to Beginning of Life, Middle of Life, End of Life stages
Chapter 4—Defining Asset Health Indicators (AHI) to Support Complex Assets Maintenance and Replacement Strategies. A Generic Procedure to Assess Assets Deterioration	Likelihood of generalized failures resulting from long-term degradation requiring a capital expenditure for assets refurbishment or replacement	Network or System level	Applicable to Middle of Life, End of Life stages
Chapter 5—A framework to embed asset management in production companies	Assurance of generation of value from assets based on a structuring AM decision-making process	System level	Applicable to Beginning of Life, Middle of Life, End of Life stages
Chapter 6—An approach to value-based infrastructure asset management	Wide variety of stakeholders ranging from the metro operator, the city council, passengers and employees each with their set of value drivers and associated metrics	Network level	Applicable to Middle of life
Chapter 7—Exploiting EAMS, GIS and Dispatching Systems Data for Criticality Analysis. Improving Maintenance and Felling and Pruning Management in Power Lines	Efficiency of the felling and pruning process while keeping the risk of the network under control	Network level	Mainly applicable to Middle of Life stage

(continued)

Table 1 (continued)

	Driven by stakeholders' requirements, value drivers/metrics	Value realization at asset system/network level	Whole-life assessment of value realization
Chapter 8—A CPS for Condition Based Maintenance Based on a Multi-Agent System for Failure Modes Prediction in Grid Connected PV Systems	Solar power prediction, and efficiency of energy generation for changing weather conditions	Network level	Mainly applicable to Middle of Life stage
Chapter 9 –Failure Mode Prediction and Energy Forecasting of PV Plants to Assist Maintenance Tasks by ANN Based Models	Solar power and reliability predictions, to properly release dynamic preventive maintenance activities	Network level	Mainly applicable to Middle of Life stage
Chapter 10—Condition-based maintenance for systems with aging and cumulative damage based on proportional hazards model	Maintenance costs and risks	Asset level	Applicable to Middle of life
Chapter 11—Implementation of a Condition Monitoring System on an Electric Arc Furnace through a risk-based methodology	Safety of the Electric Arc Furnace operation/Asset Integrity (focus on the burning system)	Asset level	Mainly applicable to Middle of Life stage
Chapter 12—A Dynamic Opportunistic Maintenance Model to Maximize Energy-based Availability while Reducing the Life Cycle Cost of Wind Farms	Wind energy generation efficiency by releasing preventive maintenance activities with a dynamic approach based on reliability thresholds	Network level	Mainly applicable to Middle of Life stage

(continued)

Table 1 (continued)

	Driven by stakeholders' requirements, value drivers/metrics	Value realization at asset system/network level	Whole-life assessment of value realization
Chapter 13—A review of the roles of digital twin in CPS-based production systems	Efficient and effective decision-making support based on real time monitoring, optimization and prediction	System, asset level	Applicable to Beginning of Life, Middle of Life, End of Life stages
Chapter 14—A Social Network of Collaborating Industrial Assets	Maintenance and operational costs and risks, and operational performance; meeting the needs of asset owners and their customers	Fleet level	Mainly applicable to Middle of Life stage
Chapter 15—Collaborative Prognostics in Social Asset Networks	Failure prediction, and maintenance costs	Fleet, asset level	Mainly applicable to Middle of Life stage
Chapter 16—The perceived value of additively manufactured digital spare parts in the industry: an empirical investigation	Value generated by digital spare parts management process measurable in different terms (i.e. reduced repair time, delivery time and costs, emissions, and inventory, and better qualities)	Network level	Mainly applicable to Middle of Life stage

Table 2 Map of book chapters with respect to the asset management "tools"

Chapter	Organization	Business process and asset-control activities	Skills and competences	Information	Technologies	Engineering methodologies
Chapter 2—A Conceptual Framework for the Alignment of Infrastructure Assets to Citizen Requirements in Smart Cities	City/nation	Strategic control, applicable from Beginning of Life, through Middle of Life until End of Life	Understanding user requirements, translating into functional specifications	Relationship from user requirements to asset specifications	Technology agnostic	Requirements engineering, Building information modeling, Data modelling
Chapter 3—Application of a performance-driven TCO evaluation model for physical asset management	Chemical company (producing basic chemicals, fertilizers, nitrogen compounds, plastics and synthetic rubber in primary forms)	Strategic control, applicable from Beginning of Life, through Middle of Life until End of Life	Technologist understanding of production processes, Modeling and simulation, Management (skills in Asset lifecycle management, in Operations and Maintenance)	Cost modeling (CAPEX and OPEX), Technical performance (e.g. OEE)	APM and CMMS/EAM/ERP	Performance measurement, Reliability modelling, Cost modelling, Simulation

(continued)

Table 2 (continued)

Chapter	Organization	Business process and asset-control activities	Skills and competences	Information	Technologies	Engineering methodologies
Chapter 4—Defining Asset Health Indicators (AHI) to Support Complex Assets Maintenance and Replacement Strategies. A Generic Procedure to Assess Assets Deterioration	Big utility company (for LNG regassification and gas transportation)	Strategic control through Middle of Life until End of Life (in the frame of assets renewal and life extension analysis)	Technologist understanding of assets' deterioration processes, Modelling assets deterioration, Criticality analysis	Asset expected life and location, load over time, health and reliability modifiers over time	PI systems, CMMS, and DCS CBM	Condition monitoring (applied to recommend replacement investment and maintenance programs in complex assets)
Chapter 5—A framework to embed asset management in production companies	Manufacturing companies (multiple case study: chemical, appliances, steel, tyre, petrochemical, food and beverage, machine tools manufacturers)	Strategic, tactical and operational control, applicable from Beginning of Life, through Middle of Life until End of Life	Management (skills in asset lifecycle management)		Technology agnostic	Business process engineering

(continued)

Table 2 (continued)

Chapter	Organization	Business process and asset-control activities	Skills and competences	Information	Technologies	Engineering methodologies
Chapter 6—An approach to value-based infrastructure asset management	Metro transportation company	Repair of tunnels	Engineering (Failure mode and effect analysis), Stakeholder analysis, Optimization	Asset condition, GIS, Dispatching systems info concerning impact of failures in service quality	CMMS, Asset register	Value-mapping techniques, Failure mode and effect analysis (FMEA), Stochastic and deterministic optimization approaches
Chapter 7—Exploiting EAMS, GIS and Dispatching Systems Data for Criticality Analysis. Improving Maintenance and Felling and Pruning Management in Power Lines	Electrical distribution company	Strategic control through the Middle of Life (in the frame of Maintenance activities)	Management (skills in Operations and Maintenance), Risk analysis, Electrical engineering	Vegetation cartography. High-precision gradient maps, GIS, Dispatching systems info concerning impact of failures in service quality	CMMS, GIS, PNOA orthophotos, Dispatching systems and their simulation capabilities, Criticality analysis software	Risk analysis techniques (where risk is calculated as the consequence of the failure for the business times the registered frequency of failure in the systems i.e. CMMS)

(continued)

Table 2 (continued)

Chapter	Organization	Business process and asset-control activities	Skills and competences	Information	Technologies	Engineering methodologies
Chapter 8—A CPS for Condition Based Maintenance Based on a Multi-Agent System for Failure Modes Prediction in Grid Connected PV Systems	PV Solar Plant O&M Company	Operational control through the Middle of Life (in the frame of Operations and Maintenance activities)	Management (skills in Operations and Maintenance), ICT systems together with ubiquitous embedded cyber physical applications	Physical structure of the assets, their functional logic and level of intelligence	PV Systems, Cyber-Physical systems	Multi-Agent Systems architecture (to serve as a framework to support the implementation of CPS and ANN for supervisory control and data interpretation)
Chapter 9 –Failure Mode Prediction and Energy Forecasting of PV Plants to Assist Maintenance Tasks by ANN Based Models	PV Solar Plant O&M Company	Operational control through the Middle of Life (in the frame of Operations and Maintenance activities)	Management skills (in Operations and Maintenance), Data science (Artificial intelligence and Machine Learning)	Operations and maintenance data besides environmental conditions of the assets	PV Systems, SCADA systems	CBM, ANN and PHM models for supervisory control data interpretation
Chapter 10—Condition-based maintenance for systems with aging and cumulative damage based on proportional hazards model	Bridge owners	Operational control through the Middle of Life (in the frame of Maintenance activities)	Failure modes of bridge components, maintenance of bridges, deterioration models	Health indicators, deterioration models, environmental condition of the assets	Condition monitoring systems	Proportional hazards model

(continued)

Table 2 (continued)

Chapter	Organization	Business process and asset-control activities	Skills and competences	Information	Technologies	Engineering methodologies
Chapter 11—Implementation of a Condition Monitoring System on an Electric Arc Furnace through a risk-based methodology	Steel making company	Operational control through the Middle of Life (in the frame of Operations and Maintenance activities)	Technical knowledge (including production process and maintenance knowledge, as well as plant automation), Data science (Machine Learning/Statistical modeling)	Process -control variables (to make state detection and health assessment)	Condition monitoring system, SCADA, PLC	CBM, and PHM models for supervisory control data interpretation
Chapter 12—A Dynamic Opportunistic Maintenance Model to Maximize Energy-based Availability while Reducing the Life Cycle Cost of Wind Farms	Wind Turbines Park O&M Company	Operational control through the Middle of Life (in the frame of Operations and Maintenance activities)	Management skills (in Operations and Maintenance), Simulation and optimization	Reliability data and environmental conditions of the assets	PV Systems. SCADA or condition monitoring systems	Dynamic Reliability modeling. Agents simulation modeling. Optimization algorithms and heuristics

(continued)

Table 2 (continued)

Chapter	Organization	Business process and asset-control activities	Skills and competences	Information	Technologies	Engineering methodologies
Chapter 13—A review of the roles of digital twin in CPS-based production systems	Manufacturing companies	Strategic, tactical and operational control, applicable from Beginning of Life, through Middle of Life until End of Life	Technologist understanding of monitoring processes, Modeling and simulation, Optimization, Management (skills in Asset lifecycle management)	Monitored and simulated information dependent on the purpose of the Digital Twin	Asset Performance Management and SCADA, PLC/CNC, condition monitoring systems, RTU	Data modelling, Machine state modelling (for monitoring), Simulation
Chapter 14—A Social Network of Collaborating Industrial Assets	Manufacturing companies	Operational control through the Middle of Life (in the frame of Operations and Maintenance activities)	O&M of industrial assets, Simulation and optimization	Health indicators, condition monitoring information, production schedules	Condition monitoring systems, demand forecasting systems, CMMS	Regression and other machine learning-based prediction models, Stochastic optimization techniques, Simulation
Chapter 15—Collaborative Prognostics in Social Asset Networks	Organisations providing through-life engineering services to a fleet of high-value assets	Operational control through the Middle of Life (in the frame of Operations and Maintenance activities)	Data science (Artificial intelligence, Machine Learning), Data management architectures and protocols	Asset Health Indicator	Condition monitoring sensors, Cyber-physical systems	Regression and other machine learning-based prediction models

(continued)

Table 2 (continued)

Chapter	Organization	Business process and asset-control activities	Skills and competences	Information	Technologies	Engineering methodologies
Chapter 16—The perceived value of additively manufactured digital spare parts in the industry: an empirical investigation	Companies within the supply network –additive manufacturing service provider and software developer, Manufacturing subcontractor, OEMs and others	Strategic control through the Middle of Life (in the frame of Spare parts supply network/Spare parts management activities)	Management (in logistics, supply chain, spare parts management), Technologist understanding of Additive Manufacturing	Logistics support data, including repair time, delivery time and costs, emissions, and inventory, and material qualities	ERP/EAM/SCM/Additive Manufacturing	

in this book, a good balance can be pointed out within such a variety of models, even if a prevalence of models dedicated to the Middle of Life stage is notable.

- Besides the evidence of the spread of stakeholders' requirements—depending on the different business contexts under consideration—justifying the different concepts of value, it is notable the fact that value as a driver concerns not only the assets under management (at individual, system or network level), but also the asset management system. In a few cases in our collection, it is apparent that asset management decisions should be made using a value-driven process in order to enable the assurance of value generation. A caveat of the evidences collected in the book is also a confirmation of the implication that decision-makers have to take a multi-objective approach, not only weighing different value drivers, but also weighing the preferences of different stakeholders during their decision-making process.
- Business processes and engineering methodologies developed in this research are technology-agnostic. Technology is simply a "tool" that, if effectively deployed, can support the implementation of these processes and methodologies. However, the ability to apply the methodologies appropriately, or the establishment of a suited business process, against the problem at hand, is the key to success.
- The skills and competencies required for asset management is wide-ranging. On the one hand, 'soft' skills such as those regarding requirements engineering and stakeholder analysis were found to be as important as technical skills such as those regarding sensor technologies and data science.
- A systematic identification of data and information requirements is absolutely critical to effective asset management decision-making. Note that the data required for asset management is not limited to those collected from/about the assets themselves. Contextual data such as those that allow us to understand the logistics and wider supply chain, physical environments, and the business context are all critical.
- The world of asset management is turning digital. Although great strides have been made in the academic research, we are only scratching the surface in terms of reaping the true benefits of the digital revolution in practice. The case studies presented in this book is a testament to its potential.

Printed in the United States
By Bookmasters